SCATTERING IN VOLUMES AND SURFACES

North-Holland
Delta Series

NORTH-HOLLAND
AMSTERDAM · OXFORD · NEW YORK · TOKYO

Scattering in Volumes and Surfaces

Edited by

M. Nieto-Vesperinas
Instituto de Optica
National Research Council (C.S.I.C.)
Madrid, Spain

J. C. Dainty
Applied Optics Section
Blackett Laboratory
Imperial College
London, U.K.

1990

NORTH-HOLLAND
AMSTERDAM · OXFORD · NEW YORK · TOKYO

ISBN: 0 444 88529 3

Published by:

North-Holland
Elsevier Science Publishers B.V.
P.O. Box 211
1000 AE Amsterdam
The Netherlands

Sole distributors for the U.S.A. and Canada:

Elsevier Science Publishing Company, Inc.
655 Avenue of the Americas
New York, N.Y. 10010
U.S.A.

Library of Congress Cataloging-in-Publication Data

Scattering in volumes and surfaces / edited by M. Nieto-Vesperinas, J.C. Dainty.
p. cm. -- (North-Holland delta series)
ISBN 0-444-88529-3
1. Surfaces (Physics) 2. Scattering (Physics) 3. Electromagnetic waves--Scattering. I. Nieto-Vesperinas, M., 1950- II. Dainty, J. C. III. Series.
QC173.4.S94S36 1990
530.4'17--dc20 89-28343
CIP

Printed in The Netherlands

PREFACE

In recent years there has been a renewed interest in the subject of electromagnetic scattering by volume media and by surfaces. There are several reasons for this, based on recent advances. First, the link between multiple scattering of light in dense media and the concept of localization, as encountered by electrons in random structures, has led to a re-examination of optical multiple scattering using concepts of solid state physics. Second, a number of measurements of light scattering from well characterized random rough surfaces has been reported; the light scattered from these surfaces also involves multiple scattering and, like in volume scattering, exhibits the phenomenon on enhanced backscattering. Also, the availability of faster computers with larger memories has allowed numerical Monte Carlo simulations both in volumes and surfaces. Finally, new results on non-linear interactions of light with surfaces and thin films have contributed, together with the aforementioned advances, to the general increase in understanding of electromagnetic scattering.

In response to this renewed interest, a workshop on Volume and Surface Scattering was held in Madrid, Spain in September 1988, organized by M. Nieto-Vesperinas, J.C. Dainty and R. Petit and sponsored by the European Research Office of the U.S. Army and the National Research Council of Spain (C.S.I.C.). Virtually all of the twenty-seven invited attendees gave informal presentations of their work over the three days of the workshop. As a result of this lively and rewarding meeting, we asked those who presented talks if they would like to contribute to a book on Volume and Surface Scattering. The nineteen articles that follow in this edited volume are therefore the outcome of activity of this workshop, although their content often goes beyond that presented informally a year ago.

The aim of this book is twofold. It provides a snapshot of the subject as perceived today by those involved in it and we hope, more importantly, that it will stimulate others to look again into this fascinating area.

We appreciate the help of Professor A.A. Maradudin paving the first stages of the publication of this volume and of Dr. M. Eligh of Elsevier who solved several aspects that arose in the course of editing this material.

Finally, we also wish to thank the authors for their time and care in preparing their works and the sponsors of the Madrid workshop for providing the means which gave rise to this book.

September 1989
The Editors

TABLE OF CONTENTS

Scattering in Volumes and Surfaces
M. Nieto-Vesperinas and J.C. Dainty (Editors)

EXPERIMENTAL AND THEORETICAL STUDIES ON ENHANCED BACKSCATTERING FROM SCATTERERS AND ROUGH SURFACES

Akira ISHIMARU

Department of Electrical Engineering, University of Washington, Seattle, Washington 98195, USA

This paper presents experimental and theoretical studies on the backscattering enhancement from discrete scatterers. The theory is based on the diffusion approximation and shows that the angular width of the peak is related to the transport length of the medium. The enhancement due to the rough surface scattering is divided into the case of large height variations and slopes and the case of small height variations with surface wave mode contributions.

1. INTRODUCTION

Backscattering enhancement phenomena have been observed for many years. It has been sometimes called the "retroreflectance" or the "opposition effect[1]." An example is "glory" which appears around the shadow of an airplane cast on a cloud underneath the airplane. The moon is brighter at full moon than at other times. Many materials such as $BaSO_4$ or $MgCO_3$ are known to cause the enhancement of scattered light in the backward direction. Some soils and vegetation are also observed to cause backscattering enhancement[2]. In spite of many observations, the several theories proposed to explain these phenomena are inadequate [1]. For example, the shadowing theory is based on the fact that in the scattering direction other than in the back direction, less light may be observed because of the shadowing. Also the scattering pattern of the surface or the scatterers such as Mie scattering may be often peaked in the back direction. The lens hypothesis is that the material may often act as corner reflectors or lenses resulting in peaked backscattering.

Recently, more quantitative experimental and theoretical studies of the enhancement have been reported. Watson[3] noted that the backscattered intensity is twice the multiple scattering and the first-order scattering. de Wolf[4] showed that the backscattered intensity from turbulence is proportional to the fourth-order moment and approximately twice the multiple scattered intensity. An excellent review is given by Kravtsov and Saichev[5]. Kuga and Ishimaru reported on an experiment which showed that the scattering from latex microspheres is enhanced in the backward direction with a sharp angular width of a fraction of a degree[6]. It was explained theoretically by Tsang and Ishimaru that the enhanced peak is caused by the constructive interference of two waves traversing through the same particles in opposite directions[7]. This idea is consistent with our earlier crude explanation[8]. We have also reported on the depolarization experiment and further theoretical studies[9-11]. Meanwhile, physicists have recognized that the transport of electrons in a strongly disordered material is governed by multiple scattering and that multiple scattering leads to "weak Anderson localization" caused by "coherent backscattering"[12,13]. It is then shown that both electron localization in disordered material and photon localization in disordered dielectrics are governed by coherent backscattering which is caused by the constructive interference of two waves traversing in opposite directions. Our experimental work in 1984 was followed by several independent optical experiments showing that the backscattering enhancement is a weak localization phenomenon[14-17]. The enhanced peak value is close to 2 and the angular width is governed by the diffusion length in the medium.

The experimental data on enhanced backscattering from rough surfaces was shown by O'Donnell and Mendez[18,19] and Gu et al.[20] and numerical studies were conducted by Nieto-Vesperinas et al.[21,22]. The theoretical explanation of the enhancement was presented by Maradudin et al.[23,24] and Bahar[25]. The backscattering enhancement is important in several applications. For example, lidar target calibration requires consideration of the enhancement[26]. Also recent studies show that the ultrasound backscattering from liquid-solid interfaces is enhanced due to the surface roughness and leaky Rayleigh waves[27,28]. This paper presents our recent studies on the backscattering enhancement from discrete scatterers and rough surfaces.

2. EXPERIMENTAL RESULTS ON BACKSCATTERING ENHANCEMENT FROM DISCRETE SCATTERERS

Experiments on backscattering enhancement are conducted using latex microspheres. The experimental setup is shown in Fig. 1. A He-Ne laser with an expanded beam diameter of 30 mm is used as the light source. The beam splitter (BS) reflects approximately 25 percent of the incident beam, and this reflected beam has fine interference patterns because of the surface coating of the BS. In order to eliminate the effect of the interference patterns, the incident beam diameter and the input aperture of the detector are adjusted to be much larger than the spacing of the interference patterns.

The transmittance of the BS varies at different angles if the beam is linearly polarized. However, if the incident beam is circularly polarized, the transmittance becomes almost constant within about 60° measured from the normal direction of the BS. Therefore, the phase of the incident beam is adjusted such that the beam at the spectrophotometer cell (SC) is circularly polarized. This ensures that the backscattered light, which may retain

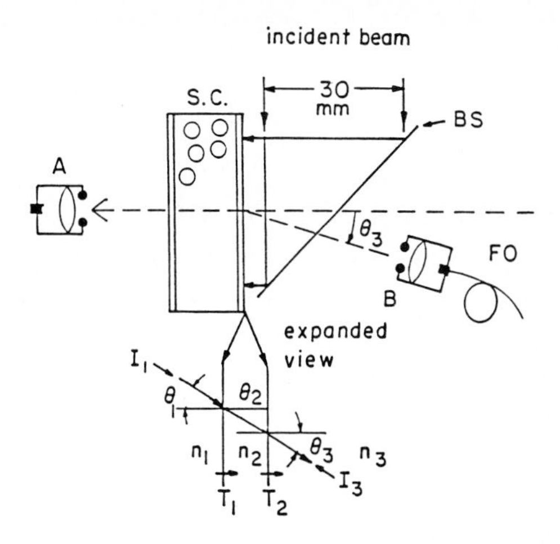

Fig. 1 Schematic diagram of the experimental apparatus (Ref. 6).

The detector, which consists of a 5-mm input aperture, a lens, and a 100-μm core optical fiber cable, has a field of view (FOV) of 0.195 °, and it is mounted on the computer-controlled rotational stage. The light output of the optical fiber cable is focused onto a low noise photodiode.

Uniform latex microspheres manufactured by Dow Chemical are used as scatterers. They are suspended in an SC with a 50-mm diameter. The path length of the SC is either 10 or 20 mm, depending on the density. In order to minimize the reflection from the SC glass wall, the SC is tilted about 5° from the vertical direction. The noise level from the water- filled SC is less than -75 dB at all angles.

At the beginning of the experiment, the SC is filled with deionized water, and then the detector is moved to position A in Fig. 1. This output is used as a reference, and each measurement is normalized by this value. Then the detector is moved to $\theta_3 = -10°$, and the backscattered intensity is measured from $\theta_3 = -10°$ to 50° at 0.1°, 0.2°, or 0.5° intervals, depending on the position of the detector. Because of the difference between the indices of refraction of water and of air, the angle θ_1 inside the SC is less than the scan angle θ_3. Therefore, the scan range from $\theta_3 = -10°$ to 50° corresponds to θ_1 from $-7.5°$ to 35°.

Since the incident beam diameter is larger than the input aperture of the detector, the area that the detector sees changes by $1/\cos(\theta_3)$ as the detector moves. Therefore, if a medium is a perfect diffuser and if the incident beam is uniform, the output of the detector should be constant. Except for the sharp peak around $\theta_3 = 0°$, the backscattered intensity from a dense solution of small particles is close to a constant within the angle of $-7.5°$ to 8°.

Experimental results are shown within the angle of $-7.5°$ to $8°$. Fig. 2 shows the experimental data for a particle size of 1.101 μm. The data are obtained for four different particle sizes ranging from an average diameter of 0.091 to 5.7 μm and at least seven different densities, up to 40 percent in some cases. In all cases, the peak of the reflectivity at 0° is observed at high densities, and we found the following results: (1) when the particle size is smaller than the wavelength ($\lambda_\omega = 0.475\mu m$ in water), the peak at 0° is small and appears at densities higher than a few percent; (2) when the particle size is 2-4 times greater than λ_ω, a sharp peak appears at 0°, and it becomes larger as the density increases; and (3) when the particle size is many times greater than λ_ω, the effect of the Mie scattering pattern becomes apparent even in a dense medium, and the sharp peak at 0° is superimposed upon the Mie scattering pattern.

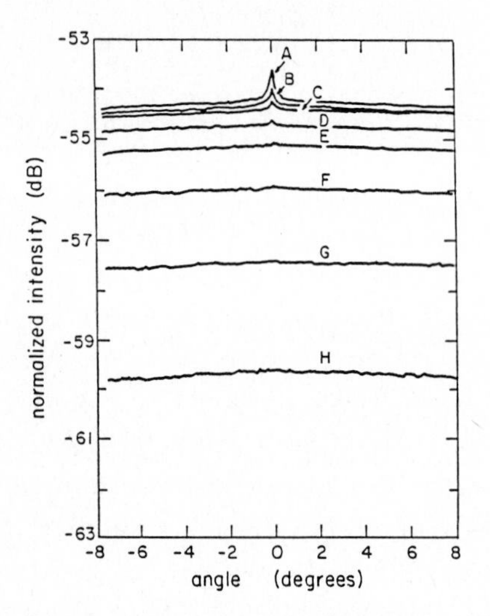

Fig. 2. Backscattered intensity versus angle. Particle size is 1.101 µm. A, experimental data for the volume density 9.55%; B, 4.78%; C, 2.39%; D, 1.19%, E, 0.597%; F, 0.299%; G, 0.149%; H, 0.075%.

3. THEORY OF BACKSCATTERING ENHANCEMENT BASED ON DIFFUSION THEORY

Let us consider a plane wave normally incident on a semi-infinite random medium as shown in Fig. 3. We start with the equation of transfer:

$$\frac{dI}{ds} = -\gamma_t I + \frac{\gamma_t}{4\pi} \int p(\hat{s}, \hat{s}')I(\hat{s}')\, d\omega', \tag{1}$$

where γ_t is the extinction coefficient and $p(\hat{s}, \hat{s}')$ is the phase function. The specific intensity I consists of the coherent intensity I_c and the diffuse intensity I_d

$$I = I_c + I_d, \tag{2}$$

$I_c = I_o \exp(-\gamma_t z)\delta(\hat{\omega} - \hat{\omega}_z)$.

The coherent intensity is a delta function pointed in the +z direction.

The diffuse intensity I_d is obtained from (1) by integration.

$$I_d = \int_o^\infty ds \exp[-\gamma_t s] \frac{\gamma_t}{4\pi} \int p(\hat{s}, \hat{s}') I(\hat{s}')\, d\omega'. \tag{3}$$

Now we substitute (2) into (3) and obtain

$$I_d = I_1 + I_{m\ell}. \tag{4}$$

The intensity I_1 is obtained by substituting I_c in the right side of (3) and represents the first-order scattering. The intensity $I_{m\ell}$ is caused by I_d in the right side of (3) and represents all the multiple scattering.

The first-order scattering specific intensity I_1 and the bistatic cross section σ_1^o per unit area of the random medium (bistatic scattering coefficient) is given by

$$\sigma_1^o = \frac{4\pi I_1}{I_o} = p(\theta, 0)\frac{1}{\mu + 1}, \quad \mu = \cos\theta, \tag{5}$$

where $p(\theta, 0)$ is the phase function in the direction of θ measured from the backward direction when the incident wave is in the direction $\hat{z}$. (See Fig. 3.)

Next consider the multiple scattering I_m. This consists of two terms $I_{m\ell}$ and I_{mc}. One corresponds to the wave multiply-scattered through many particles. This is also called the ladder term. We describe this wave by the diffusion approximation. We write the bistatic cross section per unit area of the random medium (bistatic scattering cross coefficient) $\sigma_{m\ell}^o$ as follows:

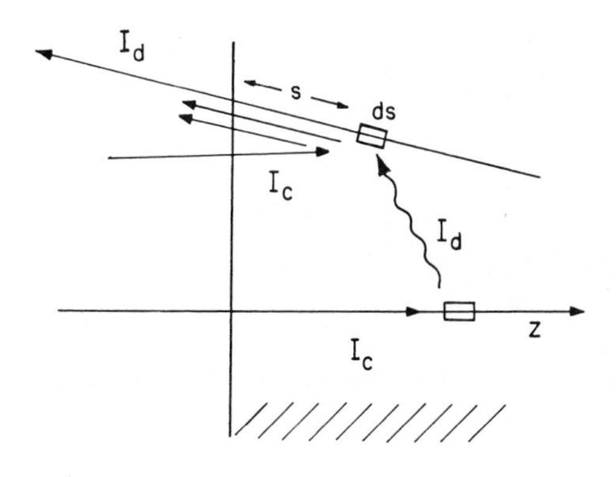

Fig. 3: First-order scattering and multiple scattering.

$$\sigma^o_{m\ell} = \int dV_2 F, \tag{6}$$

$$F = \rho\sigma_s \left[U - \frac{\bar{\mu}}{\rho\sigma_{tr}} \nabla U \cdot \hat{s}\right] e^{-\rho\sigma_t z/u},$$

where $\bar{\mu}$ is the mean cosine of the scattering angle, σ_{tr} is the transport cross section of a single particle, and $\hat{s}$ is the unit vector in the scattering direction.

$$\sigma_{tr} = \sigma_t(1 - W_o\bar{\mu}), \tag{7}$$

where $W_o = \sigma_s/\sigma_t$ is the albedo of a single particle.

The function U is given by[8]

$$U = \int_o^\infty dz' G(\rho, z; z') Q(z') + \frac{1}{2\pi h} G(\rho, z; o) Q_1(o), \tag{8}$$

$$Q(z') = 3\rho\sigma_s[\rho\sigma_{tr} + \rho\sigma_t\bar{\mu}] \frac{I_o}{4\pi} \exp(-\rho\sigma_t z'),$$

$$Q_1(o) = \frac{\sigma_s\bar{\mu}}{\sigma_{tr}} I_o, \quad h = \frac{2}{3\rho\sigma_{tr}}.$$

Green's function G is given by (Reference 8, page 184)

$$G = \frac{1}{4\pi} \int_o^\infty \frac{\lambda d\lambda}{\gamma} J_o(\lambda\rho) \left[e^{-\gamma|z-z'|} + \frac{h\gamma - 1}{h\gamma + 1} e^{-\gamma(z+z')}\right], \tag{9}$$

where $\gamma^2 = \lambda^2 + k_d^2$. Now we note that the integral in (6) is

$$\int dV_2 = 2\pi \int_o^\infty \rho d\rho \int_o^\infty dz. \tag{10}$$

Therefore, noting that

$$\int_o^\infty \rho d\rho J_o(\lambda\rho) = \frac{\delta(\lambda)}{\lambda}, \tag{11}$$

we can perform the integration in (6) by evaluating the integrand at $\lambda = 0$. The term $I_1 + I_{m\ell}$ is the diffusion approximation to the radiative transfer solution.

In addition to the radiative transfer term $I_{m\ell}$, we have the term corresponding to two waves traversing through the same particles in opposite directions (Fig. 4). This is also called the "cyclical term" or the "maximally crossed term." To express this term, we note that the difference between this term I_{mc} and the previous term $I_{m\ell}$ is that in this case, there is a phase difference given by

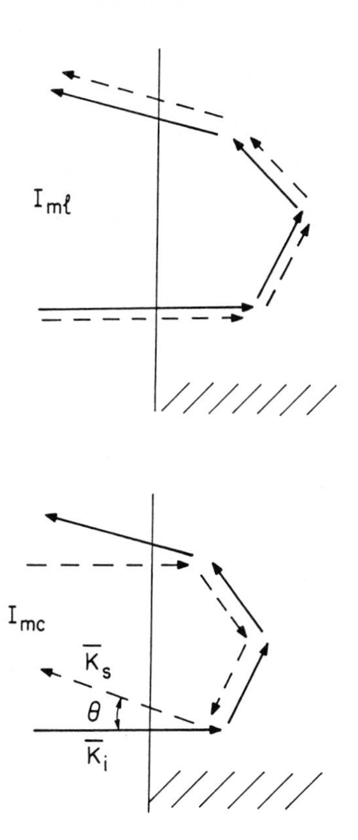

Fig. 4: Ladder term $I_{m\ell}$ and cyclic term I_{mc}.

$$\left[e^{i\bar{K}_i\cdot\bar{\rho}_1 - i\bar{K}_s\cdot\bar{\rho}_2}\right]\left[e^{i\bar{K}_i\cdot\bar{\rho}_2 - i\bar{K}_s\cdot\bar{\rho}_1}\right]^* = e^{i(\bar{K}_i+\bar{K}_s)\cdot(\bar{\rho}_1-\bar{\rho}_2)}. \tag{12}$$

Therefore, the bistatic scattering coefficient is given by

$$\sigma^o_{mc} = \int dV_2 e^{-i(\bar{K}_i+\bar{K}_s)\cdot(\bar{\rho})} F = (2\pi)\int_o^\infty \rho d\rho J_o(K_\rho)F, \tag{13}$$

where $K = |\bar{K}_i + \bar{K}_s| = k2\sin(\theta/2)$.

Now noting that

$$\int_o^\infty \rho d\rho J_o(K_\rho)J_o(\lambda\rho) = \frac{\delta(\lambda - K)}{\lambda}, \tag{14}$$

we can easily evaluate the integral (13) with the Fourier Bessel transform (9).

Noting that $K << \rho\sigma_t$, we get approximately

$$\sigma^o_{m\ell} = Af[\theta, \gamma = K_d] \cos\theta,$$

$$\sigma^o_{mc} = Af\left[\theta, \gamma = (K_d^2 + K^2)^{1/2}\right] \cos\theta, \tag{15}$$

where

$A = \frac{\sigma_s^2}{2\pi\sigma_t\sigma_{tr}}\left[\frac{\sigma_{tr}}{\sigma_t} + 2\bar{\mu}\right]$,

$f(\theta, \gamma) = \frac{\rho\sigma_{tr} - \bar{\mu}\gamma\cos\theta}{\rho\sigma_{tr} + (2\gamma)/3}$,

$K_d^2 = 3\rho\sigma_a\rho\sigma_{tr}$,

$K = k2\sin(\theta/2)$.

The second term in the numerator of $f(\theta, \gamma)$ represents the gradient term of the diffusion approximation and Equation (15) is valid when this second term is smaller than the first term. Equation (15) shows that σ^o_{mc} has a sharp peak in the backward direction $\theta = 0$. For lossless scatterers, $K_d = 0$ and σ^o_{mc} becomes a half of $\sigma^o_{m\ell}$ when the angle is approximately equal to

$$\theta = \frac{3\rho\sigma_{tr}\lambda}{4\pi(1 + 3\bar{\mu})}. \tag{16}$$

This angular width is consistent with our experimental data and it shows that the above diffusion theory correctly explains the backscattering enhancement phenomena.

The intensity near the back direction $\theta = 0$ is therefore given by (Fig. 5)

$$\sigma^\circ = \sigma_1^\circ + \sigma^\circ_{m\ell} + \sigma^\circ_{mc}. \tag{17}$$

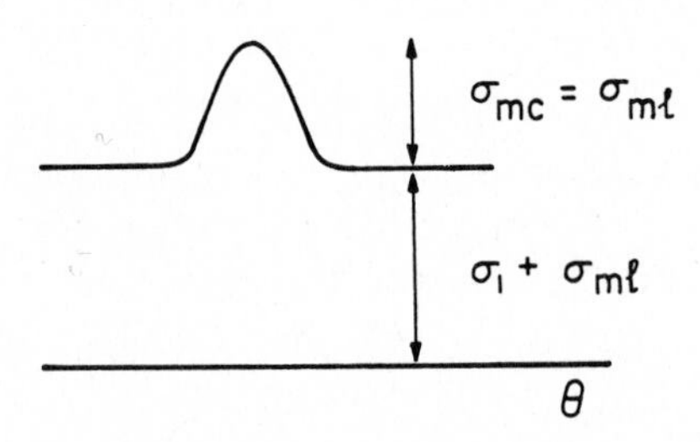

Fig. 5: Intensity near θ = 0 (back).

The enhancement factor is, therefore, close to 2 and is given by

$$\frac{\sigma_1^o + 2\sigma_{m\ell}^o}{\sigma_1^o + \sigma_{m\ell}^o} \tag{18}$$

where σ_1^o and $\sigma_{m\ell}^o$ are evaluated at $\theta = 0$. Note that σ_1^o and $\sigma_{m\ell}^o$ is the radiative transfer solution σ_{rt} and, therefore, the enhancement factor can also be expressed as

$$\frac{2\sigma_{rt} - \sigma_1^o}{\sigma_{rt}}, \quad \theta = 0.$$

The depolarization effects of the backscattering enhancement are discussed in Reference 9, and the effects of the pair correlation of the particles in a dense medium are shown in Reference 10. The approach based on the diagram method and Green's function is given in Reference 11.

4. ENHANCEMENT OF ROUGH SURFACE SCATTERING

The backscattering enhancement from rough surfaces has been observed experimentally and shown numerically[18–22]. These are two distinct enhancement phenomena. One type is for a surface with a large height and slope. The enhancement seems to occur in the range

$$\sigma \gtrsim 0.5\lambda,$$

$$slope = \sqrt{2}\,\frac{\sigma}{\ell} \gtrsim 0.5, \tag{19}$$

where σ^2 is the variance of the height variation and ℓ is the correlation distance. The surface and height correlation function is expressed by a gaussian function

$$< h_1 h_2 >= \sigma^2 \exp\left[-\frac{x^2}{\ell^2}\right]. \tag{20}$$

The slope is the root mean square slope. The enhanced peak is relatively broad and of the order of $5 \sim 20$ degrees.

In Fig. 6, the points marked with 0 represent the backscattering enhancement taken from References 18–24, and they are generally within the range indicated in (19). These points are shown approximately as they also depend on the angle of incidence and the boundary conditions. Also noted in Fig. 6 are the ranges of validity for the Kirchhoff approximation (KA), the phase perturbation (PP), and the field perturbation (FP). It should be observed that all these theories are not applicable to the range where the enhancement occurs, except the full wave theory by Bahar[25].

The other type is caused by the surface wave and, therefore, is normally associated with the rough interface between two media. They occur even when the height variations are small. In Fig. 6, these are shown with OS. For example, in optical scattering by metallic rough surfaces, surface wave modes can exist for the p-polarization when the real part of the dielectric constant of metal is negative. The surface waves cause constructive interference between two waves traversing in opposite directions (Fig. 7). Theoretical studies on this enhancement have been reported in References 23 and 24. An interesting case of backscattering enhancement of ultrasound due to leaky Rayleigh waves on the interface between a liquid and a solid has been reported recently[27,28].

In this section, we propose the diagram method to explain the physical mechanism of enhanced backscattering. Consider the one-dimensional rough surface with impedance boundary conditions (Fig. 8). The field ψ satisfies the wave equation

$$\left(\frac{\partial^2}{\partial x^2} + \frac{\partial^2}{\partial z^2} + k^2\right)\psi = 0. \tag{21}$$

At the surface $z = h(x)$, the field satisfies the impedance boundary condition

$$\frac{\partial \psi}{\partial n} + \alpha_o \psi = 0. \tag{22}$$

We assume that α_o is constant. Now let us convert (22) into the equivalent boundary conditions at $z = 0$. First note that

$$\frac{\partial}{\partial n} = \left[1 + (\frac{\partial h}{\partial x})^2\right]^{-1}\left[-\frac{\partial h}{\partial x}\frac{\partial}{\partial x} + \frac{\partial}{\partial z}\right]. \tag{23}$$

Also expand ψ about $z = 0$.

$$\psi = \sum_n (h^n/n!)\frac{\partial^n}{\partial z^n}\psi. \tag{24}$$

Substituting (23) and (24) into (22), we obtain the following boundary conditions at $z = 0$.

$$\frac{\partial}{\partial z}\psi + \alpha_o\psi + V\psi = 0, \tag{25}$$

where

$$V = \left(h\frac{\partial^2}{\partial z^2} - \frac{\partial h}{\partial x}\frac{\partial}{\partial x}\right) + \alpha_o h\frac{\partial}{\partial z} + \left[\frac{h^2}{2}\frac{\partial^3}{\partial z^3} - h\frac{\partial h}{\partial x}(\frac{\partial^2}{\partial x \partial z})\right] + \alpha_o\left[\frac{h^2}{2}\frac{\partial^2}{\partial z^2} + \frac{1}{2}(\frac{\partial h}{\partial x})^2\right] + \cdots.$$

Here h is considered a small quantity (ϵh) and shown above are the terms up to ϵ^2.

Now we let G be Green's function satisfying the boundary condition at $z = 0$.

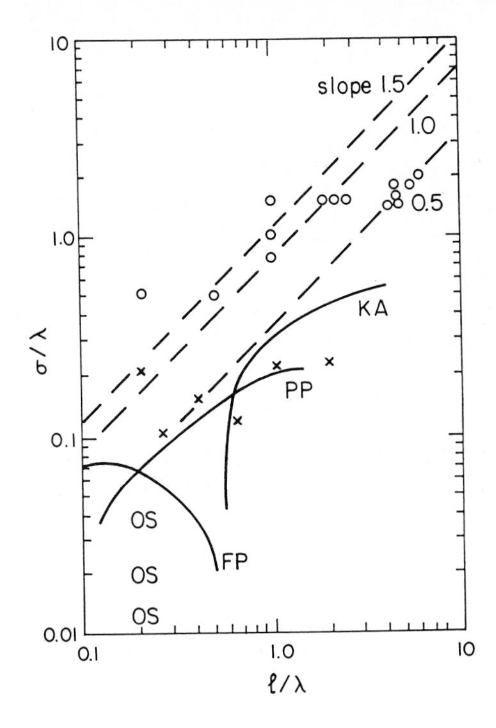

Fig. 6: Backscattering enhancement observed experimentally or found numerically at the points marked 0. Those marked OS are the enhancement due to surface wave modes. Those marked x are where the enhancement is non-existent or very small.

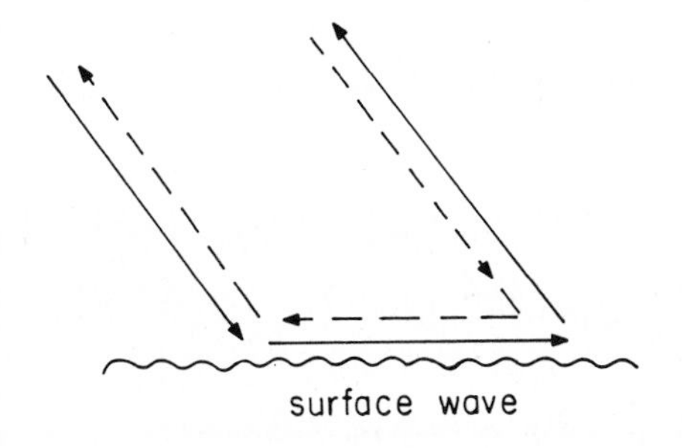

Fig. 7: Enhancement due to the surface wave.

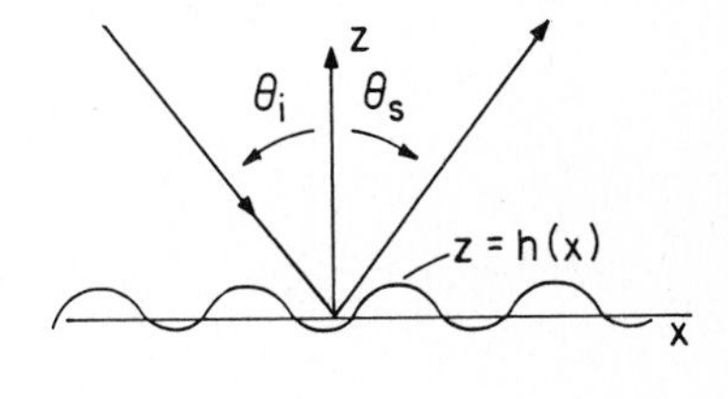

Fig. 8: One-dimensional rough surface.

$$\frac{\partial}{\partial z}G + \alpha_o G + VG = 0. \tag{26}$$

We make use of Green's function G_o which satisfies the boundary condition at $z = 0$

$$\frac{\partial}{\partial z}G_o + \alpha_o G_o = 0. \tag{27}$$

Then we apply Green's theorem to obtain

$$G(\bar{r}, \bar{r}_o) = G_o(\bar{r}, \bar{r}_o) + \int G_o(\bar{r}, \bar{r}_1)V(\bar{r}_1)G(\bar{r}_1, \bar{r}_o)\, dx_1. \tag{28}$$

Starting with (28), we can now obtain Dyson's equation for the average Green's function $< G >$.

$$< G(\bar{r}, \bar{r}_o) >= G_o(\bar{r}, \bar{r}_o) + \int G_o(\bar{r}, \bar{r}_2)M(\bar{r}_2, \bar{r}_1) < G(\bar{r}_1, \bar{r}_o) > dx_1 dx_2. \tag{29}$$

The mass operator M under the "bilocal approximation" is given by

$$M =< V(\bar{r}_2)V(\bar{r}_1) > G_o(\bar{r}_2, \bar{r}_1). \tag{30}$$

This is sometimes called the "Bourret approximation" or the "first-order smoothing." The nonlinear approximation is given by

$$M =< V(\bar{r}_2)V(\bar{r}_1) >< G(\bar{r}_2, \bar{r}_1) > . \tag{31}$$

In this paper, we make use of the "bilocal approximation." The first-order intensity is obtained from the Bethe-Salpeter equation by taking the first term of its iteration.

$$I_1 = \int\int < G(\bar{r}, \bar{r}_1) >< G^*(\bar{r}', \bar{r}_1') >< V(\bar{r}_1)V(\bar{r}_1') >< \psi_i(\bar{r}_1) >< \psi_i^*(\bar{r}_1') > dx_1 dx_1'. \tag{32}$$

The second-order intensity consists of the ladder term $I_{2\ell}$ and the cross (cyclical) term I_{2c}. The second-order ladder term is given by

$$\begin{aligned} I_{2\ell} = \int & < G(\bar{r}, \bar{r}_2) >< G^*(\bar{r}, \bar{r}_2') >< V(x_2)V(x_2') >< G(x_2, x_1) >< G^*(x_2', x_1') > \\ & < V(x_1)V(x_1') >< \psi_i(x_1) >< \psi_i^*(x_1') > dx_1\, dx_2\, dx_1'\, dx_2'. \end{aligned} \tag{33}$$

Similarly the second-order cyclical term is given by

$$I_{2c} = \int < G(\bar{r}, \bar{r}_2) >< G^*(\bar{r}, \bar{r}_1') >< V(x_2)V(x_2') >< G(x_2, x_1) >< G^*(x_1', x_2') >$$
$$< V(x_1)V(x_1') >< \psi_i(x_1) >< \psi_i^*(x_2') > dx_1\, dx_2\, dx_1'\, dx_2'. \qquad (34)$$

The physical meaning of these terms is shown in Figures 9, 10, and 11[29]. The above formulations are applicable to the case of small height variations and are similar to those discussed by Maradudin et al.[23,24]. The cyclical term (34) exhibits enhanced backscattering in the range $\ell \lesssim \lambda$ and $\sigma << \lambda$.

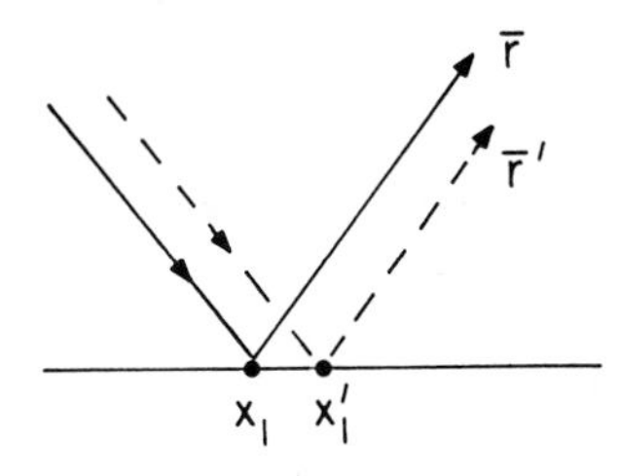

Fig. 9: First-order scattering.
(Conjugate wave is shown in dotted lines.)

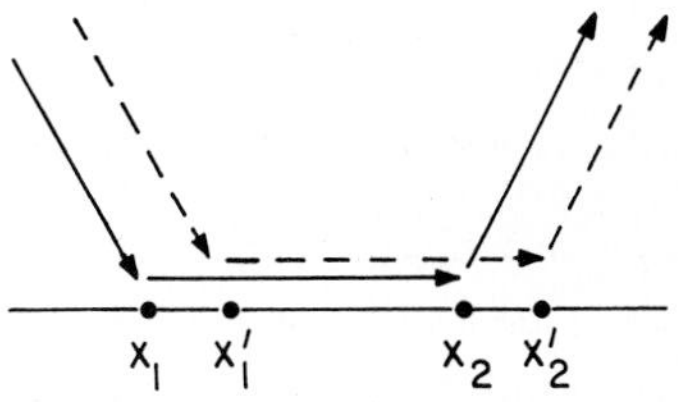

Fig. 10: Second-order ladder term.

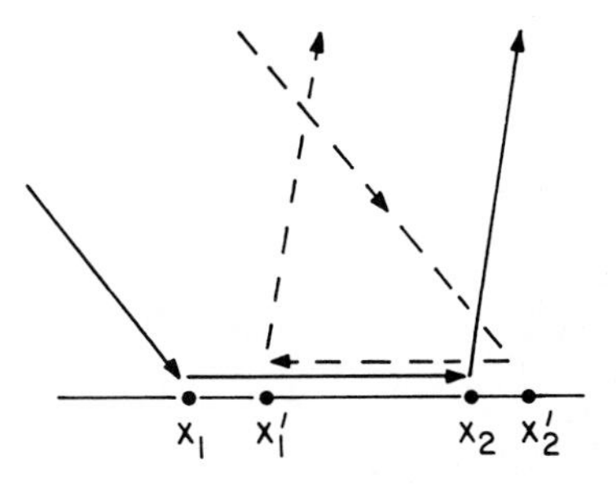

Fig. 11: Second-order cyclic term.

5. CONCLUSIONS

In this paper, we discussed the backscattering enhancement from discrete scatterers based on the diffusion approximation. The theoretical predictions are in agreement with the experimental data. There is, however, a need for more investigation on higher order moments, dense media, and polarizations. The enhancement due to rough surface scattering may be divided into the case of large height variations and slopes and the case of small height variations with surface wave mode contributions. The latter enhancement can be explained theoretically, but the former requires additional investigation.

ACKNOWLEDGEMENT

This work was supported by the U.S. Army Research Office, the National Science Foundation, the U.S. Army Engineer Waterways Experiment Station, and the Office of Naval Research.

REFERENCES

1) W. G. Egan and T. W. Hilgeman, Optical Properties of Inhomogeneous Materials, (Academic Press, New York, 1979).

2) F. Becker, P. Ramanantsizehena, and M.-P. Stoll, Appl. Opt. 24, (1985) 365.

3) K. Watson, J. Math. Phys. 10 (1969) 688.

4) D. de Wolf, IEEE Trans. Antennas Propag. AP-19 (1971) 254.

5) Y. A. Kravtsov and A. I. Saichev, Sov. Phys. Usp. (1982) 494.

6) Y. Kuga and A. Ishimaru, J. Opt. Soc. Am. A 1 (1984) 831.

7) L. Tsang and A. Ishimaru, J. Opt. Soc. Am. A 1 (1984) 836.

8) A. Ishimaru, Wave Propagation and Scattering in Random Media, (Academic Press, New York, 1978).

9) Y. Kuga, L. Tsang, and A. Ishimaru, J. Opt. Soc. Am. A 2 (1985) 616.

10) L. Tsang and A. Ishimaru, J. Opt. Soc. Am. A 2 (1985) 2187.

11) A. Ishimaru and L. Tsang, J. Opt. Soc. Am. A 5 (1988) 228.

12) D. E. Khmel'nitskii, Physica 126B (1985) 235.

13) P. W. Anderson, Philosophical B 52 (magazine) (1985) 505.

14) M. P. Van Albada and A. Lagendijk, Phys. Rev. Lett. 55 (1985) 2692.

15) P. E. Wolf and G. Maret, Phys. Rev. Lett. 55 (1985) 2696.

16) P. E. Wolf, et al., J. Phys. France 49 (1988) 63.

17) E. Akkermans, et al., J. Phys. France 49 (1988) 77.

18) E. R. Mendez and K. A. O'Donnell, Opt. Commun. 61 (1987) 91.

19) K. A. O'Donnell and E. R. Mendez, J. Opt. Soc. Am. A 4 (1987) 1194.

20) Z.-H. Gu, et al., Appl. Opt. 28 (1989) 537.

21) M. Nieto-Vesperinas and J. M. Soto-Crespo, Opt. Lett. 12 (1987) 979.

22) J. M. Soto-Crespo and M. Nieto-Vesperinas, J. Opt. Soc. Am. A 6 (1989) 367.

23) A. Maradudin, et al., Opt. Lett. 14 (1989) 151.

24) V. Celli, et al., J. Opt. Soc. Am. A 2 (1985) 2225.

25) E. Bahar and M. A. Fitzwater, J. Opt. Soc. Am. A 5 (1988) 89.

26) D. A. Haner and R. T. Menzies, Appl. Opt. 28 (1989) 857.

27) M. de Billy and L. Adler, J. Acoust. Soc. Am. 72 (1982) 1018.

28) P. B. Nagy and L. Adler, J. Acoust. Soc. Am. 85 (1989) 1355.

29) A. Ishimaru, "The diagram and smoothing methods for rough surface scattering," URSI International Symposium on Electromagnetic Theory, Stockholm, Sweden, August 14-17, 1989.

Scattering in Volumes and Surfaces
M. Nieto-Vesperinas and J.C. Dainty (Editors)

MULTIPLE SCATTERING IN DYNAMIC SYSTEMS

Isaac Freund

Department of Physics, Bar-Ilan University

Ramat-Gan, Israel

New, universal, multiple scattering correlation functions are derived for a dynamic system with strong interparticle interactions. This new theory is used to extract the particle dynamics in an interacting colloid from recent photon correlation data. We also consider the approach to the ensemble average in a dynamic multiple scattering medium, and derive an exceedingly slow rate of decrease of the optical contrast of integrated (blurred) speckle in such media.

1. INTRODUCTION

Over the past 25 years, quasi-elastic light scattering (QELS) has developed into an important tool in many areas of physics, chemistry, and biology[1,2]. An important restriction has always been that only single scattering systems could be studied, since there did not exist an appropriate theory which would permit quantitative analysis of light which was inelasticly scattered a large number of times. In practice, this meant that QELS was restricted to dilute, nearly transparent systems, so that a wide range of significant problems remained inaccessible. In recognition of the importance of multiply scattering systems, numerous attempts were made to solve the problem of dynamic multiple scattering, principally by considering successive low orders of scattering, such as double or triple scattering[4]. Unfortunately, such an approach did not lead to a definitive solution of the problem, since as is now understood, the intermediate regime of low orders of scattering is the most complicated one, and only in the limit of strong multiple scattering does a new level of simplicity finally emerge from an apparently extreme level of complexity.

As is often the case, the key to the solution of the problem of dynamic multiple scattering had to await developments in other areas, in this case the solution of the problem of electron transport in random media. Following the recognition that many of the insights developed in electron transport were also applicable to (classical) optical waves[4,5], the phenomenon of weak localization of light, manifest as a coherent backscattering peak, was demonstrated by Ishimaru and coworkers,[6] van Albada and Lagendijk[7], and Wolf and Maret[8]. Akkermans et. al.[9], and also Stephen and Cwillich,[10] subsequently presented a quantitative theory of the coherent backscattering which serves as a basis for much of the recent progress in the treatment of multiple scattering. In the initial experimental studies[6-8], the scattering system was a fluid suspension of microscopic polystyrene spheres in water. It was soon realized that in a solid, however, the coherent backscattered peak would be swamped by the large-scale intensity fluctuations known as speckle, but could be recovered by ensemble averaging. This was demonstrated by Etemad and co-workers[11] and by Kaveh et al.[12]. These latter authors also presented a quantitative theory both of the multiply-scattered speckle, and of the ensemble averaging which yields the coherent backscattering peak. Thus, it became apparent that the pioneering studies of the coherent backscattering[6-8] were successful because the required ensemble averaging was automatically provided by the rapid diffusive motion of the polystyrene spheres which comprised the fluid scattering system. This quite naturally led to a reconsideration of the problem of dynamic multiple scattering, and to its definitive solution using methods similar to those developed for the treatment of the ensemble averaging of the coherent backscattering peak[12]. This solution was given by Maret and Wolf[13] and by Rosenbluh et al.[14]. These latter workers also presented a detailed quantitative comparison between their theoretical expressions and their experimental results, finding good agreement between the two. Further important results were provided by Stephen[15], by Edrei and Kaveh[16], by Pine et. al.[17], by Freund et. al.[18], by Rimberg and Westervelt[19], by Edrei and Freund[20], and by Qu and Dainty[21].

In all of the forgoing, a system of non-interacting scatterers was assumed. However, in the dense systems for which multiple scattering is of importance, interparticle interactions are almost always present, so that a theory which includes these is required. The problem here is not the construction of a formal theory involving many unknowns, but rather the development of suitable approximations which yield simple, useful expressions that can be employed in the interpretation of experimental data. We develop such a theory here, and obtain very simple expressions which involve only the relative displacements of the particles. Comparison of these expressions with photon correlation data permits the time evolution of the relative particle displacements to be obtained directly, and thus yields a simple, fundamental characterization of the particle dynamics in dense, interacting, multiple-scattering media.

2. MULTIPLE SCATTERING AND INTENSITY FLUCTUATIONS

Following our earlier treatment of optical fluctuations in multiply-scattering media[12], we treat the multiple scattering via a model in which incident photons are considered to be successively scattered from one particle to another, so that they undergo a quasi-random walk and thus traverse a diffusive-like path before ultimately emerging from the medium with an amplitude and phase which is dependent on the total optical path length L. There are, of course, many different possible paths through the medium, each of which yields its own amplitude and phase, and so the net optical field at a given point on the output face of the sample is obtained by summing over all paths which terminate at that point. As the positions of the particles change, the optical path lengths (phases) of the individual paths also change, so that the net optical field at the output face fluctuates with the motions of the particles. The corresponding intensity fluctuations, which are the quantity of primary interest here, may be measured in one of two equivalent ways. Imaging optics can be used to collect the light emitted at a single point on the output face of the sample, thereby yielding an experiment which corresponds directly to the theoretical calculations[17,19]. Alternatively, and more conveniently from an experimental point of view, one may measure the fluctuating intensity at some point in a far-field (or free-space) speckle pattern[13,14,18,19]. In this case, each speckle spot receives light emitted by all exit points on the sample. Far above the optical Anderson transition[4,5], it is usually assumed that there are no important long-range correlations between the optical fields at sample points which are separated by distances larger than the optical mean free path[22-25]. Under these circumstances, the statistical properties of the intensity fluctuations at a single point on the output face of the sample are identical to those at a point in the far-field (or free-space) speckle pattern, so that no distinction need be made between the two methods of measurement. This is the situation assumed here.

In single scattering experiments, the form factor of individual scatterers, and the structure factor of correlated scatterers, can be measured directly. In multiple scattering, however, where the random walk of the photon results in an averaging over all scattering angles, such detailed information is lost, and the only scatterer property which survives is a simple average quantity, the transport mean free path l^*. This is defined[26] in terms of the extinction depth τ as $l^* = m\tau$, where m, the average number of scattering events required to randomize the initial photon direction, is given by

$$1/m = \langle 1 - \cos\theta \rangle, \qquad (1)$$

with θ the angle of scattering for a single scattering event. In performing the average in Eq. (1), both the single-scatterer form factor and the structure factor of the medium are used as weighting factors. Although the structure factor is often unknown, this presents no special problem since l^* is easily measured experimentally. Thus, for a static system, there are no essential differences between multiple scattering by randomly positioned uncorrelated scatterers as opposed to correlated scatterers, since the effects of correlations in the scatterer positions are automatically absorbed into the definition of the transport mean free path l^*, the only microscopic parameter to survive the extensive averaging inherent in multiple scattering. We find here that a very similar situation exists as regards a dynamic system, and that the inter-particle interactions can be absorbed into a single average parameter which describes the motion of a particle relative to its neighbors.

3. MULTIPLE SCATTERING BY MOVING SCATTERERS

As the particles defining a given path move in time, the optical length L of the path changes by some amount ΔL, and the resulting random phase fluctuations $k\Delta L$ yield rapid intensity fluctuations in the scattered light. The characteristic time scale of these fluctuations may be measured by means of the half width of the (normalized) intensity-intensity correlation function (cross correlation coefficients) C(t), defined as

$$C(t) = [\langle I(0)I(t)\rangle - \langle I\rangle^2]/[\langle I^2\rangle - \langle I\rangle^2], \qquad (2)$$

where <> denotes an ensemble average. Assuming that different photon paths have statistically independent random phases, and calculating the optical field as a weighted sum over all possible paths, we have

$$C(t) = |\sum_a w_a \langle \exp[ik\Delta L_a(t)]\rangle|^2. \quad (3)$$

Here $k = 2\pi/\lambda$ is the wavevector of the light, the weight of a path is $w_a = \langle I_a\rangle/\sum_a\langle I_a\rangle$, where I_a is the intensity contributed by photon path a, and in accordance with our model of statistically independent paths, $\langle I^2\rangle = 2\langle I\rangle^2$. We now assume that the range of the interparticle interactions is much shorter than the photon transport mean free path l^*. In typical systems l^* is 50 - 200 times the interparticle spacing, so that a volume of order $(l^*)^3$ contains 10^5-10^7 particles. Our basic restriction is thus that the number of particles whose motions are strongly correlated is very much less than this, so that flow, large-scale turbulence, etc. are excluded. Under the above circumstance, ΔL is composed of a large number of independent contributions, and may be considered to be a zero-mean, random, gaussian process. Accordingly, we immediately obtain the important result that

$$\langle \exp[ik\Delta L]\rangle = \exp[-k^2\langle(\Delta L)^2\rangle/2]. \quad (4)$$

Our problem thus reduces to the geometrical one of calculating $\langle(\Delta L)^2\rangle$ in terms of the displacements of the individual particles comprising the path.

4. PATH GEOMETRY AND THE CORRELATION FUNCTIONS

We describe the geometry of the system using a set of vectors **r** which describe the particle displacements, and a set of unit vectors **B** which describe the path of the multiply-scattered photon. These vectors are shown in Fig. 1, where the circles correspond to the scatterers, the lines joining them (the bonds) to the photon path, the angle θ is the change in photon direction upon scattering, and the subscripts label successive scattering events along the path. The bond lengths, themselves, or mean spacing between scattering events, determines the extinction depth, but is otherwise not a parameter directly relevant to our problem. We note that $\mathbf{B}_{i,i-1} = -\mathbf{B}_{i-1,i}$, and that $|\mathbf{B}| = 1$.

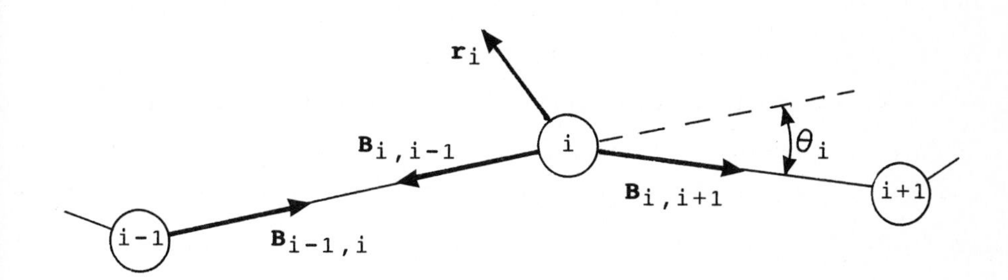

FIGURE 1

We now define two new sets of vectors,

$$\mathbf{b}_i = \mathbf{B}_{i,i+1} + \mathbf{B}_{i,i-1} = \mathbf{B}_{i,i+1} - \mathbf{B}_{i-1,i}, \tag{5a}$$

where

$$b_i = |\mathbf{b}_i| = 2\sin(\theta_i/2), \tag{5b}$$

and

$$\mathbf{d}_i = (6)^{-1/2}[(\mathbf{r}_i - \mathbf{r}_{i+1}) + (\mathbf{r}_i - \mathbf{r}_{i-1})]. \tag{6}$$

The vectors $\mathbf{b}$ contain the information about the geometry of the path which we will require, while the vectors $\mathbf{d}$ describe the motion of a particle relative to its neighbors along the specified path. The normalization of $\mathbf{d}$ is chosen such that if the scatterer motions are uncorrelated, then $\langle|\mathbf{r}|^2\rangle = \langle|\mathbf{d}|^2\rangle$. The change ΔL in the length of the path due to the displacements $\mathbf{r}$ can be written in one of two equivalent ways:

$$\Delta L = \sum_i \mathbf{r}_i \cdot \mathbf{b}_i \,, \tag{7a}$$

or

$$\Delta L = \sum_i (\mathbf{r}_i - \mathbf{r}_{i+1}) \cdot \mathbf{B}_{i,i+1}. \tag{7b}$$

Each of these two forms displays explicitly only one of the two important aspects of the problem. Eq. (7a) explicitly shows the dependence upon the geometry of the path, since if all $\theta = 0$, then to first order in the particle displacements (the only order required here) $\Delta L = 0$. On the other hand, the effects of correlations in the motions of neighboring particles is not immediately apparent from this form. Such correlations are made explicit in Eq. (7b), which gives ΔL as a sum over changes in the lengths of the bonds connecting neighboring particles, so that if two neighbors have exactly the same displacement **r**, then the connecting bond does not change its length, and it therefore makes no contribution to ΔL.

We now seek a form for ΔL which explicitly displays both aspects of the problem, so we attempt to write ΔL in terms of **d** and **b**, rather than **r** and **b**. In doing this, we make the approximation that the correlations which are of primary importance in our problem are correlations in the particle displacements. Such correlations are contained in the magnitude of **d**. There are also angular correlations which depend upon the direction of **d** relative to the path. These we neglect on the basis that in the presence of strong multiple scattering, there are many different, essentially independent, paths passing through any pair of particles, and an effective angular averaging takes place which washes out all angular correlations. This is not strictly true, of course, since the angular correlations derive from particular local arrangements of the particles which give rise to different weights for different paths. This effect, however, is clearly a secondary one. Accordingly, we attempt to retain correlations which directly arise from the relative displacements of the particles, but neglect correlations which arise indirectly from the positions of the particles. We thus adopt the philosophy that with the magnitude of **d** as our guarantor that the major correlations are accounted for, a somewhat liberal treatment of angular terms, which includes uncorrelated averaging, becomes permissable.

Returning to Eq. (7a), and assuming only that the path contains a sufficient number of scatterers so that end-point effects can be neglected, we obtain without approximation

$$3\Delta L = 6^{1/2}\sum_i \mathbf{d}_i \cdot \mathbf{b}_i + \sum_i \mathbf{r}_i \cdot (\mathbf{B}_{i+1,i+2} - \mathbf{B}_{i-2,i-1}). \quad (8)$$

The vector $\mathbf{B}_{i+1,i+2} - \mathbf{B}_{i-2,i-1} = \mathbf{b}_i(3)$ describes the change in the photon direction after <u>three</u> successive scattering events starting at particle (i-2) and ending at particle (i+2), i.e. --> (i-1) -> (i) -> (i+1) -->. $\mathbf{b}(3)$ has the magnitude $b(3) = 2\sin[\theta(3)/2]$, where $\theta(3)$ is the change in angle due to the three successive scattering events. The value of $<|\mathbf{b}(3)|^2> = 4<\sin^2[\theta(3)/2]>$ we approximate by assuming that the number of scattering events within a single transport mean free path l^* is relatively large, and thus that the average angle θ for a single scattering event is relatively small. This approximation is well satisfied when the particles have a diameter of order the wavelength of light, and improves as the particles become more correlated in position. Under these circumstances, we can write

$$<\sin^2[\theta(3)/2]> = 3<\sin^2[\theta/2]>, \quad (9)$$

since the multiply scattered photon executes a random walk in θ-space.

Returning now to Eq. (8), we obtain for $<(\Delta L)^2>$,

$$9<(\Delta L)^2> = 6<(\sum_i \mathbf{d}_i \cdot \mathbf{b}_i)^2> + 3<(\sum_i \mathbf{r}_i \cdot \mathbf{b}_i)^2>, \quad (10)$$

where in accordance with our philosophy of uncorrelated angular averages, we have dropped both off-diagonal and cross-terms, and where we have also used Eq. (9). The second sum in Eq. (10) will be recognized from Eq. (7a) to be simply ΔL. Writing N_s for the number of scattering events in the path, neglecting again off-diagonal terms, and performing an uncorrelated angular average, we obtain

$$<(\Delta L)^2> = (4/3)N_s<|\mathbf{d}|^2><\sin^2(\theta/2)>. \quad (11)$$

Using the elementary trigonometric identity $2\sin^2(\theta/2) = 1 - \cos(\theta)$ together with Eq. (1) for the average number of scattering events in a transport mean free path l^*, and defining N the number of transport mean free paths l^* in the total path L by

$$N = L/l^* = N_s/m, \quad (12)$$

we obtain

$$<(\Delta L)^2> = (2/3)N<|\mathbf{d}|^2>. \quad (13)$$

We note that for the two limits in which $<(\Delta L)^2>$ is definitely known, Eq. (13) gives exactly the right result. When all the scatterers are locked together so that they move as a single rigid body, then **d**, and therefore $<(\Delta L)^2>$ are zero. This is clearly correct, since under such a circumstance the scattered intensity is independent of time, and $C(t) = 1$ for all times. At the other extreme, when the scatterers are totally uncorrelated, Eq. (7a) yields without any approximations, $<(\Delta L)^2> = (2/3)N<|\mathbf{r}|^2>$, in full accord with Eq. (13), since $<|\mathbf{d}|^2> = <|\mathbf{r}|^2>$ for uncorrelated scatterers. We are thus led to expect that since Eq. (13) is exactly correct for no correlation and for total correlation, it will also provide a reasonable approximation for intermediate levels of correlation.

It is convenient to define an effective average displacement per particle

$$r_{eff}^2 = <|\mathbf{d}|^2>, \quad (14)$$

which permits a simple, transparent interpretation of data in terms of the effective motion of a single particle. When, for example, the particle is trapped in a cage formed by its neighbors, and cannot move relative to the cage walls, r_{eff} is zero, while the fact that the cage as a whole, including its occupant, might be moving relative to the laboratory is of no interest when we are concerned with the microscopic dynamics of the particles.

Returning to Eq. (3), we reorder the sum by collecting together all the terms corresponding to all the different paths which have a given value of N, and replace the sum over individual paths by a weighted sum over N. This yields our central result

$$C(r_{eff}) = |\sum_N W_N \exp[-(N/3)k^2 r_{eff}^2]|^2 , \qquad (15)$$

where the W_N are normalized probabilities for finding a path with the given value of N. We emphasize that Eq. (15) makes no statement about the particle dynamics. This latter quantity enters when we specify how r_{eff} evolves with time. We also emphasize that Eq. (15) and the results to be derived from it are very general, and describe how the correlation function for the intensity fluctuations depends upon the scatterer displacements without regard to how these displacements come about. Finally, we note in passing that when photon absorption is important, then $(k^2 r_{eff}^2)$ in Eq. (15) and in the results to be derived from it, are replaced by $(k^2 r_{eff}^2 + 3l^*/l_{abs})$, where l_{abs} is the photon absorption length.

The W_N depend upon dimensionality and sample geometry, and are usually computed from a diffusion equation upon assuming that N is large. Recently, appropriate solutions of this equation within various contexts have been discussed by a number of authors[14-20]. With the aid of these results, we give $C(r_{eff})$ explicitly for three important configurations:

(i) the classic point source in an infinite medium

$$C(r_{eff}) = \exp[-2(S/l^*)kr_{eff}], \qquad (16a)$$

where S is the source-receiver separation;

(ii) diffuse transmission through a slab of thickness S

$$C(r_{eff}) = \{[S/(1+\Delta)l^*]\sinh[(1+\Delta)kr_{eff}]/\sinh[(S/l^*)kr_{eff}]\}^2, \qquad (16b)$$

where the Milne equation value $\Delta = 0.71$ determines the boundary conditions for the photon diffusion equation[26];

(iii) diffuse reflection from an infinite half-space

$$C(r_{eff}) = \exp[-2(1+\Delta)kr_{eff}]. \qquad (16c)$$

Our treatment thus far has been for a scalar wave, i.e. in common with all previous treatments[13-20] we have neglected the vector nature of the optical field. When this is included, the weights W_N in Eq. (15) are dependent on the states of polarization of the input and output fields. For a linearly polarized input field, for example, short paths (small N) are more probable if the output has the same polarization as the input (copolarization), while long paths (large N) are favored if the output polarization is orthogonal to the input one (cross-polarization). The effects of this are that the correlation function for copolarized light is broader than the one calculated for a scalar wave, while the correlation function for the crosspolarized case is narrower. Accordingly, we may view the scalar theory as yielding the average of these two, thereby corresponding very nearly to the case in which the output light is not polarization selected. We note in passing that theoretical results for a vector field for a system of point scatterers, the only model for which reasonably simple results could be expected, will not be useful in most instances, since the depolarization predicted by this model deviates markedly from that observed in most real systems.

In noninteracting systems where r_{eff} is the actual root-mean-square particle displacement, and where the particle dynamics are known , C(t) is obtained from Eq. (16) by specifying how r_{eff} evolves with time. When this is done, Eq. (16), as well as Eq. (15), then become identical to the already known results for noninteracting systems[13-20].

For a Maxwell-Boltzman velocity distribution, $r_{eff}(t) = (3k_BT/M)^{1/2}t$, where k_B is the Boltzman constant, T the temperature, and M the particle mass. Such a distribution usually corresponds to very short time scales which are much less than the mean time between collisions, or the time over which the particle's initial motion is damped out. For noninteracting particles undergoing diffusive Brownian motion with particle diffusion constant D, $r_{eff}(t) = (6Dt)^{1/2}$, leading to a stretched exponential in the important case of diffuse reflection[17] [Eq. (16c)]. In this latter instance, the half width of C(t) is predicted to be narrower than the typical (i.e. 90°) half width for single scattering by a universal factor of $6(1+\Delta)^2/\ln(2) = 25$, in full accord with the experimental results of Rosenbluh et. al.[14] who found a 25-fold acceleration in the rate at which the multiply-scattered light fluctuated in time, as compared with the rate of single-scattering fluctuations for the same system.

Because of the universality and simplicity of Eq. (16c), and the convenience and general applicability of measuring the diffuse reflectance, this configuration will likely prove to be the most important one experimentally. In this instance, when studying strongly interacting systems for which the particle dynamics are not known, r_{eff} may be obtained simply and directly from photon correlation measurements of C(t) by means of

$$r_{eff}(t) = -\ln[C(t)]/2k(1+\Delta). \qquad (17)$$

5. COMPARISON WITH EXPERIMENT

We now use these new theoretical results to extract r_{eff} from the diffuse reflectance photon correlation data of Pine et. al[17] for a multiply-scattering, interacting colloid. These authors studied a polydisperse system comprised principally of 0.5 μm charged polystyrene spheres in water. Data for this strongly interacting system (colloidal glass), and for the same system with the interactions screened out by a high electrolyte concentration, were presented (their Fig. 3) but not analyzed quantitatively, presumably due to lack of a theory which includes interparticle interactions. In Fig. 2, we present r_{eff} as obtained from these data for the interacting colloidal glass, curve (a), and for the noninteracting fluid system, curve (b). In curve (c), we present the results of a calculation for monodisperse, noninteracting 0.5 μm spheres using the free particle diffusion constant. The good agreement between this calculation and the measurements, curve (b), indicates that the effects of polydispersity are not severe, and that the r_{eff} values obtained here reflect principally the motion of the 0.5 μm spheres. We note that finite size effects are minimized by restricting the analysis to t > 50 μsec.

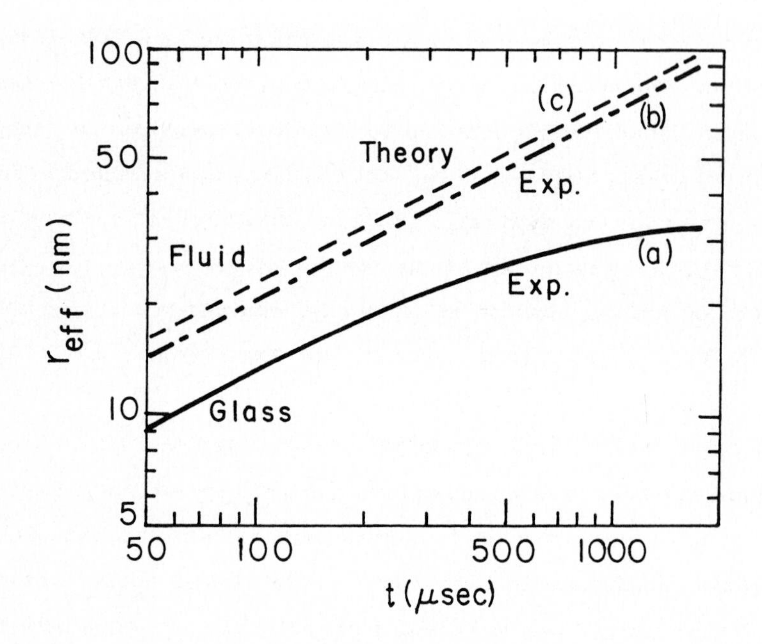

FIGURE 2

Curve (a) for the colloidal glass shows that up to about 20 nm (300 μsec), r_{eff} grows as $t^{1/2}$, indicating that in this region the interacting particles undergo normal diffusive motion, albeit with a reduced diffusion constant. At longer times, a fairly rapid crossover to a very much slower rate of increase occurs, and within the scale of the experiment r_{eff} appears to saturate at about 35 nm. All of this is in striking agreement with the expected behavior as discussed by Pusey[27]. Each particle may be considered as contained in a cage formed from its neighbors, and since the particles all carry the same charge and the interactions are repulsive, the central particle is constrained to diffuse within this cage. Thus, r_{eff} is expected to grow initially with time as $t^{1/2}$ until the particle nears the cage walls, where r_{eff} is of order the cage radius, which is determined by the particle charges and the solvent screening. Since the free diffusion of the particle has now ended, a crossover behavior is expected to occur at longer times, and r_{eff} is expected to saturate at a value approximating the cage radius. This is precisely what is observed, and our analysis of the data suggests an effective cage radius of some 35 nm, and an effective particle diffusion constant which is reduced to about 40% of the free particle value by hydrodynamic and other interactions. The rearrangement of the walls of the cage, which requires the correlated motion of many particles, occurs very much more slowly, and the long time behavior of r_{eff} reflects the time scale of this slow rearrangement. Unfortunately, the current data do not extend to sufficiently long time scales to permit the extraction of quantitative information about this slow cage rearrangement, but we anticipate that the new analysis given here will encourage further experimentation on a variety of systems.

6. APPROACH TO THE ENSEMBLE AVERAGE

As indicated in the Introduction, an important part of the initial motivation for studying how multiple scattering affects the time dependence of the intensity fluctuations arose out of the realization that it was these fluctuations which automatically gave rise to the ensemble averaging[11,12] required to observe the coherent back-scattering peak[6-8]. Accordingly, we now turn to a detailed consideration of how this averaging proceeds in time. We start by recalling that the large-amplitude optical fluctuations in the backscattered speckle pattern which mask the peak have an optical contrast of unity[28]. As ensemble averaging proceeds over a time interval T, the optical contrast is reduced, the speckle spots smear out, and the presence of the underlying coherent peak becomes apparent.

What is the time scale over which the contrast falls from unity to say 1/e? An intuitively reasonable, almost obviously true, estimate would be that this time scale must be the same as the time scale over which the correlation function C(t) decays to say 1/e, since C(t) measures precisely the mean time over which a speckle spot retains its integrity. In stark contrast to this estimate, Rosenbluh and coworkers[14] found experimentally that even when the averaging time T was more than an order of magnitude greater than the half-width of C(t), the optical contrast was still very high. Only when T became as large as 100 times the width of C(t), was the contrast finally reduced appreciably. This striking, counterintuitive, experimental result is the central subject of this section.

The optical contrast $\mathcal{C}$ is defined[28] by

$$\mathcal{C}^2 = \langle I^2\rangle/\langle I\rangle^2 - 1. \tag{18}$$

If the speckle pattern is integrated over a time T, then I in Eq. (18) is everywhere $I = \int_0^T I(t)dt$. This integrated intensity is what is measured experimentally at each point in the speckle pattern, and so this is the intensity inserted into Eq. (18) when computing the experimental contrast. Computing $<I^2>$ as an iterated integral, using the fact that $C(t)=C(-t)$, appropriately translating the limits of integration, and recalling that $<I^2>=2<I>^2$, we obtain for the optical contrast of multiply-scattered integrated (i.e. blurred) speckle

$$\mathcal{C}^2(T) = (2/T)\int_0^T C(t)dt - (2/T^2)\int_0^T C(t)tdt. \qquad (19)$$

It is easily verified that Eq. (19) gives the correct limiting values for the contrast for very short averaging times (a contrast of unity), and for very long averaging times (a contrast of zero). In Fig. 3, we plot the contrast calculated from Eq. (19) for the important case of diffuse reflection as a function of $(T/t_1)^{1/2}$, where t_1 is the angle-averaged single-scattering time[21], and is given in terms of the particle diffusion coefficient D and the wavevector of the light k as $t_1 = 1/4k^2D$. Since the integrals are elementary, we do not give the resulting cumbersome expressions.

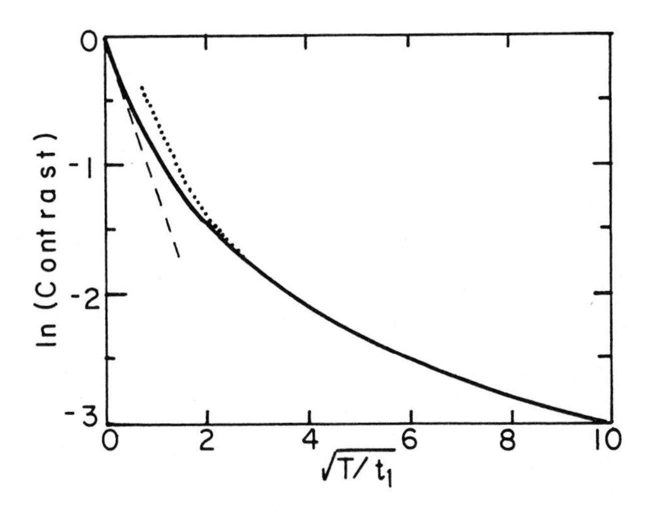

FIGURE 3

Also shown in Fig. 3 as the dashed curve is the short time limiting form

$$\mathcal{C}(T) = \exp[-(4/15)(1+\Delta)(6T/t_1)^{1/2}], \qquad (20a)$$

and as the dotted curve, the long time limiting form

$$\mathcal{C}(T) = 2^{1/2}[(1+\Delta)(3T/t_1)^{1/2}]^{-1}. \qquad (20b)$$

We note that just like the correlation function for diffuse reflection, the contrast for diffuse reflection is also a universal function of T/t_1.

The calculated contrast, in complete agreement with the experimental results of Rosenbluh et. al.[14], is 80% at the half-width of C(t) (i.e. $T = t_1/25$), and still exceeds 50% for averaging times which are ten times this half width. As indicated by Eq. (20b), for long times the contrast falls exceedingly slowly, as only $T^{-1/2}$, and thus does not reach 1%, for example, until T is 50,000 times the half width of C(t)! This very slow falloff of the optical contrast arises from the very slow rate of loss of correlation by the short multiple-scattering photon paths, in marked contrast to the fast initial drop of C(t) which arises from the very rapid loss of correlation by the long photon paths.

7. SUMMARY

We have presented new, universal, multiple-scattering correlation functions which include most of the important effects of interparticle interactions and correlations. We believe that while it will prove possible to relax many of the (sometimes sweeping) approximations we have made, and therefore to provide a firmer basis for our theoretical expressions, the essential forms of these expressions are likely to be substantially preserved. We have used these new theoretical results to analyze recent photon correlation data on an interacting colloidal glass, and we have succeeded in extracting from these data significant information about the time-dependent motions of the colloidal particles. The results of this analysis have been found to be in striking agreement with theoretical predictions, thereby holding out the expectation that our theoretical expressions will prove useful in the study of a wide variety of dynamic multiple scattering systems, systems which were hitherto unaccessible because of the lack of a suitable theory. We have also derived theoretical expressions for the optical contrast of integrated, or blurred, multiple scattering speckle patterns, and have found these to be in full agreement with recent experimental results. We conclude by noting that the enormous progress made over the past few years in our understanding of all aspects of multiple scattering has substantially removed the traditional impediments which have hindered the application of optical methods to the study of highly random media.

ACKNOWLEDGEMENT

I am pleased to acknowledge important discussions with my colleagues Profs. N. Wiser, M. Kaveh, and M. Rosenbluh. This work has been supported by the U.S.-Israel Binational Foundation (BSF), Jerusalem, and by the Israel Academy of Sciences.

REFERENCES

1) B. J. Berne and R. Pecora, Dynamic Light Scattering With Applications to Chemistry, Biology and Physics (Wiley, New York, 1976).

2) H. Z. Cummins and E. R. Pike eds. Photon Correlation Spectroscopy and Velocimetry (Plenum, New York, 1977).

3) P. C. Colby, L. M. Narducci, V. Bluemel, and J. Baer, Phys. Rev. A 12 (1975) 1530; F. C. van Rijswijk and U. L. Smith, Physica (Amsterdam) 83A (1976) 121; C. M. Sorensen, R. C. Mockler, and W. J. O'Sullivan, Phys. Rev. A 14 (1976) 1520; A. Boe and O. Lohne, Phys. Rev. A 17 (1978) 2023; F. Gruner and W. Lehman, J. Phys. A 13 (1980) 2155.

4) S. John, Phys. Rev. Lett. 53 (1984) 2169.

5) P. W. Anderson, Philos. Mag. B 52 (1985) 505.

6) Y. Kuga and A. Ishimaru, J. Opt. Soc. Am. A 1 (1984) 831; L. Tsang and A. Ishimaru, J. Opt. Soc. Am. A 2 (1985) 2187.

7) M. P. van Albada and A. Lagendijk, Phys. Rev. Lett. 55 (1985) 2692.

8) P. E. Wolf and G. Maret, Phys. Rev. Lett. 55 (1985) 2696.

9) E. Akkermans, P. E. Wolf, and R. Maynard, Phys. Rev. Lett. 56 (1986) 1471.

10) M. J. Stephen and G. Cwillich, Phys. Rev. B 34 (1986) 7564.

11) S. Etemad, R. Thompson, and M. J. Andrejco, Phys. Rev. Lett. 57 (1986) 575.

12) M. Kaveh, M. Rosenbluh, I. Edrei, and I. Freund, Phys. Rev. Lett. 57 (1986) 2049.

13) G. Maret and P. E. Wolf, Z. Phys. B 65 (1987) 409.

14) M. Rosenbluh, M. Hoshen, I. Freund, and M. Kaveh, Phys. Rev. Lett. 58 1987) 2754; M. Kaveh, M. Rosenbluh, and I. Freund Nature (London) 326 (1987) 778.

15) M. J. Stephen, Phys. Rev. B 37 (1988) 1; M. J. Stephen Phys. Lett. A 127 (1988) 371.

16) I. Edrei and M. Kaveh, J. Phys. C: Solid State Phys. 21 (1988) L971; I. Edrei and M. Kaveh, Phys. Rev. B 38 (1988) RC950.

17) D. J. Pine, D. A. Weitz, P. M. Chaiken, and E. Herbolzheimer, Phys. Rev. Lett. 60 (1988) 1134.

18) I. Freund, M. Kaveh, and M. Rosenbluh, Phys. Rev. Lett. 60 (1988) 1130.

19) A. J. Rimberg and R. M. Westervelt, Phys. Rev. B 38 (1988) RC5073.

20) I. Edrei and I. Freund, Phys. Rev. B 39 (1989) 9660.

21) D. N. Qu and J. C. Dainty, Optics Lett. 13 (1988) 1066.

22) B. Shapiro, Phys. Rev. Lett. 56 (1986) 1809.

23) M. J. Stephen and G. Cwillich, Phys. Rev. Lett. 59 (1987) 285.

24) S. Feng, C. Kane, P. A. Lee, and A. D. Stone, Phys. Rev. Lett. 61 (1988) 834.

25) P. A. Mello, E. Akkermans, and B. Shapiro, Phys. Rev. Lett. 61 (1988) 459.

26) P. M. Morse and H. Feschbach, Methods of Theoretical Physics (McGraw-Hill, New York, 1953).

27) P. N. Pusey and W. van Megan, Phys. Rev. Lett. 59 (1987) 2083; P. N. Pusey, J. Phys. A: Math. Gen. 8 (1975) 1433; J. C. Brown, P. N. Pusey, J. W. Goodman, and R. H. Ottewill, J. Phys. A: Math. Gen. 8 (1975) 825.

28) J. W. Goodman, Statistical Properties of Laser Speckle Patterns, in: Laser Speckle and Related Phenomena, ed. J. C. Dainty (Springer-Verlag, New York, 1974).

Scattering in Volumes and Surfaces
M. Nieto-Vesperinas and J.C. Dainty (Editors)

DYNAMICS OF BROWNIAN PARTICLES FROM STRONGLY MULTIPLE LIGHT SCATTERING

Pierre Etienne WOLF * and Georg MARET +

* Centre de Recherches sur les Très Basses Températures, CNRS
+ Hochfeld Magnetlabor, Max Planck Institut für Festkörperforschung
166x, *F*-38042 Grenoble-Cedex, France

The time autocorrelation function of the light intensity multiply back-scattered from optically dense colloidal suspensions is found to decay as $-t^{1/2}$ at small correlation times t, in agreement with transport theory. The analogy of this dynamic behaviour to the shape of the static coherent backscattering cone and to the absorption dependence of the total scattered intensity is outlined and quantitatively verified. A method is described to determine the particle's diffusion constant from a combined measurement of these effects, or by simple comparison with a calibrated sample.

1. INTRODUCTION

Quasielastic light scattering is widely used for measuring dynamic properties such as diffusion constants in colloidal suspensions. In its standard form[1] mostly used for particle sizing, this technique is restricted to (dilute) suspensions where only single scattering occurs. Even a small amount of double or triple scattering makes the interpretation of data quite intricate[2]. This technique thus appears inappropriate for studying optically dense suspensions. However, it has been realized recently[3-8] that also in the regime of *very* multiple scattering, where the transport theory of light applies, useful information may be obtained from the decay of the time autocorrelation function of the light intensity. In a former article[3] we have shown that this decay is strongly nonexponential at short times revealing the very fast decay of coherence in extended scattering paths containing many scatterers. We have suggested a quantitative interpretation of our results in terms of a relation between the field autocorrelation function $G_1(t)$ and the Green's function describing the diffusive transport of the intensity. Qualitatively similar results have been discussed by Rosenbluh *et. al.*[4]. Subsequently Stephen[5] has made analytic predictions for $G_1(t)$ for the backscattering and transmission geometries. Corresponding combined measurements have been recently carried out by Pine *et. al.*[6] showing convincingly that "*diffusing light spectroscopy*[6] " is a powerful technique.

In this article we emphasize that in the strong multiple scattering limit the self diffusion constant D can be reliably obtained from the *short time* decay of $G_1(t)$. This is because the short time decay is controlled by contributions of the large paths to the intensity fluctuations for which the transport theory of light applies. We corroborate this idea by measurements on aqueous suspensions of polystyrene microspheres.

2. THEORY

2.1 *Dynamic correlation function $G_1(t)$*

Let us first consider independent scatterers undergoing brownian motion. We assume that the average distance l between successive scattering events along a scattering path is much larger than the optical wavelength λ (far field approximation) and that there are no interferences on average between the different random paths of the diffusing light. Then the average autocorrelation function of the optical field E (*one* polarisation) along paths with n scattering events is simply given by

$$< E_n(0)\, E_n^*(t) > = I(n) < \prod_i^n \exp^{-i\mathbf{q}_i \delta \mathbf{r}_i(t)} >$$

$\mathbf{q}_i$ is the scattering vector at the ith scatterer and $\delta\mathbf{r}_i(t)$ its displacement. $I(n)$ denotes the average total intensity scattered per solid angle through the nth order path and < > represents the average over the particle motion and over the configurations of paths. The $\mathbf{q}_i\delta\mathbf{r}_i(t)$ are independent variables with a gaussian distribution of variance Dq_i^2t. Therefore

$$< E_n(0)\, E_n^*(t) > = I(n) < \prod_i^n \exp^{-Dq_i^2t} >$$

Let us assume that the successive q_i's along a path are independent which may be considered correct for large paths where the condition $\Sigma\mathbf{q}_i=\mathbf{k}_s-\mathbf{k}_0$ is not restrictive ($\mathbf{k}_s$ and $\mathbf{k}_0$ are the incident and emerging wavevector, respectively). Then

$$< E_n(0)\, E_n^*(t) > = I(n) < \exp^{-Dq^2t} >^n_q$$

$< >_q$ is an average over q weighted by the form factor $F(q)$ of the individual scattering events. The total field correlation function is obtained by summing over paths with different n,

$$< E(0)\, E^*(t) > = \sum_n I(n)\, \langle e^{-Dq^2t}\rangle_q^n \tag{1}$$

Because of the assumption of independent q_i's, the contributions of large paths, but not of short paths, are properly described in the sum in eq.(1). For times smaller than the single backscattering relaxation time $\tau = 1/D(2n_0k_0)^2$, where n_0 is the refractive index of the scattering medium, we retain only the first cumulant. Then, to leading order in $D\langle q^2\rangle t$, the *un*normalized correlation function $G_1(t)$ becomes

$$G_1(t) = < E(0)\, E^*(t) > \simeq \sum_n I(n)\, e^{-D\langle q^2\rangle nt} \tag{2}$$

Finally we note[3] that $\langle q^2\rangle = (2n_0k_0)^2\, l/(2l^*)$, where l and l^* are the scattering and transport mean free paths, respectively. Introducing the average length of paths of n scatterers, $L=nl$, we have

$$G_1(t) = \sum_L I(L)\, e^{-t/\tau\, L/2l^*} \tag{3}$$

which is the expression used in our former work[3] and by Pine *et. al.*[6]. In this first order cumulant expansion the correlations associated to the longer paths decay faster.

In the following we consider the particular case of backscattering. For this geometry, in simple transport theory, $I(L)$ is only a function of the reduced path length L/l^*, but not of $1/l^*$. Then $G_1(t)$ is a sole function of t/τ independent of the other characteristics of the scatterers such as l or l^*. This is the physical basis for the possibility of extracting information about *single* particle dynamics.

In principle eq.(2) is true only for time scales much shorter than τ. For longer times Stephen[5] has pointed out that, in the case of point-like particles, $\langle e^{-Dq^2t}\rangle = \int e^{-Dq^2t} q dq / \int q dq \propto 1/t$. Hence $G_1(t)$ should decay as $1/t$ rather than exponentially. Experiments in the long time domain[3,6], however, show that $G_1(t)$ decays much faster than $1/t$. This is probably due to the fact that for $t \geq \tau$ the contributions of small paths become so large that eq.(1) itself is no longer valid. In addition it appears experimentally[3] that for large scatterers (of diameter $d \simeq \lambda$) the scaling law $G_1(t,\tau,l^*) = G_1(t/\tau)$ remains approximately valid up to $t \simeq 2 - 3\,\tau$. This could originate from two facts: (*i*) The contribution of low order scattering ($l \leq L << l^*$) is small for these large scatterers[9], so that $I(L)$ in eq.(3) is properly described by the transport theory. (*ii*) For well behaved form factors $F(q)$ the first cumulant expansion $\langle e^{-Dq^2t}\rangle \simeq e^{-D\langle q^2\rangle t}$ is valid up to $t \simeq 1/D\langle q^2\rangle \simeq \tau\, 2l^*/l >> \tau$ rather than up to $\tau = 1/D(2n_0k_0)^2$. A detailed analytical prediction for $G_1(t)$ in the long time regime seems a difficult task. It appears therefore a priori more reliable to use only the short time ($t << \tau$) behavior of $G_1(t)$ in the data analysis.

We shall show that a direct quantitative estimate of D can be obtained from the very short time decay of $G_1(t)$ with no assumption other than the validity of the transport theory at large distance scales and, again, independent scatterers. As mentionned above, for $t << \tau$

$$G_1(t) = \sum_L I(L/l^*)\, e^{-t/\tau\, L/2l^*} \tag{4}$$

For the large paths which contribute to the short time decay the light is fully depolarized[9] and we may use for $I(L/l^*)$ the prediction for a scalar wave corresponding to diffusive transport of the light intensity. The functional form of $I(L/l^*)$ then depends on the boundary conditions for the diffusion equation, i.e. that the diffusive intensity vanishes on a plane at a distance $z_0 = 2l^*/3$ from the interface[10].

2.2. *Relation of $G_1(t)$ to static coherent backscattering*

We can compare eq.(4) to the angular variation of the static intensity scattered per solid angle near backscattering or, more precisely to the coherent albedo $\alpha_c(Q)$ which arises from constructive interference between time reversed scattering paths[9].

$$\alpha_c(Q) = \sum_L I(L/l^*)\, e^{-L/l^*\,(Ql^*)^2/3} \tag{5}$$

$\mathbf{Q} = \mathbf{k_s} + \boldsymbol{k_0}$ is related to the scattering angle off backscattering by $Q = 2\pi\Theta/\lambda$ for $\Theta << \pi$. Here Θ and λ are both measured outside the sample.

Expression (4) can be further compared to the variation of the incoherently scattered intensity $\alpha_i(l_a)$ with the concentration of an absorbing dye. i.e. in the presence of a finite absorption mean free path l_a [9]. Per solid angle

$$\alpha_i(l_a) = \sum_L I(L/l^*)\, e^{-L/l^*\, l^*/l_a} \tag{6}$$

Thus we have a direct correspondence between eq.(4), eq.(5) and eq.(6) provided the identification $t/(2\tau) \longleftrightarrow (Ql^*)^2/3 \longleftrightarrow l^*/l_a$ is made. This correspondence holds for any given direction of observation, in transmission or reflection experiments. The sample geometry only enters via the functional form of $I(L/l^*)$. Note that since $I(L/l^*)$ is given by the classical transport theory eq.(4) and eq.(5) are only valid outside the coherent backscattering cone[9]. Inside the cone $I(L/l^*)$ should be replaced by $I(L/l^*)\,(1+e^{[-L/l^*\,(Ql^*)^2/3]}$. This accounts for the constructive interference between time reversed paths[3,9].

In the near backscattering direction and for plane wave illumination $I(L)$ decays as $1/L^{3/2}$. This results in a triangular singularity of $\alpha_c(Q)$ near $Q = 0$ [9]

$$\alpha_c(Q) = \alpha_c(0) - \beta\, Ql^* + ...$$

and therefore also implies

$$G_1(t) = G_1(0) - \beta\,\sqrt{3t/2\tau} + ... \tag{7}$$

$$\alpha_i(l_a) = \alpha_i(\infty) - \beta\,\sqrt{3l^*/l_a} + ...$$

where β is given by the transport theory. These equations show that the triangular singularity of the coherent albedo[9] is translated into a square root singularity of the time autocorrelation function $G_1(t)$ or into a square root singularity of the albedo under absorption. The above equations should be exact since only large paths contribute to the singular part of the decay. Consequently β should not depend on the parameters (l, l^*, τ) of the sample. However, as $\alpha_c(Q)$, $\alpha_i(l_a)$ and $G_1(t)$ are unnormalized quantities, β depends on the incident flux F_0 and on the characteristics of the detection. In the context of the discussion[9] of $\alpha_c(Q)$ we have found that $\beta = 3F_0/8\pi\,(1 + z_0/l^*)^2$ in the diffusion approximation[10] independent of l/l^*. It should be noted that the calculation of $\alpha_c(Q)$ by Tsang and Ishimaru[11] using a modified version of the diffusion approximation results in $\beta = 3F_0/8\pi\,(l/l^* + z_0/l^*)^2$. In the latter case β depends on the size dependent anisotropic scattering of the particles, i.e. on l/l^*.

One usually measures *normalized* quantities, i.e. $G_1(t)/G_1(0)$, $\alpha_i(l_a)/\alpha_i(\infty)$, $\alpha_c(\Theta)/\alpha_c(0)$. Since the total average intensity equals $G_1(0) = \alpha_i(\infty) = \alpha_c(0)$ when single scattering is negligible, we find

$$\frac{G_1(t)}{G_1(0)} = 1 - \delta\,\sqrt{3t/2\tau} \tag{8}$$

$$\frac{\alpha_i(l_a)}{\alpha_i(\infty)} = 1 - \delta\,\sqrt{3l^*/l_a} \tag{9}$$

$$\frac{\alpha_c(\Theta)}{\alpha_c(0)} = 1 - \delta\,Ql^* \tag{10}$$

where the coefficient $\delta = \beta/G_1(0)$ is, for a given system, the same in eq.(8)-(10). Because, for a given flux F_0, $G_1(0)$ depends on the contribution of paths of any size, it is *not* given by the diffusion theory. Hence, unlike β, δ depends on the sample studied and generally cannot be precisely evaluated. This is illustrated by the different theoretical evaluations of δ in eq.(10), which range from 1.5 to 2.5 depending on the low L behaviour of $I(L/l^*)$ [12]. Thus, measurements of $G_1(t)$ alone may give τ to within a factor of, say, 4 only. This may be sufficient if only an order of magnitude estimate of the particle size is desired. More precise values of τ can be obtained by simultaneous measurements of $\alpha_c(Q)$ and $\alpha_i(l_a)$. Combination of the latter quantities gives both δ and l^*. (Only one of these additional measurements is necessary, of course, if l^* is known). From the initial slope $dG_1/dt^{1/2}$ the characteristic time τ can then be deduced.

2.3 *Interparticle Correlations*

Up to now we have considered the case of independent scatterers. However, in the dense media where multiple scattering usually cannot be avoided the scatterers are most probably spatially correlated. We therefore sketch the corresponding modifications to eq.(8) in the following approximate picture. The static effect of correlations between scatterers is accounted for by the single scattering structure factor $S(q)$ of the suspension. For short range correlations with correlation length much shorter than l, the current approach to the problem of time correlations is then to divide the suspension into small uncorrelated regions with locally defined $S(q)$, which act as new elementary scatterers. The time autocorrelation function of the field single-scattered from such a region is, for $t \ll \tau$ [13]

$$< E(0)\,E^*(t) > \;\propto\; S(q) - Dq^2t \simeq S(q)\,e^{-Dq^2t/S(q)} \tag{11}$$

The physical meaning[13] of eq.(11) is that for small times ($t \ll \tau$) the absolute decay of correlations is the same than for noninteracting particles because the scatterers do essentially not yet feel each other. Only the initial static value at t=0, i.e. $S(q)$, is affected resulting in a faster decay in the normalized autocorrelation function. The term $e^{-D\langle q^2\rangle nt}$ in eq.(2) must then be replaced by $e^{-D\langle q^2/S(q)\rangle nt}$ where the average is now weighted by $S(q)F(q)$. We further express n as $n = L/l' = L/l^{*'}\ l^{*'}/l'$ where l' and $l^{*'}$ are the scattering and transport mean free paths, respectively, as modified by interactions.

$$l^{*'}/l' = \int_0^{2n_0k_0} S(q)F(q)qdq \Big/ \int \frac{q^2}{2n_0^2k_0^2} S(q)F(q)qdq$$

We finally obtain

$$G_1(t) = \sum I(L/l^{*'}) \exp\left[-\frac{L}{l^{*'}} \frac{t}{2\tau} \frac{\langle q^2/S(q)\rangle}{2n_0^2k_0^2} \frac{l^{*'}}{l'} \right]$$

where we make explicit that the relevant quantity in the transport theory is $L/l^{*'}$. We therefore have in the short time domain ($t \ll \tau$)

$$\frac{G_1(t)}{G_1(0)} = 1 - \delta\sqrt{3t/2\tau'}\ . \tag{12}$$

The square root cusp in $G_1(t)$ remains and the effect of correlations is entirely contained in the new characteristic time τ'

$$\tau'/\tau = 2n_0^2k_0^2 \frac{l'/l^{*'}}{\langle q^2/S(q)\rangle} = \int_0^{2n_0k_0} q^2S(q)F(q)qdq \Big/ \int q^2F(q)qdq \tag{13}$$

which is equal to $l^*/l^{*'}$. This means that the correlations should modify τ and $(l^*)^{-1}$ in exactly the same way. Eq.(13) allows to estimate the corrections of the short time decay of $G_1(t)$ due to correlations if $S(q)$ is known. At later correlation times the situation becomes more complex not only because of the reasons outlined above but also since in interacting systems the self diffusion constant becomes itself a function of time.

3. EXPERIMENTAL

In order to test the method outlined above we have carried out measurements on monodisperse aqueous suspensions of polystyrene spheres. We briefly recall our experimental set-up [3]. The vertically polarized beam of a mono-mode Ar^+ laser (λ = 514.5 nm) incident from a direction a few degrees off the normal of the sample cell has a waist (FWHM) of about 5 *mm*. The intensity scattered at Θ = 171° from a 125μm diameter area at the sample is imaged onto the photomultiplier tube using a f-number of 0.01. A second polarizer is used to select the vertical (VV) or horizontal (VH) component of the scattered intensity. From the photomultiplier output $I_0(t)$ the normalized intensity autocorrelation function $g_2(t) = \langle I_0(0)I_0(t)\rangle/\langle I_0(0)\rangle^2$ is determined using a digital multibit correlator (ALV 3000) with up to 1024 channels. For the observed[9] gaussian distribution of scattered fields, $g_2(t)$ is related to the normalized field autocorrelation function $g_1(t)$ by [1]

$$g_2(t) - 1 = |\ g_1(t)\ |^2$$

Because the area of observation cannot be infinitesimaly small compared to the coherence area[1], $g_2(0)$

and $g_1(0)$ are experimentally always somewhat smaller than 2, and 1, respectively. The quantitative comparison of $g_1(t)$ with the theory is then made through $g_1(t)/g_1(0) \equiv G_1(t)/G_1(0)$.

We have used aqueous suspensions of d = 0.102, 0.180, 0.305, 0.460, 0.797μm diameter polystyrene spheres and solid fractions ρ of 10% as purchased from *Sigma* or *Polysciences*. The spheres were gently centrifugated and sonicated before the experiment in order to eliminate the few large aggregates which tend to form after some time. The sample volume was 1cm * 1cm * 1cm. We have previously calculated[9] $S(q)$ and the corresponding ratio $l^*/l^{*\prime}$ ($= \tau'/\tau$) of these suspensions for pure hard core interactions. Even at ρ = 10% the expected corrections to τ are small except for the smallest beads used (for which we found $l^*/l^{*\prime} \simeq 0.66$). Data obtained at ρ=3% using d=0.46μm agree, within the experimental accuracy, with data at ρ=10% corroborating the weakness of correlation effects under our conditions. We have therefore directly plotted the measured quantity $g_1(t)$ as a function of $(t/2\tau)^{1/2}$ - where $\tau=1/D(2n_0k_0)^2$ is the separately measured[3] (heterodyne) decay time for very dilute suspensions of the same spheres in the backscattering direction at the same temperature (20°C). The obtained values of τ are in 1% agreement with those calculated from the Stokes-Einstein formula. Actually measurements were taken at temperatures between 18°C and 23°C, and all time scales were corrected to 20°C according to the temperature dependence of the viscosity of water.

4. RESULTS AND DISCUSSION

4.1 *Short time behavior of $g_1(t)$*

Fig.1 shows $g_1(t)$ plotted *vs.* t and $t^{1/2}$ for a stock suspension of d=0.46μm beads at ρ=10%. Obviously the decay for $t \leq 10\mu sec$ (or $t/\tau \leq 0.01$ with τ = 1.03 msec) follows a $t^{1/2}$ law. We emphasize the sharpness of this $t^{1/2}$ singularity: After 0.25 μsec (at the second reliable measurement channel, $\simeq \tau/4000$) $g_1(t)$ has decreased by 0.024 corresponding to an average decay rate about *100 times* larger than in the single scattered regime also shown for comparison.

4.2. *Comparison between $g_1(t)$, $\alpha_c(Q)$ and $\alpha_i(l_a)$*

Since $g_1(t)$ varies like $t^{1/2}$ at small times we can estimate τ from the short time slope of $g_1(t)/g_1(0)$ by the use of eq.(8). From a fit of a straight line through the data in Fig.1 up to $t^{1/2}$ = 0.003 we obtain $1/g_1(0)\ \ dg_1(t)/dt^{1/2}$ = -54 sec$^{-1/2}$ at T=20°C. The parameter δ has been previously measured[9] from both the absorption dependence of the incoherently scattered intensity $\alpha_i(l_a)$ and from the small angle slope of the coherent backscattering cone $\alpha_c(Q)$. We obtained δ = 1.35 ± 0.15 from both measurements. Inserting these values into eq.(8) we obtain τ = 0.94 ± 0.17 msec in fair agreement with the free particle value τ = 1.03 msec. This agreement and the quantitative analogy between $\alpha_i(l_a)/\alpha_i(0)$, $\alpha_c(Q)/\alpha_c(0)$ and $g_1(t)/g_1(0)$ is directly illustrated in Fig.2 where we plot the three sets of normalized data against the variables $(l^*/l_a)^{1/2}$, $Ql^*/3^{1/2}$ and $(t/2\tau)^{1/2}$, respectively, scaled as suggested by eq.(8,9,10). Within experimental error these different physical quantities do indeed vary in a very similar way corroborating the combined validity of eq.(8,9,10).

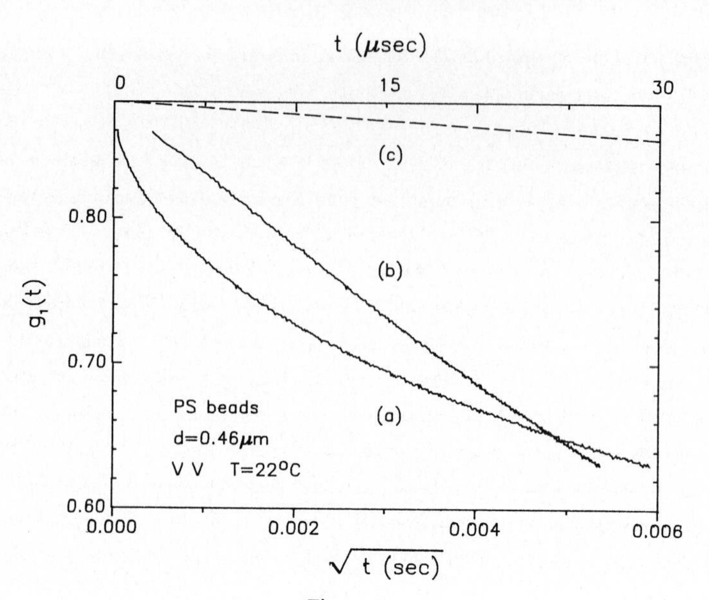

Figure 1
The short time autocorrelation function $g_1(t)$ in the multiple scattering regime (ρ=10%) plotted *vs.* t (*a*) and *vs.* $t^{1/2}$ (*b*) in comparison to single scattering ($\rho=10^{-5}$), curve (*c*).

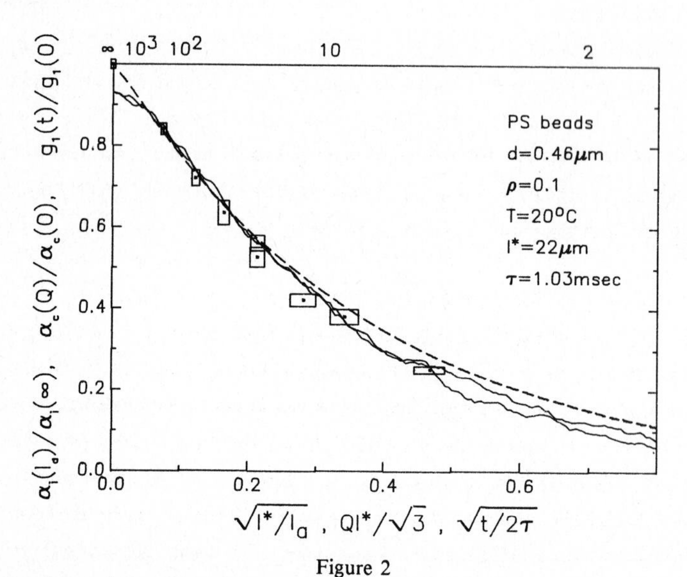

Figure 2
Comparison of the dynamic correlation function $g_1(t)/g_1(0)$ *vs.* $(t/2\tau)^{1/2}$, dashed line, with the absorption dependence of the static intensity $\alpha_i(l_a)/\alpha_i(\infty)$ *vs.* $(l^*/l_a)^{1/2}$, rectangles, and the angular dependence of both wings of the coherent backscattering cone $\alpha_c(Q)/\alpha_c(0)$ *vs.* $Ql^*/3^{1/2}$, continuous lines. The size (in units of l^*) of the typical largest contributing loop is indicated in the upper abscissa.

The quantitative agreement actually extends well beyond the linear regime, which must be viewed as a direct test of eqs.(4,5,6), independently of the precise form of $I(L)$. We believe that the extended range of validity of eq.(4,5,6) has the same physical origin than the scaling $G_1(t) = G_1(t/\tau)$ discussed in section 4.1.3, that is the small weight of the short paths ($L \leq l^*$) for the large scatterers investigated. These results demonstrate that, in backscattering, both dynamic and static experiments can be interpreted in terms of the transport theory, which was questionned recently for the case of transmission[14]. More precisely, our data do not reveal any evidence for a *dynamic correlation length* $\Lambda = 2mm$ [14], since its existence would imply a difference between absorption and correlation data below $l_a=\Lambda=2$mm , i.e.$(l^*/l_a)^{1/2} = 0.1$, which is not observed.

4.2.2 *Comparison with theory*

We now compare the observed short time slope of $g_1(t)/g_1(0)$ with the theoretical expression, eq.(8), $-\beta/G_1(0)\ (3/2\tau)^{1/2}$. We have previously measured[9] that $G_1(0) = (3.1 \pm 0.3)\ F_0/4\pi$ per solid angle for the same samples (d=0.46μm, ρ=10%). Inserting this value and τ = 1.03 msec into eq.(8) we find $\beta = (4.5 \pm 0.5)\ F_0/4\pi$. This is in good agreement with $\beta = 3F_0/8\pi\ (1 + z_0/l^*)^2 \simeq 4.2\ F_0/4\pi$ for $z_0/l^* = 2/3$. The β value obtained from the expression[11] $\beta = 3F_0/8\pi\ (l/l^* + z_0/l^*)^2$ would be much smaller due to the large value of l/l^* ($\simeq$5) for the 0.46 μm diameter beads. This suggests that the simplest version of the diffusion theory[12] gives better quantitative results than the more elaborate theory[11]. We shall see in the following section that this point is confirmed by the analysis of the size dependence of $g_1(t)$.

4.2.3 *Particle sizing*

Fig.3 shows $g_1(t)$ measurments for five different sphere sizes, all at ρ = 10% and at a temperature of 20°*C* plotted against $(t/2\tau)^{1/2}$. With the exception of the smallest spheres (d=0.102 μm), the short time slopes are nearly identical. This combined with the observation[9] of a nearly size independent static intensity $G_1(0)$ again corroborates eq.(8). The same conclusion can be drawn from data taken in the VH configuration of the polarizers (data not shown). In Fig.4 we plot the data of Fig.3 as a function of $1/G_1(0)\ (t/\tau)^{1/2}$ for the four larger diameter spheres studied. $G_1(0)$ was measured simultaneously with $g_1(t)$ at constant incident laser power for all samples. We find that now all normalized curves superimpose at small times. This clearly demonstrates that β is indeed independent of the particle size in this range of l/l^* ($\simeq$ 2 to $\simeq$ 10), again in agreement with the simplest version[12] of the diffusion theory, but not with ref.11. It can also be seen from Fig.4 that the similarity of all curves extends well beyond the linear regime. Such a scaling does not follow in a straightforward manner from the diffusion theory.

4.2.4 *Polarisation*

For the smallest scatterers (d=0.102μm) the observed slope is, in parallel polarization (VV), about 0.68 times smaller and, in crossed polarization (VH), about 1.36 times larger, respectively, than the

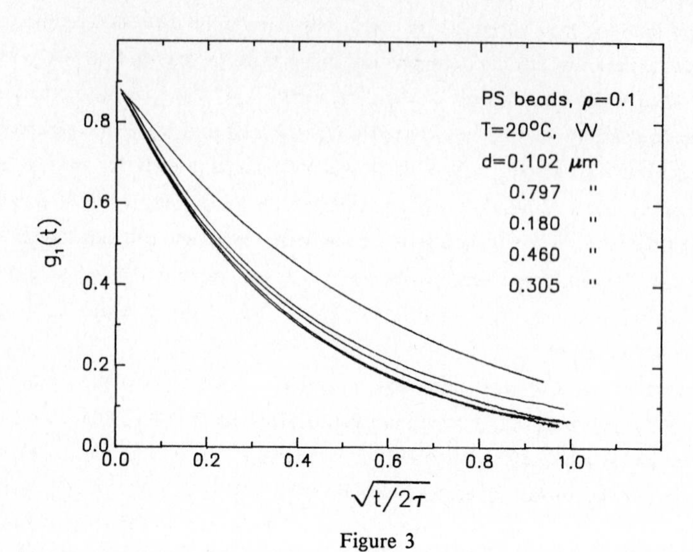

Figure 3
$g_1(t)$ for suspensions of monodisperse polystyrene spheres at various different sizes. The time scales are reduced by τ as obtained from the free particle diffusion.

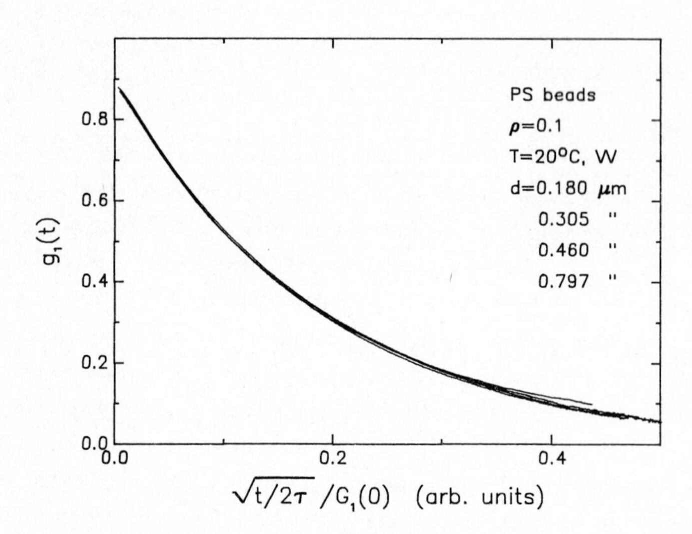

Figure 4
The data as given in Fig.3, but the $(t/2\tau)^{1/2}$ scale has been normalized to the static scattered intensity $G_1(0)$.

above values. Theses differences can be essentially[15] accounted for by the static intensity $G_1(0)$ which is 1.35 times larger in (VV) and 0.63 times smaller in (VH) as a result of the lower degree of depolarisation for these small beads [9]. This is illustrated by the data in Fig.5 where we plot $g_1(t)$ *vs.* $1/G_1(0)$ $(t/2\tau)^{1/2}$ for VV and VH for the 0.102 μm beads in comparison with the largest beads (d=0.797μm) studied. The agreement may seem less good than in Fig.4. But considering that, when plotted *vs.* $t^{1/2}$, the slope for the 0.102μm beads in VH differs from the slope for the 0.797μm beads in VV by a factor 6, the observed scaling appears satisfying. It suggests a simple alternative to the sizing method outlined above: One measures, with the same optics, both the decay of $g_1(t)$ and $G_1(0)$ for a sample of known τ such as calibrated polystyrene spheres. The unknown value τ_u can then be determined to within $\simeq$ 20% by adjusting τ_u to superimpose both short time slopes in a plot of $g_1(t)$ *vs.* $1/G_1(0)$ $(t/2\tau)^{1/2}$.

In order to fully exploit the particle sizing methods outlined in this section, attention must be payed to a number of experimental factors such as absorption, finite size effects and lack of laser coherence. They are discussed in the following section.

4.2 *Experimental corrections*

4.2.1 *Absorption*

In the presence of absorption $I(L/l^*)$ must be multiplied by a damping factor e^{-L/l_a}. Then eq.(4) implies that $G_1^{abs}(t) = G_1(t+\tau_a)$ with $\tau_a/\tau = 2l^*/l_a$. Because the initial decay of $G_1(t)$ is determined by the long loops, even a small amount of absorption may significantly round off the the $t^{1/2}$ singularity at short times. For example, the $t^{1/2}$ decay shown in Fig.1 which extends up to $t/\tau \simeq 0.01$ is remarkably affected even for l_a/l^* as large as 1000.

4.2.2 *Sample and beam geometry*

A finite thicknes Δ of the sample will cut on average paths longer than Δ^2/l^*, and correspondingly round off the decay of $g_1(t)$ for times smaller than $t/\tau \simeq (l^*/\Delta)^2$. This can be avoided by using a large sample cell. For the relatively large l^* value of 100μm and a cell with Δ=1cm the initial decay will be altered only for $t/\tau << 10^{-3}$. However, the finite size of the illuminated area - which is usually smaller than the sample - , or a finite area of observation will also introduce a cut-off of long paths. This effect can be easily evaluated for a gaussian beam profile of $1/e$ intensity radius W and a circular area of observation of radius R. The contribution $I(L)$ of paths of length L for plane wave illumination must now be replaced by $I'(L) = I(L)\,[1-e^{-R^2/(4Ll^*+3W^2)}]$. For a small ($R<<W$) or large ($R>>W$) area of observtion, respectively, those paths are affected whose average lateral extension is larger than the beam diameter ($Ll^*>W^2$) or than the area of observation ($Ll^*>R^2$). We see that rounding occurs at a time $t/\tau \simeq (l^*/W)^2$ or $(l^*/R)^2$, respectively. We have checked this by numerical calculations of eq.(4) using the above expression for $I'(L)$ and by various experiments. It turns out that a

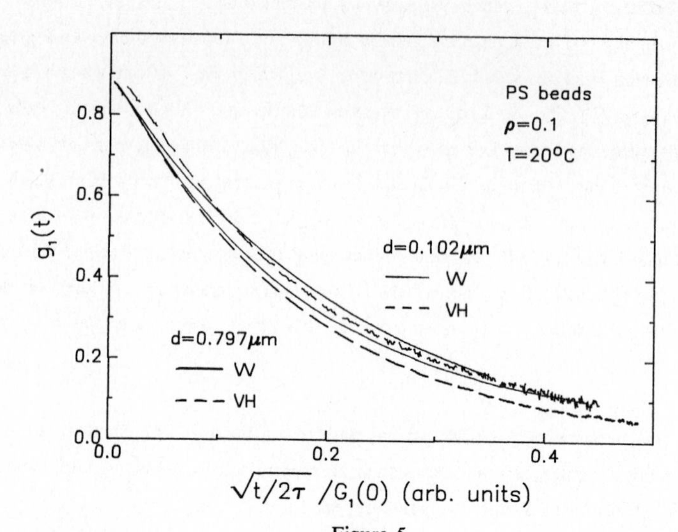

Figure 5
Polarisation dependence of $g_1(t)$. The ratio of the decay rates in VH polarisation and in VV polarisation are much larger ($\simeq$ 2.14) for the small spheres than for the large spheres ($\simeq$1.15), but can be accounted for by the differences in the static intensity $G_1(0)$ $(G_1(0)_{VH}/G_1(0)_{VV}$ = 0.46 and 0.89, respectively)

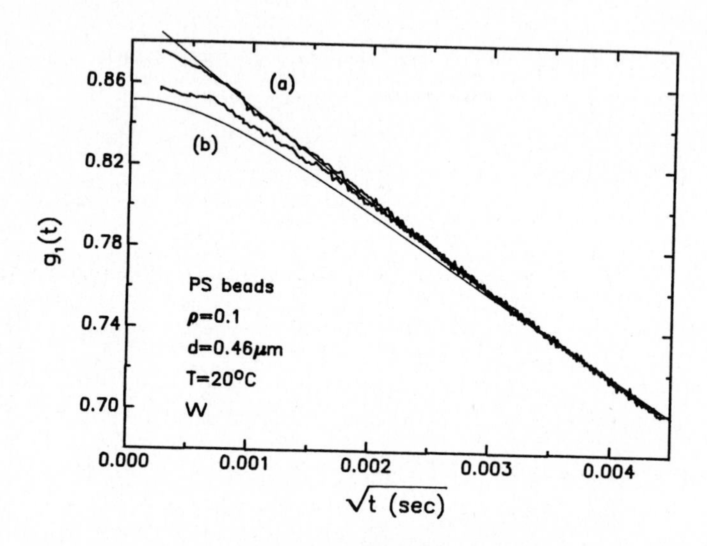

Figure 6
Effect of the laser coherence length ξ on the short time decay of $g_1(t)$. (*a*) $\xi \simeq$ 2m, (*b*) $\xi \simeq$ 40mm. The numerical calculations (continuous lines) are discussed in the text.

linear regime of $G_1(t)$ *vs.* $t^{1/2}$ still exists at times larger than the rounding time, but with too small a slope. This effect may easily become sizeable. For example, in the case of the suspensions used in Fig.1 ($l^* \simeq 22\mu m$), $W = R = 500\mu m \simeq 23\ l^*$ leads to a rounding for $t/\tau \leq 0.01$ which is roughly the time domain over which the true linear initial decay occurs. For $t/\tau > 0.01$ we still could observe a $t^{1/2}$ decay but with a slope about 25% smaller than the ideal case. This leads to a $\simeq$ 50% error in τ. Thus, as the area of observation is usually kept as small as possible, care must be taken to ensure $W >> l^*$. It may be worth noting, however, that even in the case where long paths are cut either by absorption or geometry, τ can still be measured by comparison of $\alpha_i(l_a)$ and $g_1(t)$ provided the same parameters are used in both cases. Then $I'(L)$ is the same in eq.(8) and eq.(9) so that the analogy between $\alpha_i(l_a)$ and $g_1(t)$ remains valid.

4.2.3 *Laser coherence length*

The random interferences between light emerging from different scattering paths result in rapid angular variations of the scattered intensity (speckle pattern), if a coherent light source is used. Dynamic speckle fluctuations (typically on a μsec to msec time scale) then arise from the brownian motion of the scatterers. If the coherence lenght ξ of the incident radiation is finite, interferences between paths differing in length by more than ξ will be averaged out on a time scale ξ/c (typically nsec) and so will be the corresponding contributions to the fluctuating speckle pattern. In single scattering this effect is usually negligible since ξ is typically much larger than the dimensions of the scattering volume. In strong multiple scattering, however, the presence of long paths ($L >> \xi$) implies also long path length differences ($|L-L'| > \xi$) and hence short time rounding of $g_1(t)$ will occur. One can show that then, averaged on a time scale $\geq \xi/c$

$$|G_1(t)|^2 = \sum_{L,L'} I(L)I(L')\, e^{-|(L+L')|\, t/2\tau l^*}\, |\Gamma(L-L')|^2 \tag{14}$$

where $\Gamma(t) = \Gamma(L/c)$ is the correlation function of the incident field and has a temporal width of order ξ/c. For finite ξ the $t^{1/2}$ singularity is suppressed. For times longer than $t/\tau \simeq l^*/\xi$ essentially only paths with lengths $L < \xi$ contribute to eq.(4) so that $\Gamma(L-L') \simeq 1$ and $|G_1(t)|^2$ is not affected by the finite coherence length. $|G_1(t)|^2$ should be rounded for times $t/\tau \simeq l^*/\xi$. We have confirmed this by a direct numerical calculation of eq.(14). The result is plotted in Fig.6 in comparison with the experiment. The experimental curve (*a*) was obtained, like all other data reported here, using a mono-mode *Ar*-laser with intra cavity etalon ($\xi \simeq$ 2m), while (*b*) was obtained without etalon. In the latter case ξ was directly measured interferometrically, $\xi \simeq 40\ mm \simeq 1800\ l^*$. The rounding is clearly observable for $(t/\tau)^{1/2} \leq 0.05$. The continuous line (*a*) shows a fit of the data set (*a*) to eq.(15), which is discussed in section 4.3. We obtain $g_1(0) = 0.898$ and $\gamma = 1.46$. This curve corresponds to the behavior at $\xi = \infty$ and fits the data with the exception of the very shortest times. The continuous line (*b*) is 0.898 $G_1(t)/G_1(0)$ where $G_1(t)$ results from eq.(14) using $I(L) \propto 1/L^{-3/2}\ e^{-3\gamma^2 l^*/4L}$. This choice of $I(L)$ in eq.(4) yields eq.(15). γ is the same as in curve (*a*). The curve (*b*) is in qualitative agreement with the data set (*b*). Both calculation and experiment directly demonstrate the substantial effect of a finite coherence length which is noticeable up to $t/\tau \simeq 2\, l^*/\xi$. Unlike the rounding effects due to finite absorption or finite geometry, this rounding only appears in the dynamic data, but not in $\alpha_i(l_a)$ or $\alpha_c(Q)$. Therefore when measuring τ by comparing eq.(8) to eq.(9) or eq.(10) it is essential to operate with a coherence length larger than $\simeq 10^4\, l^*$.

4.3 *Long time behavior of g_1(t)*

We have shown in section 4.1 that the diffusion constant D of scatterers can be measured from the *short* time decay of $g_1(t)$. This is because this decay is dominated by the long scattering paths for which the various approximations involved in the derivation of eq.(4) apply. However, as outlined in section 4.2, the short time decay may be most sensitive to effects cutting predominantly long paths. This is not so for the *long* time decay which is controlled by shorter paths. Thus the described method would be more widely applicable, if D could be determined at longer times. The first observation by Pine *et. al.*[6] that in various polystyrene sphere suspensions the decay of $g_1(t)$ closely follows

$$g_1(t) = g_1(0)\exp^{-\gamma\sqrt{3t/2\tau}} \qquad (15)$$

up to times $t \simeq \tau$ suggests this possibility. Our data (Fig.7) where we plot ln $g_1(t)$ *vs.* $(t/2\tau)^{1/2}$ for the polystyrene spheres with d=0.46μm and ρ=10% also follow this stretched exponential decay over the time domain studied. Fitting a straight line through the full data set up to $t/\tau \simeq 1$ gives $\gamma = 1.59$. For the data set below $t/\tau = 0.01$ we find $\gamma = 1.50$ which corresponds to an absolute slope of 57.3 sec^{-1}, in good agreement with the value obtained from the data shown in Fig.1. Comparison of eq.(15) and eq.(4) shows that $\gamma \equiv \delta \equiv \beta/G_1(0)$ so that γ should scale like $1/G_1(0)$ for a constant flux F_0. We have verified this behavior experimentally. For instance, γ values only slightly differ between VH and VV for the 0.46μm diameter beads, whereas they differ by a factor 2.0 for the 0.102μm diameter beads, as does $G_1(0)$. Hence, it appears possible to roughly infer the short time slope from the long time decay according to eq.(15).

Pine *et. al.*[6] have originally derived an expression for $G_1(t)$ corresponding to eq.(15) under the assumption that the conversion from propagating to diffusing light takes place at a single distance γl^* from the interface. This assumption appears a priori physically less appropriate than a conversion decaying exponentially with distance which, in turn, gives rise to a power law decay[5] of $G_1(t)$. The more satisfying condition that $I(L)$ is given by the transport theory above the length scale l^*, and equals zero below, leads[16] to $g_1(t) = e^{-\gamma\sqrt{3t/2\tau}}/(1+(2t/3\tau)^{1/2})$. Recently another (nonexponential) decay of $G_1(t^{1/2})$ has been derived[17] by accounting for a smooth crossover from ballistic to diffusive propagation at distances $\simeq l^*$. However the dynamic range over which $g_1(t)$ has been measured is large enough to preclude a curvature consistent with either of these calculations[16,17].

Thus a theoretical justification of the stretched exponential behaviour (eq.15) is still missing. The fact that for long times ($t \geq \tau$) not only short paths, not described by the transport theory, but also higher cumulants of eq.(1) contribute significantly to $G_1(t)$ make this a difficult problem. For instance, the statement explicit in eq.(2) that longer paths decay faster can only be considered true on average[18]. In addition data obtained under somewhat other conditions than those in Fig.7 seem to reveal small but systematic deviations from eq.(15). This may indicate that a behaviour as shown in Fig.7 could be somewhat accidental.

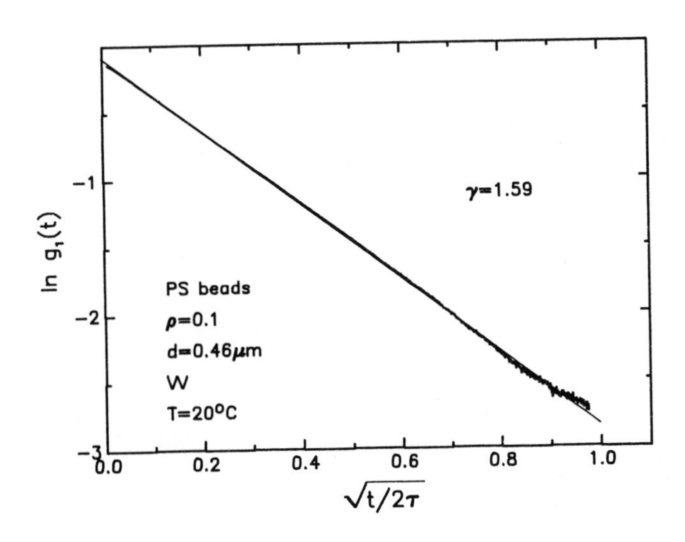

Figure 7

$g_1(t)$ for polystyrene suspensions plotted on the same scales as in ref.6.

5. CONCLUSIONS

Our experimental study shows that, for non-absorbing systems *(i)* the initial decay of $g_1(t)$ is proportional to $t^{1/2}$ as predicted by the transport theory, *(ii)* the initial slope $dg_1/dt^{1/2}$ provides a measure of τ to within about 20%, if the coefficient δ is known from an independent measurement, and *(iii)* the $t^{1/2}$ - dependence corresponds to the Θ - dependence of the backscattering cone and to the $(1/l_a)^{1/2}$ - dependence of the intensity on absorption, well beyond the initial linear regime.

The quantitative reproducibility of our data was generally of about 10%. Therefore these results suggest that diffusion constants can be measured within $\simeq$ 10% for any brownian suspension by comparing the initial decay of $g_1(t)$ to that of a calibration sample, say of polystyrene beads, when both measurements are performed under identical experimental conditions. The ratio of the initial slopes $dg_1/dt^{1/2}$ times the ratio of the static intensities provides the ratio of $\tau^{1/2}$ for the two different systems and hence the dynamics of concentrated suspensions could be probed. It appears particularly interesting to study suspension under conditions of strong and controlled correlations (for example at various ionic strenghts) and to examine the expected sensitivity of $dg_1/dt^{1/2}$ to $S(q)$ by comparison with an independent determination of $S(q)$ via, for example, neutron scattering.

ACKNOWLEDGEMENTS

We kindly acknowledge support from the Deutsche Forschungsgemeinschaft and discussions with D.Pine and D.Weitz.

REFERENCES

1) B.J.Berne and R.Pecora, Dynamic Light Scattering (Wiley, New York, 1976)

2) C.M. Sorensen *et. al.*, Phys.Rev.A, *14*, 1520, (1976)
R.C.Colby *et. al.*, Phys.Rev.A, *12*, 1530 (1975)

3) G.Maret and P.E.Wolf, Z.Phys.B *65*, 409, (1987)

4) M.Rosenbluh, M.Hoshen, I.Freund and M.Kaveh, Phys.Rev.Lett *58*, 2754 (1987)

5) M.J.Stephen, Phys.Rev.B 37,1 (1988)

6) D.J.Pine, D.A.Weitz, P.M.Chaikin and E.Herbolzheimer, Phys.Rev.Lett *60*, 1134 (1988)

7) A.J.Rimberg and R.M.Westerfeld, Phys.Rev.B *38*, 5073 (1988)

8) D.N.Qu and J.C.Dainty, Opt. Lett. (USA), *13*, 1066 (1988)

9) P.E.Wolf, G.Maret, E.Akkermans and R.Maynard, J.Phys.(Paris) *49*, 63 (1988)

10) See e.g. Ishimaru "Multiple Light Scattering in Random Media", Vol.1 (Academic Press, New York, 1978)

11) A.Ishimaru and L.Tsang, JOSA, *5*, 228 (1988)

12) E.Akkermans, P.E.Wolf, R.Maynard and G.Maret, J.Phys.(Paris) *49*, 77 (1988)

13) P.Pusey, J.Phys.Math.Gen.A, *8*, 1433 (1975)

14) I.Freund, M.Kaveh and M.Rosenbluh, Phys.Rev.Lett. *60*, 1130 (1988)

15) At our level of experimental accurary (typically 10%) the $\simeq$30% decrease in τ, expected due to hard core correlations in the above approximation for the smallest spheres, could not be detected. But whether $S(q)$ is properly described by such correlations remains an open question.

16) D.J.Pine, D.A.Weitz, G.Maret, P.E.Wolf, E.Herbolzheimer and P.Chaikin, In "Classical Wave Localisation" Ed. P.Sheng, World Scientific, London, Singapore (1989)

17) F.MacKintosh and S.John, preprint

18) For example, for a semicircular path of n steps with identical q_i $(=2n_0k_0 \sin\pi/2n)$ the decay rate is $\pi^2/4n\tau$ which *de*creases with n. Although the weight of such paths decreases exponentially with n, the total contribution to $g_1(t) \propto (\Sigma e^{-n}\, e^{-t/\tau n})$ is just of order $e^{-(t/\tau)^{1/2}}$.

Scattering in Volumes and Surfaces
M. Nieto-Vesperinas and J.C. Dainty (Editors)

MULTIPLE SCATTERING IN DENSE MEDIA

P. Bruscaglioni, G. Zaccanti

Department of Physics - University of Florence, Via Santa Marta, 3
50139 - Florence - Italy

This paper presents numerical methods, based on particular versions of the Monte Carlo scheme. They can be employed to deal with the propagation of optical radiation in turbid media. A large range of optical depths, and a variety of geometries of the radiation beam and of the medium are allowed to be dealt with by the numerical procedure. Indications are given of the procedures suitable to the different cases to be examined. Examples of comparison with experimental results are presented.

1. INTRODUCTION

Propagation of electromagnetic radiation in turbid media, when multiple scattering is important, is a complex phenomenon, whose characteristics depend on several factors: geometry of the problem (beam and receiver geometry, extension of the medium), scattering properties of the particulate responsible for the medium turbidity, homogeneity or non-homogeneity of the continuous or non continuous medium. This paper presents the methods used by our group at the Department of Physics, University of Florence, Italy (DPUF) to deal with the propagation of beams having a small or substantial divergence, in turbid media constituted by a homogeneous fluid in which particulate is suspended. The case of monochromatic radiation is considered in this paper.

As is known, in a medium whose properties are described by an extinction coefficient σ, a single scattering albedo w_0 and a phase function p, generally functions of the position, propagation of specific intensity can be described by the radiative transfer equation (transport equation) which, for the time independent case, can be written as:

$$1) \qquad \frac{dI\,(\mathbf{r},\mathbf{s})}{ds} = -\,\sigma I\,(\mathbf{r},\mathbf{s}) + \sigma\, w_0 \int_{4\pi} p\,(\mathbf{s},\mathbf{s}')\; I\,(\mathbf{r},\mathbf{s}')\; d\omega' + \epsilon\,(\mathbf{r},\mathbf{s})$$

where: $I(\mathbf{r},\mathbf{s})$ is specific intensity, function of position $\mathbf{r}$ and direction $\mathbf{s}$, $p(\mathbf{s},\mathbf{s}')$ is considered normalized to one for integration over the complete solid angle 4π and $\epsilon(\mathbf{r},\mathbf{s})$ is a term due to possible sources in the medium. Analytic methods can be employed to deal with this equation if certain assumptions are made. Typically, for the case of the propagation of a collimated or small divergence light beam, the assumption is often made that even multiply scattered radiation propagates with a small angle with respect to the original beam direction. This is the small angle assumption (SA) which, necessarily, presumes that the phase function has negligible values outside a small interval about the zero value of its argument, i.e. that the sizes of the diffusing particulate are larger than (or of the order of) the radiation wavelength. Thus the integral at second member of Eq.(1) can be put in the form of a convolution integral and Fourier techniques become applicable. The SA assumption is generally also accompanied by other assumptions concerning the phase function and the beam, both often modelled with Gaussian functions of the relevant arguments [1, 2, 3].

Other simplifyng assumptions are based on considering the contribution of scattered radiation to specific intensity in the integral at second member of Eq. 1 as a truncated power expansion of the angular variables (Fante Dolin approximation[4, 5]). The diffusion approximation[6, 7] is considered in cases of beams impinging on very dense media,within which the radiation field, after a large number of scattering events have occurred, is assumed to be represented by an isotropic term plus a small correction term connected to a residual anisotropy.

Even in cases of collimated beams and phase functions with high forward peaks, the SA assumption loses its validity as the optical depth of the medium (passed by the beam) becomes larger. There is always an interval between the optical depths where the SA assumption (with certain limits) can be invoked, and those at which one can rely on the diffusion approximation. Generally, but especially in this interval, or anyway when the SA approximation fails because the diffusing particles are not large in comparison with the radiation wavelength, numerical methods are appropriate.

The numerical methods, developed at DPUF particularly for transmission of laser beams, are based on particular versions of the Monte Carlo (MC) technique. In addition to their conceptual simplicity, the main positive features of MC schemes are connected to the possibility of simulating propagation in media with any boundary shape and general scattering characteristics (phase function, extinction coefficient, albedo). Thus, for instance, it is not necessary, unlike many analytic treatments of the transport equation, to rely on the SA assumption mentioned above. Any distribution of radiance of a beam incident on the medium, and any characteristics of a receiving apparatus, are allowed to be dealt with. Thus, a general experimental situation can be simulated. Moreover, as it is implicit with the development of an MC procedure, in addition to the received power, one can have statistical information on the positions within the medium of the scattering events, and on the relative lengths of the paths followed by radiation: consequently on the time spread of received radiation, if pulses are transmitted. However, especially if there is a small probability for emitted radiation to be received by the considered apparatus, or more generally to reach a particular area of interest, calculation time necessary to obtain reliable statistical results can be excessively long, if one does not use suitable procedures of variance reduction.

Our MC codes can be divided in three groups: 1) Elementary MC procedure, 2) Semi Monte Carlo, 3) SEMOC. In the following sections we shall outline their main characteristics in order to show their applicability to different situations of moderate or large optical depths. The results of calculations in representative situations will be presented. They will be also compared to results of laboratory experiments, aiming at a validation of the numerical techniques. We premise that our codes, in the present version, do not take into account polarization of the radiation. We denote as σ the extinction coefficient of the medium (which is assumed to be homogeneous or composed of homogeneous layers with plane boundaries). σ_S indicates the scattering coefficient and $w_0 = \sigma_S/\sigma$ the single scattering albedo. A phase function only depending on the scattering angle : $p(\theta)$ is considered. However, the introduction of Stokes vectors and scattering matrices, which can be done in a simple way if it is permitted by memory capacity of the computer, would allow one to modify our codes in order to deal with polarization, with a limited increase of calculation time. τ indicates the optical depth of the medium interposed between a source and a receiver.

2. ELEMENTARY MC PROCEDURE

As is well known, the basic MC procedure consists in drawing by lot, according to laws of probability relevant to the particular medium and geometry of the considered beam, the lengths of the paths followed by a series of "photons" between each successive point of scattering, and the angles of scattering. If particles in the medium causing scattering have a non unitary albedo (w_o) the weight of a photon after each scattering event has to be reduced by a factor equal to the albedo. Or otherwise a probability is assigned (equal to w_o) to the photon for continuing its trajectory after the scattering point. In the case of a simulated transmission experiment the trajectories followed by photons are then computed positively if they arrive on the receiver area with an angle comprised in its field of view. Or more generally photons are counted if their trajectories eventually have a prefixed positive termination. Such is the case for instance for the simulated experiment presented by Bucher[8], where photons, originally coming from a collimated thin beam, were counted if they passed across a plane parallel slab of diffusing particles representing a cloud. The elementary MC procedure proves to be very effective when the number of "positive" photons is not very small in comparison with that of the photons emitted by a source.

3. SEMI MONTE CARLO CODE

This is the name given to our Monte Carlo based code, where the modifications specified below were made to the elementary procedure[9]. We have to mention that Poole et al.[10] used the same type of MC code for studies of light beam propagation in sea water . Further modifications were considered of utility for certain circumstances of calculations and will be explained in the next section, where our third type of code (named SEMOC) will be presented.

The principal point of the Semi Monte Carlo (SM) code consists in using the elementary MC method to evaluate, by means of the conventional lotting procedures, the positions of the scattering points for each considered trajectory of a photon. Figure 1 explains the principle of the calculation. Once the code has determined a trajectory (index i) O P_{1i} ...P_{ki}... for a photon originally emitted in the z direction, the probability q_{ki} is evaluated that the photon scattered at the position P_{ki} where kth event of scattering occurs is received by an element of the receiver area ΔA, with the assumption that no further scattering event happens between P_{ki} and ΔA. One has for this probability :

$$q_{ki} = \frac{\Delta A}{r_{ki}^2} \cos \varphi_{ki} \, p(\theta_{ki}) \exp(-\sigma r_{ki}) \qquad \varphi_{ki} \leq \alpha$$

$$q_{ki} = 0 \qquad \varphi_{ki} > \alpha \qquad (2)$$

where α indicates the field of view angle of the receiver, **n** the direction of its optical axis and θ_{ki} is the scattering angle, as shown in Fig.1. For the validity of Eq.(2) it was assumed that ΔA is so small that $\Delta A/r^2_{ki} \cos\varphi_{ki}$ is the

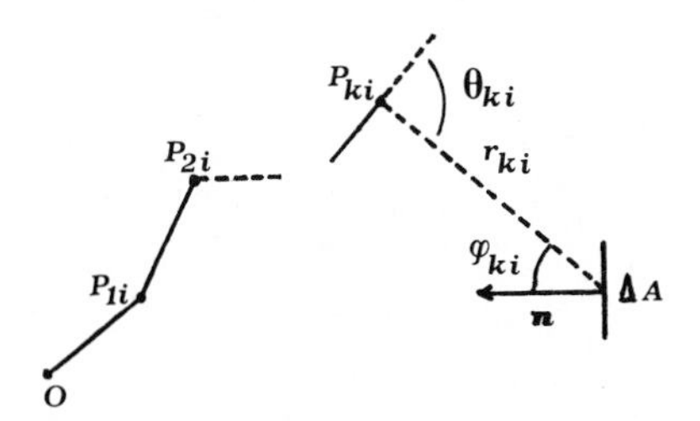

FIGURE 1
Principle of basic calculation for the SM code. A generic trajectory, labelled i, is considered: a photon emitted at O is subsequently scattered at $P_{1i}, P_{2i}....P_{ki}$. The probability of receiving the photon scattered at P_{ki} (position of kth scattering event) by an element of area ΔA of the receiver, without any further scattering event between P_{ki} and ΔA, is denoted as q_{ki}.

solid angle subtended by the element of area, seen from point P_{ki}. The quantity δ_{ki} is also calculated, where:

$$3) \qquad \delta_{ki} = \frac{\ell_{ki} - L}{L} ,$$

ℓ_{ki} is the length of the trajectory O $P_{1i}....P_{ki}$ ΔA, and L is the distance O- ΔA.

One then considers the density of received power at the position of ΔA per unitary power emitted , as a sum of the contributions of the different orders of scattering. Indicating as S_k the contribution of the kth order of scattering , one has:

$$4) \qquad S_k = \frac{w_o^k}{\Delta A\ N} \sum_{i=1}^{N} q_{ki}$$

where the sum is extended to the total number N of the photons considered to be emitted. The factor w_o^k takes into account the reduction in the photon weight, when the medium albedo w_o is not unitary. For a limited number of values of the integer n the average value of the nth power of δ_{ki} is also calculated:

$$5) \qquad <\delta_k^n> = \sum_{i=1}^{N} w_o^k\, q_{ki}\, (\delta_{ki})^n \Big/ \sum_{i=1}^{N} q_{ki}$$

The ratio between the total received scattered power (P_s) and the emitted power (P_e) is then given as:

$$6) \qquad \frac{P_s}{P_e} = \sum_k \int_A S_k \, dA$$

where the integration has to be performed over the area A of the receiver. In the practical use of SM code, the numerical integration, requires the knowledge of S_k for a very limited number of points. SM code is to be employed for large values of τ (see Sect.5), at which one can expect that S_k is almost constant on the area of a receiver. A correct procedure of computation obviously restricts the upper limit of the summation in Eq.6 to a sufficient number of orders of scattering, which can be decided from an examination of the partial results.

The further average of $< \delta_k{}^n >$ over the orders of scattering and over the receiver area is also calculated:

$$7) \qquad < \delta^n > = \sum_k \int_A < \delta_k^n > S_k \, dA \Big/ \sum_k \int_A S_k \, dA$$

Let now us indicate as δ the ratio $(\ell\text{-}L)/L$, when ℓ is the length of a generic trajectory from the source to the receiver. An estimation of the probability of finding a certain value of δ can be furnished by a histogram: the code considers all the sorted trajectories, for every order of scattering and every necessary element of the receiver. The found range of values of δ is divided in intervals. The level h_m of the histogram in the mth interval is:

$$h_m = \int_A \sum_{i,k} q_{ki} \, w_o^k \, dA$$

where the sum is now extended to the trajectories for which δ_{ki} is comprised in the mth interval. From the histogram, whose area is normalized to one, one can have an estimate of the function $f(\delta)$, such that $f(\delta)\,d\delta$ represents the probability of finding a trajectory in the interval δ - $\delta+d\delta$. $f(\delta)$ is therefore related to the distribution in time of scattered radiation. For very dense media when received power is practically all due to scattered radiation, $f(\delta)$ represents the time response of the system when impulses are transmitted.

A comparison of SM procedure to compute the ratio P_s/P_e with that of the elementary MC method, mentioned in the preceding section, shows that a substantial reduction of calculation time is achieved when the ratio P_s/P_e is a small number. In fact , although each non zero term of the sum in Eq.4 can be much smaller than one, their number is much larger than the number of the non zero terms obtained in the case of the elementary MC procedure. The convergence of the statistical results thus is much faster, even if the time spent in calculating the trajectory of each photon and the series of probabilities given by Eq.(2) for the subsequent scattering points is larger than time spent for calculating a trajectory with the elementary MC procedure.

Examples of histograms, calculated and measured $< \delta >$, and of higher moments $< \delta^n >$ will be given in Sect.s 5 and 6. The importance of calculating $< \delta_k^n >$ will be also shown in Sect. 4, where SEMOC code will be described. The calculation of the moments $< \delta_k^n >$ is an essential step for the use of the scaling relationship employed with that code.

4. SEMOC CODE

SEMOC code was developed from the SM code to deal with cases of moderate optical depths (τ up to the order of 10). Its main characteristic is the use of scaling formulas allowing one to use the results pertaining to a reference optical depth τ to obtain the data relevant to an other optical depth τ'. By the SEMOC code, following the line of the SM code the quantities q_{ki} of Eq.2, and the average quantities $< \delta_k^n >$ considered in Sect.3, are evaluated. They are calculated for an opportune reference value of τ different for each order k of scattering. Typically τ = k for kth order of scattering, since it was seen that this value corresponds to a relatively faster convergence of the statistical results. Thus the quantity S_k indicated in Sect.3 are calculated from Eq.4 , at a different value of τ for each scattering order. But the results relevant to the particular real situation can be deduced from these data by using scaling relationships. These latter are here briefly recalled (for more details see [11,12]).

Let us consider the probability that an element of the receiver area ΔA receives the photon after a certain trajectory O $P_{1i}...P_{ki}$ ΔA, of length ℓ_{ki}, with a number k of scattering events, when the optical depth of the medium between the source and the receiver is τ . Let us now keep the geometrical parameters of the trajectory fixed (lengths of the tracts and angles), and change the scattering coefficient σ_s and albedo w_o to σ'_s and w'_o. τ changes from $\tau = \sigma_s L/w_o$ to $\tau' = \sigma'_s L/w'_o$. The probability changes by a factor:

$$\left(\frac{\sigma'_s}{\sigma_s}\right)^k \exp\left[-\ (c-1)\ \tau\,\ell_{ki}\ /\ L\right]$$

where $c = \tau'/\ \tau$. Therefore the quantity $S_k(\tau)$ of Eq.4 changes into:

$$8)\quad S_k(\tau') = \frac{w_o^k}{\Delta AN} \sum_{i=1}^{N} q_{ki}(\tau) \left(\frac{\sigma'_s}{\sigma_s}\right)^k \exp\,[-(c-1)\,\tau\,\ell_{ki}\,/L\,] =$$

$$= S_k(\tau) \left(\frac{\sigma'_s}{\sigma_s}\right)^k \exp\,[\,-(c-1)\,\tau\,] < \exp\,[\,-(c-1)\,\tau\,\delta_{ki}\,] >$$

with δ_{ki} given by Eq.3 and where:

$$9)\quad < \exp\left[-\,(c-1)\,\tau\,\delta_{ki}\right] > = \sum_{i=1}^{N} q_{ki}\,(\tau)\,\exp\left[-\,(c-1)\,\tau\,\delta_{ki}\right] \Big/ \sum_{i=1}^{N} q_{ki}(\tau)$$

The preceding equations are derived by taking into account the change in the exponential attenuation along the length of the photon path, and in the probability of scattering at the position P_{1i}, P_{2i},... P_{ki}.

The scaling relationship (8) allows one to obtain S_k, hence the scattered received power P_s, after summation on k and integration over the receiver area, at the desired value of τ common to all orders of scattering, if one is able to evaluate the term in the < > brackets of Eq.8. This can be done in an approximate way (whose adequacy is checked by the program) in two types of cases. First we can note that the scaling relationship takes a simple form if one can put $\ell_{ki} = L$, i.e. if the lengths of the trajectories in the medium giving substantial contribution to received power are not effectively different from the medium geometrical width . In this case the exponential factor within the brackets can be put equal to one. The approximate scaling relationship comes out:

$$10)\quad S_k(\tau') = S_k(\tau) \left(\frac{\sigma'_s}{\sigma_s}\right)^k \exp\left[-(c-1)\,\tau\right]$$

where $c = \tau'/\tau$.

If one cannot make the simplification considered above (which is equivalent to accepting the small angle approximation scheme), still one can use Eq.8 by means of an approximate evaluation of the exponential factor within the < > brackets. Let us consider the power expansion:

$$11)\quad < \exp\left[-\,(c-1)\,\tau\,\delta_{ki}\right] > = 1 - (c-1)\,\tau < \delta_k > + \frac{1}{2!}\,(c-1)^2\,\tau^2 < \delta_k^2 > - \frac{1}{3!}\,(c-1)^3\,\tau^3 < \delta_k^3 > + ...$$

where the $< \delta_k^n >$ are the nth order statistical moments introduced in Sect.3, Eq.(5). To use Eq.8 in this case, one has to evaluate a sufficient number of the quantities $< \delta_k^n >$. The series (11), truncated at a convenient number of terms, is then introduced in Eq.8 to have an approximate scaling relationship. The simplified form (10) in often sufficient when the phase function $p(\theta)$ has a pronounced forward peak, and the optical depth is not large. In other situations the practice of our calculations showed that for optical depths smaller than $\approx$ 10, for which the consideration of no more than $\approx$10 orders of scattering was required, the first three terms after the unity of the power expansion (11), were sufficient for a good convergence. In our code, the choice of the sufficient number of terms and of orders of scattering derives from the examination of the partial results.

We have mentioned that, before using of the scaling relationship, $S_k(\tau)$ is evaluated at $\tau \approx k$. We can add that our practice showed that the main contributions to P_s come generally from the orders of scattering nearly equal to τ for moderate values of optical depth. Thus the use of the scaling relationship is facilitated: terms S_k requiring a large c factor in Eq.8 are of small importance.

5 - COMPARISON BETWEEN THE CODES AND EXAMPLES OF RESULTS

SEMOC is our most efficient code for moderate values of τ. The set of statistical data given by the code can be used to evaluate the ratio P_s/P_o (scattered received power divided by unscattered received power) for a wide range of optical depths. Let us define as τ_s the contribution to τ due to scattering ($\tau_s = w_o\tau$). In our practice the code is effectively employed by considering 10 orders of scattering and the moments $< \delta^n >$ for n = 1,2,3. The obtained data are sufficient when $\tau_s < \tau_{max}$. τ_{max} is about 8 when the ratio R/L between the receiver radius and the source-receiver distance is $\approx$ 0.2 and α is few degrees, and is (somewhat) larger for smaller values of R/L. The code proved to be applicable for the case of diffusing particles both larger and smaller in comparison to wavelength. The increase of the significant orders of scattering with the increase of τ_s makes the code umpractical for τ_s larger than the indicated values. The moments $< \delta^n >$ are smaller for particles having a more pronounced forward peak. Thus the convergence of the power expansion (Eq.11) is faster for larger particles suspended in the medium, allowing one therefore to deal with larger optical depths. Also the effect of an albedo smaller than one is that of diminishing the moments $< \delta^n >$. In any case the actual range of τ that can be dealt with by means of the code has to be checked by the examination of the numerical results.

The examples of results of calculations using our codes, presented in this and the following sections, take Fig.2 as a reference for the definition of the geometrical parameters. In all the cases the wavelength λ = 0.6328 μm was considered. The source is placed at point O and emits a laser beam whose axis is the axis of symmetry of the system (z axis). The receiver has a circular area with radius R, coaxial with the z axis and a field of view whose semiaperture is α. The cross section radius of the beam at the scattering cell is r. The beam divergence is β. The scattering cell is cylindrical, coaxial with the z axis, with diameter D and length L. The distance between the exit plane of the scattering cell and the receiver is d. The scattering cell is filled with water with a suspension of latex spheres (polystyrene). Possibly a certain quantity of dye is added, to change the absorption of the medium. For the calculations and a comparison with the experimental results, the characteristics of the medium are considered to be determinated by the properties of the suspended spheres and the dye when present. The attenuation of water is neglected in the calculations, since it is taken into account during calibration of the measurement apparatus. All the calculations were carried out by means of an HP 1000 E series minicomputer, and all the calculation times referred to are therefore relative to this computer.

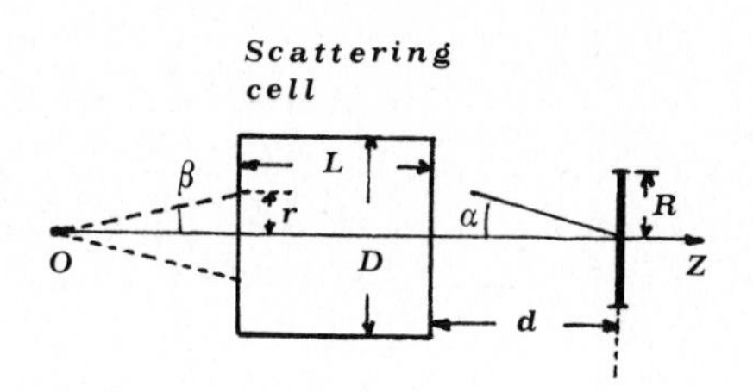

FIGURE 2
Reference scheme for calculations and presentation of numerical and experimental results. The beam source is at O. The receiver of circular area with radius R has a FOV semiaperture α .

As an example relative to the use of SEMOC code, Fig.3 shows the calculated ratio P_S/P_O plotted versus τ. The figure refers to the following situation: He-Ne laser beam with small cross section (r $\approx$1 mm) and small divergence (β = 0.3 mrad) (thin collimated beam), R = 2cm, L = 10cm, D = 14cm, α = 3°, d = 10cm. (Note that calculations made by assuming r = 0 and β = 0 gave equal results as those made by taking into account the actual beam radius and divergence). The optical depth shown on the abscissa axis

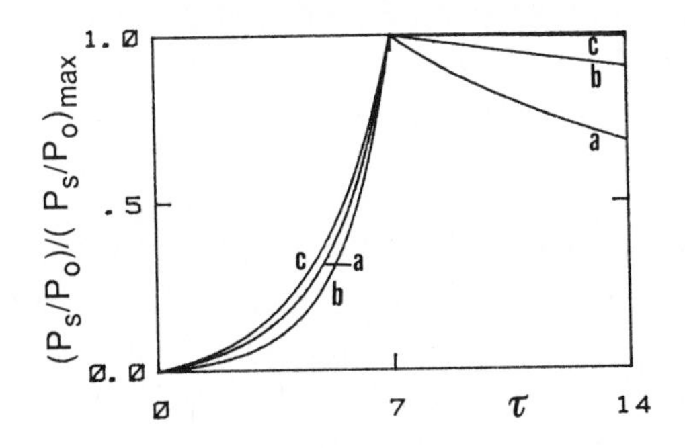

FIGURE 3
Example of SEMOC results. For three types of sphere the relative variation of P_s/P_o is plotted versus τ . Curves a,b,c: spheres with diameter 0.33, 0.995, 15.8μm respectively. $0 < \tau < 7$: effect of the suspension of non absorbing spheres. $7<\tau<14$: effect of the added dye.

is due to the suspension of purely scattering polystyrene spheres for $0<\tau < 7$. The change of τ from $\tau = 7$ to $\tau = 14$ is obtained by adding to the suspension different quantities of purely absorbing dye which changed the medium albedo without changing the medium phase function. (It was checked by calculations with Mie theory that the change of the imaginary part of the refractive index of the water due to the presence of dye did not alter the scattering properties of the spheres, for the dye quantities used in our laboratory measurements, reported in Sect.6). Curves a,b,c refer to spheres with diameter 0.33, 0.995, and 15.8 μm respectively. The curves are normalized to their maxima ($(P_s/P_o)_{max}$ is 0.13, 2.2, 20.0, for curves a,b,c respectively). All of the data for each curve came by a single set of results obtained by means of SEMOC code with a calculation time of a few hours. One can note that for $\tau > 7$ the curves have a down slope. This is due to the larger attenuation (absorption) of the scattered with respect to the unscattered radiation because of the different average path length of the two components of received radiation. The down slope is different for the three curves: particles with larger radii have a more pronounced forward peak, therefore the average path length of received radiation is smaller within the cell. This feature of the results indicated the possibility of measuring the moments $< \delta^n>$ by means of measurements of attenuation of a continuous wave beam in an absorbing medium (see Sect. 6).

The mentioned limited range of values of τ suitable to be dealt with by means of the SEMOC code is however of interest, since it extends to values where some common approximations may fail. The small angle approximation is known to be inadequate when τ is larger than a few units, especially if the forward peak of the phase function is not very high. As an example: some calculations were made with SEMOC, using a Henyey-Greenstein function as phase function:

$$p(\theta) = \frac{1-g^2}{4\pi} (1+ g^2 - 2g \cos\theta)^{-1.5}$$

where g is the asymmetry parameter. In the range of τ from $\approx$ 5 to 10 the code required the use of three terms in the power expansion (Eq.11), when g=0.7. As was mentioned, this implies the inadequacy of the small angle approximation.

For values of τ substantially larger than those mentioned above SEMOC code can not be used due to the large number of orders of scattering to be considered and the poor convergence of power expansion (11). SM code has been used for τ up to 100. As was explained in Sect.3 the SM code evaluated the ratio P_s / P_e together with the statistical moments $< \delta^n >$ and histograms of δ. For the situation of Fig.2 the quantities ℓ_{ki} and L considered by the code refer to path lengths within the scattering cell, and to the cell width.

To give examples of calculation time necessary for good statistical results: for one value of τ of the order of some tens and α of the order of the degree, one hour is a typical time necessary for our minicomputer. The histogram may however require times of the order of ten hours.

We point out that SM code can give results in the intermediate range of τ where neither the small angle approximation nor the diffusion approximation is applicable. In particular SM code could even be employed to test the limits of both approximations, and in general of an approximate procedure.

As an example of SM results, Fig.4 shows the apparent optical depth $\tau_a = -\ln (P_r/P_e)$ where P_e is received power in the absence of the spheres. τ_a is different from the true optical depth τ since P_r comprehends scattered and unscattered radiation. The simulation parameters are: R = 2cm L = 10cm, D = 14cm, β = 0.3 mrad, r= 1 mm, α = 12°. Marks +, x, □ refer to spheres with diameters 0.33, 0.995 and 15.8 μm respectively. Figure 5 shows examples of histograms: sphere diameters: 0.33 μm, τ = 20,30,50,70 for curves a,b,c,d respectively. Geometry as for Fig.4.

As for the elementary MC codes (Sect.2), one can notice that there are

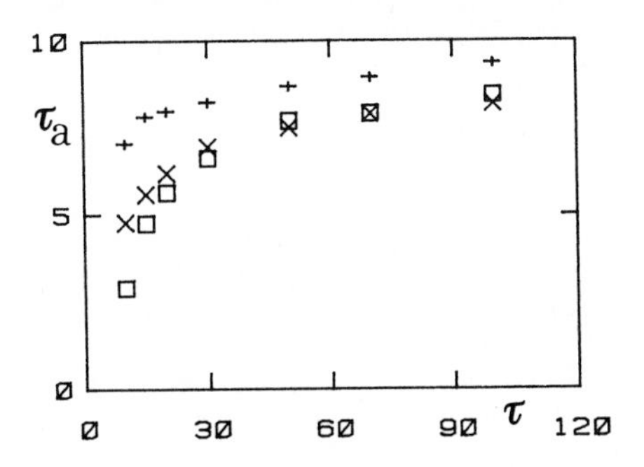

FIGURE 4
Example of SM code results. Apparent optical depth plotted versus optical depth . Spheres with diameter 0.33, 0.995 and 15.8μm for marks +, x, □ respectively.

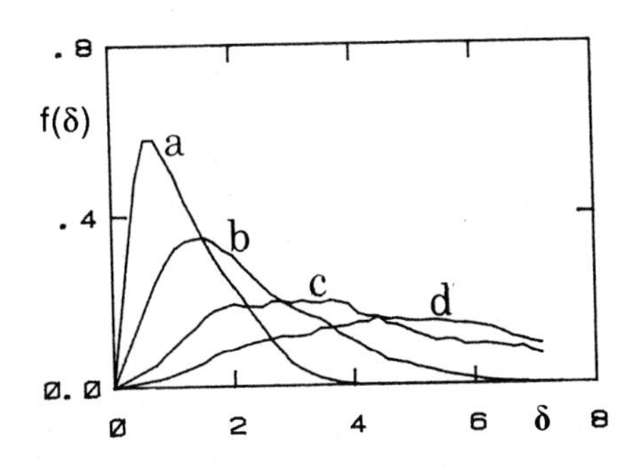

FIGURE 5
Estimation of function f (δ) by means of histograms of the quantity δ. Spheres with diameter 0.33μm . SM code. Curves a, b, c, d refer to τ = 20, 30, 50, 70 respectively.

circumstances in which this type of code is appropriate. Elementary MC code efficiency is high in the cases where the number of positive photons is a substantial fraction of the emitted photons. In particular, the presence of a simple geometry of the problem, as for the case of a plane wave impinging on a homogeneous slab of diffusers, can make the calculation of specific intensity I faster if one uses the elementary MC code. In the mentioned case of a plane geometry, every photon arriving at the plane boundary of the slab has to be considered to be "positive" for the calculation , since I is independent of the position on the boundary. As an example of elementary MC results, fig.6 shows the calculated transmitted specific intensity for the case of a plane wave of unitary density of power, normally incident on a slab

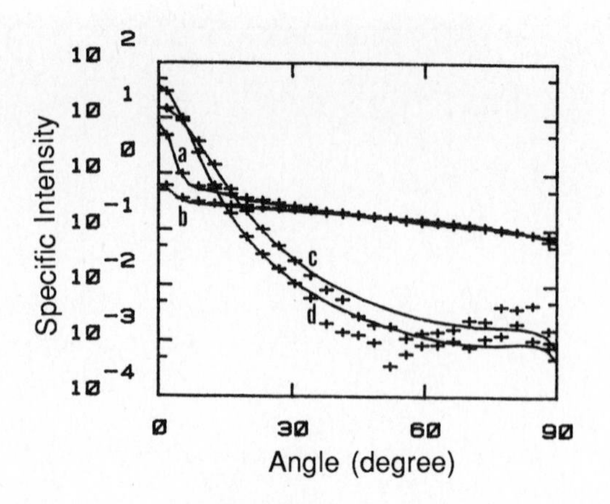

FIGURE 6
Specific intensity after the transmission of a plane wave across a slab of spherical scatterers with size parameter = 50. Crosses: results of calculation with the elementary MC code. The continuous lines show the results obtained by Kuga et al., by means of the discrete ordinate method (Ref.13, Fig.1).

of scatterers. The size parameter of the considered monodispersion of scatterers is 50. The continuous lines to which our data (crosses) are compared show the results obtained for this case by Kuga et al.[13], who used the discrete ordinate method to solve the radiative transfer equation. For curves a and b τ is equal to 5 and 10 respectively, with the real part n_r of the refractive index n equal to 1.01. For curves c and d, τ = 5 and 10 respectively and n_r = 1.3. For all the curves the imaginary part of n is equal to 10^{-9}. The crosses show that our results from the elementary MC code agree with Kuga's data. We point out that the convergence of the results was obtained with a calculation time of ≈ 30 minutes with the minicomputer used.

Another very simple case for which an elementary MC code is suitable is that of an isotropic source in a homogeneous medium. This again is due to the simple symmetry of the problem.

6 - COMPARISON WITH EXPERIMENTAL RESULTS

A series of laboratory experiments were carried out to test our numerical methods. The geometry of the test apparatus is that of Fig.2. A He-Ne laser is the beam source (λ = 0.6328μm). The scattering cell has glass windows with antireflecting coating and blackened lateral walls. It is filled with bydistilled water in which polystyrene spheres are suspended. Three types of spheres were used with diameters 0.33, 0.995 or 15.7μm. The optical receiver area is circular. A series of diaphragms is used to change its radius R. A second series of diaphragms in the focal plane of the optical system is used to change the receiver field of view semiaperture α. The measured received power is indicated as P_r and is the sum of the contributions of received unscattered P_o, and scattered power P_s. Examples of results of calculations with SEMOC code and of measurements are presented in Fig.s 7, 8 by means of the ratio P_s / P_o plotted versus the optical depth τ of the suspension. (The small optical depth of the absorbing water in the absence of spheres is accounted for during the calibration procedure). It was then necessary to evaluate P_s and P_o from the measurements of total received power. In addition, it was worthwhile to make measurements of τ, without relying on data taken from nominal figures given by the supplier of the latex spheres. The evaluation of P_o and P_s from P_r was accomplished by measuring received power at different values of α and by taking into account the different behaviours with α of P_o (whose variation with α depends on the geometry of the measurement and the given beam divergence) and P_s (whose dependence on α can be calculated from the geometry of the measurement and the known scattering properties of the suspended spheres). The extrapolation procedure to obtain P_o from P_r is explained in[14-16] for the cases of a "thin" beam and of a "conical" beam. Given P_o, one obtain the optical depth : $\tau = - \ell n (P_o / P_e)$ where P_e is received power in the absence of the spheres.

Figures 7 and 8 show the comparison of the measured (crosses) and calculated (continuous line) ratio P_s/P_o for the following cases. Figure 7: spheres of 0.33μm, α = 0.5° R = 2 cm (a), 1 cm (b), 0.5 cm (c) D = 8 cm,

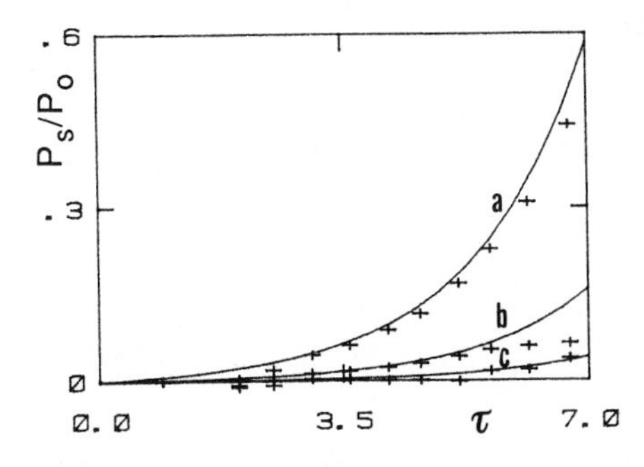

FIGURE 7

Ratio P_s/P_o plotted versus τ, for the transmission of a conical beam in the scattering cell. Crosses: experimental results. Continuous lines: results of SEMOC code. Spheres with diameter 0.33μm a: R = 2 cm, b: R = 1cm, c: R = 0.5cm, α = 0.5°.

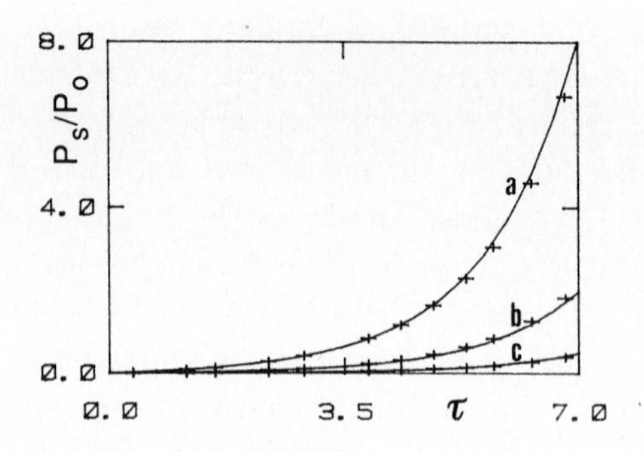

FIGURE 8
Same as Fig.7. Spheres with diameter 0.995 μm.

d = 10 cm, β = 9.3°, L = 10 cm r = 2.95 cm. Figure 8: same parameters as for Fig.7, but with spheres of 0.995 μm. As was explained in Sect.4 the calculation data were obtained at one value of τ for each order of scattering, then scaled to the complete range of τ of the figures. One has to point out that an agreement between calculations and experiments as good as that shown by the figures would not have been obtained if the nominal optical depth of the suspension had been used, instead of the one measured as indicated above.

Figure 9 shows an example of results of an experiment made to test the possibility of measuring the differences in path lengths followed by received scattered in comparison with unscattered radiation (see Fig.3 and the relevant comments in Sect.4). The ratio P_s/P_o is plotted versus τ. In the range $0< \tau < 7.1$ only an aqueous suspension of 0.33 μm spheres was in the cell. The change in τ, in the range $7.1 < \tau < 14$, was obtained by adding purely absorbing dye to the suspension . As was explained in Sect.4, the down slope of ratio P_s/P_o in this second range is due to the larger average path followed by scattered radiation within the cell. The numerical data (obtained by SEMOC code: continuous lines) agree with the experimental data (+ : α = 3°, □ : α = 2°, Δ : α = 1°) . Thin beam: r = 1 mm, β = 0.3 mrad, d = 10 cm, L= 10 cm, D= 8 cm R = 2 cm. An analysis of the down slope gave the following values of $< \delta >$ and $< \delta^2 >$. Measured data $< \delta > = 0.045$ $< \delta^2 > = 0.0016$, SEMOC data $< \delta > = 0.05$, $< \delta^2 > = 0.007$ (the values were practically the same for the three angles α). The agreement is good for $< \delta >$. No more than an order of magnitude comparison was obtained for $< \delta^2 >$. We point out that scattered radiation is only a small part of unscattered received radiation at this moderate value of τ, for this type of small spheres and this geometry.

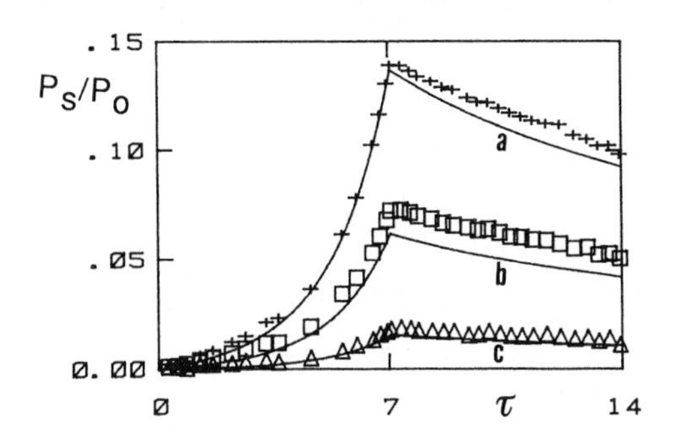

FIGURE 9
Ratio P_s/P_o versus τ. Spheres with diameter 0.33μm. R = 2cm. Thin beam. The continuous curves show calculated results (SEMOC code). For curves a,b,c α = 3°,2°,1° respectively. For $\tau \leq 7.1$ only scattering spheres are in the cell. The addition of dye then changes the optical depth up to 14. Marks represent experimental results.

Examples of results of SM code compared to laboratory experiments are shown in Fig.s 10 and 11. Figure 10 refers to spheres of diameter 0.995μm and to a case of thin beam: R= 2 , r = 1mm, β = 0.3mrad, L = 10cm, D = 16cm, d = 60cm α = 3°. The apparent optical depth: $\tau_a = - \ell n\ (P_r/P_e)$, is plotted versus the true optical depth, $\tau = - \ell n\ (P_o/P_e)$.The marks indicate the results of the SM code. The continuous line is obtained by connecting the results of about 40 laboratory measurements. For $0 < \tau < 10$ the measured values of τ were obtained by means of the extrapolation procedure indicated above for Fig.s 7,8 and 9. The procedure allowed us to establish the linear relationship between the optical depth τ and the weight of the quantity of original suspension inserted into the scattering cell. The relationship was then used to evaluate τ for values larger than 10 (Ref.14).

Figure 11 shows a comparison between measurements and calculations of the function f(δ) defined (see Sect.3) as the probability function for received radiation to have followed a path of length ℓ such that $(\ell - L)/L = \delta$. (f(δ) represents the impulse response of the system, as indicated in Sect. 3). The figure refers to a case of 0.33μm spheres in the cell with optical depth of the suspension τ = 31. The histogram b in the figure shows the results of SM code. The continuous line c indicates the results of best fitting the calculation results with a function: $F(\delta) = A\ \delta^C \exp(-B\ \delta)$, where B,C are constants determined by the best fit procedure and A is a normalization constant such that $\int_0^\infty F(\delta)\, d\delta = 1$. Curve a shows the results of best fitting the measurement results by means of the same type of function. Measurements were carried out by adding different quantities of absorbing

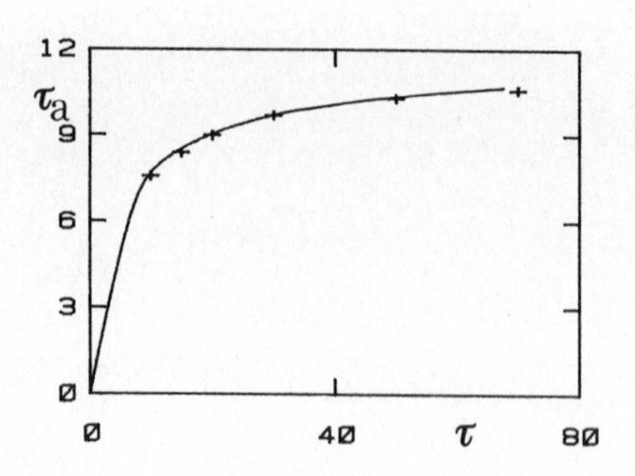

FIGURE 10
Apparent optical depth τ_a plotted versus τ. The marks show the results of calculation (SM code). The continuous lines are obtained by connecting points representing measurements. Spheres with diameter 0.995μm.

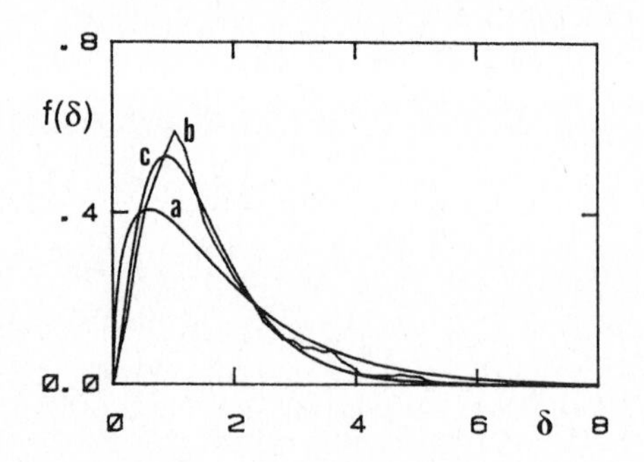

FIGURE 11
f(δ) versus δ. a: curve obtained by fitting experimental data to the function F (δ). b: histogram obtained from SM code. c: best fitting of the histogram to function F (δ). Spheres with diameter 0.33μm. Optical depth of the suspension of spheres: $\tau = 31$.

dye to the suspension in water of latex spheres, and by examining the down slope of received power as a function of the optical depth of the dye, which is known since the relationship between quantity of dye and absorption was previously measured. In this case, unlike the case to which Fig.9 refers, received power is practically all due to scattered power. Measurements and calculation results are the following; $< \delta >$ measured = 1.77, $< \delta >$ calculated = 1.46, $< \delta^2 >$ measured = 5.17, $< \delta^2 >$ calculated = 2.96.

7 - CONCLUSION

By means of a selected number of comparisons between results of calculations and laboratory experiments, we have shown the validity of computation codes used by us to deal with the propagation of light beams in turbid media. We have indicated the different situations for which each version of our Monte Carlo codes is particularly suitable.

References

1 - A. Deepak, U.O. Farrukh, and A. Zardecki: Appl. Opt. 21, 439 (1982)

2 - W.G. Tam and A. Zardecki: Appl. Opt. 21, 2405 (1982)

3 - L.B. Stotts: J. Opt. Soc. Am. 67, 815 (1977)

4 - A. Zardecki and W.G. Tam: Appl. Opt. 21, 2413 (1982)

5 - A. Zardecki, A. Deepak: Appl. Opt. 22, 2970 (1983)

6 - A.Ishimaru: Wave Propagation and Scattering in Random Media, Vol. I (Academic Press, New York 1978)

7 - A.Ishimaru, Y.Kuga, R.L.T.Cheung and K.Shimitzu: J.Opt.Soc.Am. 73, 131 (1983)

8 - E.A.Bucher: Appl.Opt. 12, 2391 (1973)

9 - P.Bruscaglioni, G.Zaccanti, L.Pantani and L.Stefanutti: Int. J.Remote Sensing 4, 399 (1983)

10- L.R.Poole, D.D.Venable and J.W.Campbell: Appl.Opt. 20, 3653 (1981)

11- E.Battistelli, P.Bruscaglioni,A.Ismaelli and G.Zaccanti: J.Opt.Soc.Am._ 2, 903 (1985)

12- P.Bruscaglioni and G.Zaccanti: Optica Acta, 33,1119, (1986)

13- Y..Kuga, A.Ishimaru, Hung-Wen Chang and L.Tsang: Appl.Opt. 25 ,3803 (1986).

14-G.Zaccanti and P.Bruscaglioni: J.mod.Optics.35,229 (1988)

15- E.Battistelli, P.Bruscaglioni, A.Ismaelli, L.Lo Porto and G.Zaccanti: Appl.Opt. 25, 420 (1986)

17- P.Bruscaglioni, G.Zaccanti and M.Olivieri: J.Phys. D: Appl. Phys. 21, S45 (1988)

Scattering in Volumes and Surfaces
M. Nieto-Vesperinas and J.C. Dainty (Editors)

ROLE OF THE INNER SCALE OF ATMOSPHERIC TURBULENCE IN OPTICAL PROPAGATION AND METHODS TO MEASURE IT

A. Consortini

Department of Physics - University of Florence, Via Santa Marta, 3
50139 Florence, Italy

After a short description of atmospheric turbulence and its models, the problem of the effect of the inner scale on laser scintillation in the case of strong turbulence is briefly described. Methods to measure inner scale by means of laser irradiance are mentioned. The second part of the paper is devoted to the effect of the inner scale on phase related quantities such as (differential) angle-of-arrival fluctuations of large beams or lateral displacements (wandering) of thin beams. The possibility of obtaining the values of the inner scale through measurements of the above quantities is examined. With reference to angle-of-arrival fluctuations a property of the correlation functions of phase related quantities that strongly depends on the inner scale is shown. A method for measuring inner scale based on this property is presented.

1. INTRODUCTION

The inner scale of atmospheric turbulence represents the dimension of the smallest turbulent inhomogeneities. Typically it is of the order of millimeters and therefore much larger than the wavelength at optical frequencies. The refractive index fluctuations are small so that the problem of optical propagation through atmospheric turbulence is essentially a forward scattering problem. Here we refer to clear atmosphere, so that additional effects can be neglected. We consider linear propagation, neglecting non linear problems connected with propagation of high power lasers, and finally we do not treat the problem of double pass through the same turbulence where enhancement of intensity (backscatter amplification effect[1]) can take place.

Due to turbulence, an initially coherent radiation, propagating in the clear atmosphere, undergoes random phase and amplitude fluctuations, that deteriorate the performance of optical systems utilizing it. A large number of systems are involved, ranging from astronomical telescopes to ground or space based laser systems with different applications such as communications, remote sensing or image formation. Recently, optical adaptive techniques and phase conjugated methods have also been introduced to correct the effect of phase deterioration.

To know the performance of all these systems and to provide criteria to construct them, the investigation of optical propagation through atmospheric turbulence has received great attention, since the invention of the laser.

Alternatively knowledge of the effect produced by turbulence allows one to use the laser as a powerful tool to probe the turbulence itself. A typical quantity utilized for this purpose is laser irradiance variance (a quantity often called intensity in the literature) which is also sensitive to the inner scale and useful to measure it. Here we are mainly interested in two other quantities, connected to phase fluctuations, which are sensitive to the inner scale, namely angle-of-arrival fluctuations and lateral displacements of thin beams.

In this paper, after a short introduction to describe atmospheric turbulence and a section devoted to the role of inner scale in irradiance fluctuations, we will describe the effects of the inner scale on the phase related quantities and the methods to measure it based on these effects.

2. ATMOSPHERIC TURBULENCE

From the point of view of optical propagation, atmospheric turbulence, at point P, is described by a randomly fluctuating refractive index n. The average value of n coincides with the value n_0 in the absence of turbulence and the fluctuation δn is a small quantity with zero average

2.1) $$n = n_0 + \delta n$$

For practical purposes sometimes the value of n_0 is assumed to be one, but this approximation should not deceive us because one is just the minimum value that the fluctuating index n can reach. The quantity δn is a random function of both point and time and, as is well known, statistical treatment is required. To have complete information one should know the probability density distribution of δn at a point and also its joint probability density distributions at all possible points.

A basic function in the statistics of n, as well as of any fluctuating quantity, is the correlation function of the fluctuations at two points P_1, P_2, $B_n(P_1, P_2)$, also called covariance of the refractive index for obvious reasons. We refer here to conditions of isotropy and homogeneity so that the correlation depends only on the distance r between the two points and is the same for any pair of points. One has

2.2) $$B_n(r) = \langle \delta_n(P_1)\delta_n(P_2) \rangle$$

where brackets denote ensemble averages or infinite time averages, due to the ergodic theorem.

Another function, that allows one to treat problems of "locally" homogeneous turbulence, is the structure function of refractive index

2.3) $$D_n(P_1,P_2) = <[\delta n(P_1) - \delta n(P_2)]^2>$$

because a "slow" varying average is removed by difference.

In the case of homogeneous and isotropic turbulence D_n is a function of r and represents an alternative choice with respect to B_n:

2.4) $$D_n(r) = 2[B_n(0) - B_n(r)]$$

In general, structure functions are more suitable than correlation functions for measurements, because they are less sensitive to systematic or random additive errors affecting the measured quantities at P_1 and P_2 in the same way.

It is well known that the theory of random functions states (Wiener-Khinchin theorem) that the correlation function has a Fourier transform $\Phi_n(\underline{\kappa})$ representing the spatial power spectrum.

In our case the power spectrum is only a function of the modulus κ of $\underline{\kappa}$ called spatial frequency and is given by:

2.5) $$\Phi_n(\kappa) = \frac{1}{2\pi^2\kappa}\int_0^\infty rB_n(r)\sin(\kappa r)dr$$

Inversely

2.6) $$B_n(r) = \frac{4\pi}{r}\int_0^\infty \kappa\Phi(\kappa)\sin(\kappa r)d\kappa$$

An explicit expression for $B_n(r)$ or $D_n(r)$ or for the spectrum $\Phi_n(\kappa)$ is said to be a model of turbulence.

Several different models are found in the theoretical treatments of optical propagation, generally based on the hypothesis that refractive index fluctuations are produced by fluctuations of temperature or wind velocity. Some of them are models easy to be handled mathematically, while others are analytic forms derived from the theory of turbulence developed by Kolmogorov. Kolmogorov's treatment is based on a dimensional analysis of the production of inhomogeneities.

Without entering into the details of these derivations and referring to specialized literature [2,3], we will limit ourselves here to recalling some useful fundamentals and definitions. As is well known an initially laminar flow becomes turbulent when its Reynolds number reaches a limiting value. In the atmosphere very high Reynolds numbers can be reached and initial inhomogeneities (eddies) produced , having a large size, say L_0 and high cinetic energy. They are not stable and break up transferring their energy to inhomogeneities of smaller and smaller size. In the large size eddies energy dissipation is negligible but it increases when the size becomes smaller until a stable dimension ℓ_0 is reached where the energy is dissipated into heat. The two quantities L_0 and ℓ_0 are called outer and inner scale of turbulence respectively and delimit different regions where the structure function behaviour is different

2.7) $$D_n(r) = C_n^2 \ell_0^{2/3} \left(\frac{r}{\ell_0} \right)^2 \qquad r << \ell_0$$

2.8) $$D_n(r) = C_n^2 r^{2/3} \qquad \ell_0 << r << L_0$$

The quantity C_n^2, called the structure constant of the refractive index, is a measure of the turbulence "strength". The region $r << \ell_0$ is the dissipation region, region ℓ_0<<r<< L_0 is the inertial subrange. Eq. 2.8 is well known as "two-thirds law".

Region $r >> L_0$ corresponds to values of r where $B_n(r) \to 0$ and where the turbulence is no longer homogeneous and isotropic. However, for computational reasons sometimes one assumes

2.9) $$D_n(P_1, P_2) = C_n^2 L_0^{2/3} \qquad r > L_0$$

In some other cases, e. g. Tatarskii[2], outer scale L_0 is assumed to be large enough to allow it to tend to infinity

2.10) $$L_0 \to \infty$$

3. INTENSITY FLUCTUATIONS AND INNER SCALE MEASUREMENT METHODS BASED ON SCINTILLATION

The theory of laser scintillation in the atmosphere has received a solution only in the case of small fluctuations. In this case there are analytical expressions of the irradiance variance, σ_I^2, in terms of the parameters of turbulence. They are well known and not described here. (sec. eg Ref.s 2 and 3). In the case of strong fluctuations much progress has been made in recent years but the problem is still under study. Here we want to indicate the point concerning the effect of the inner scale on intensity fluctuations. According to the theory of small fluctuations the variance should indefinitely increase with increasing turbulence, or path length, while experimental results have shown that after a path of some hundred meters a "saturation" effect takes place. More precisely, the variance reaches a peak and is expected to tend to a limiting value, theoretically found, but never reached in practice. Initially some theoretical asymptotic results were found without taking into account the inner scale, or better by letting it vanish. To account for experimental values of the variance much larger then theoretically expected, an inner scale contribution was suggested by Fante [4] and included by him in an asymptotic theory. The inner scale has also been introduced in numerical solutions of the parabolic equation (for the fourth-moment of the field from which σ_I^2 is obtained as a particular case) both for plane wave [5,6,7] and source point [8] and also in numerical simulations of wave propagation by large computers[9]. Unfortunately the theoretical results seem still to underestimate the values of the variance. On the other hand there are a few data where all relevant quantities including the inner scale were simultaneously measured, so that comparison is difficult. A general trend however appears from all authors. For a fixed value of the inner scale, σ_I^2 increases with C_n^2 (and distance) reaching a maximum value, then it decreases to the limiting value of one. For a fixed distance the value (see e.g. fig.2 of Ref.8) of the maximum in σ_I^2 increases for increasing ℓ_0. But the maximum is reached at larger values of C_n^2, so that in some cases an increase of ℓ_0 produces an increase of σ_I^2 and in some others a decrease.

The methods, based on intensity fluctuations, to measure inner scale, are of course developed in the case of small fluctuations and require paths of the order of no more than about a hundred meters. Here we will limit ourselves to mentioning these methods and referring to the authors' original publications and to Ref.10 for a comparison and a discussion of the methods.

Hill and Ochs[11] started measuring the inner scale by using two circular apertures of different size incoherently illuminated with near IR radiation. Subsequently[12] they developed a method based on the ratio of laser-irradiance variance of a small aperture to incoherent large aperture scintillometer variance. Azar et al.[13] developed a method based on bichromatic correlation of laser irradiance and Frehlick[14] obtained inner scale values through measurements of intensity covariance.

4. THE EFFECT OF INNER SCALE ON PHASE RELATED QUANTITIES

As is well known from optics, the effects of smooth changes in the refractive index, like those due to turbulence, are firstly felt on the phase, so that quantities related to phase seem to be suitable for very sensitive methods to investigate turbulence. Two such quantities sensitive to inner scale ℓ_0 are angle-of-arrival fluctuations and lateral displacements of thin beams.

4.1 Angle-of-arrival fluctuations

Angle-of-arrival fluctuations, first investigated by Tatarskii, describe the local deviations of a wave-front with respect to the unperturbed wave-front. In the literature they are often referred to as "differential" angle-of-arrival fluctuations in order to avoid confusion with the beam wandering that is sometimes called angle-of-arrival fluctuation. Here they will be referred to as angle-of-arrival fluctuations without the additional adjective "differential". Angle-of-arrival fluctuation at a point P is defined as the angle that the normal to the wavefront in P makes with respect to the normal in the absence of turbulence. With reference to fig.4.1, where as an example an initially coherent spherical wave is used, $SP_1P'_1$ and $SP_2P'_2$ are the unperturbed directions of two rays; let P_1Q_1 be the normal, at P_1, to the deteriorated wave front at a given time. Quantities α_1 and β_1 are the components of the angle of arrival fluctuation at P_1 in the plane of the two rays and in the normal plane, respectively. Sometimes the plane of the two rays will be referred to as horizontal plane. Due to turbulence α_1 and β_1 fluctuate in time. In the figure a spherical wave was used for simplicity. From now on we will refer to plane waves, unless otherwise explicitly stated.

Tatarskii has investigated the variances of both in plane and out of plane components as well as their correlation functions:

4.1) $$B_\alpha(\rho) = <\alpha_1\alpha_2>$$

4.2) $$B_\beta(\rho) = <\beta_1\beta_2>$$

where ρ denotes the distance between the two points. By using a simple argument, Tatarskii[2] showed that at the output of a turbulent layer one has:

4.3) $$B_\alpha(\rho) = \frac{1}{2}D''_\theta(\rho)$$

4.4) $$B_\beta(\rho) = \frac{D'_\theta(\rho)}{2\rho}$$

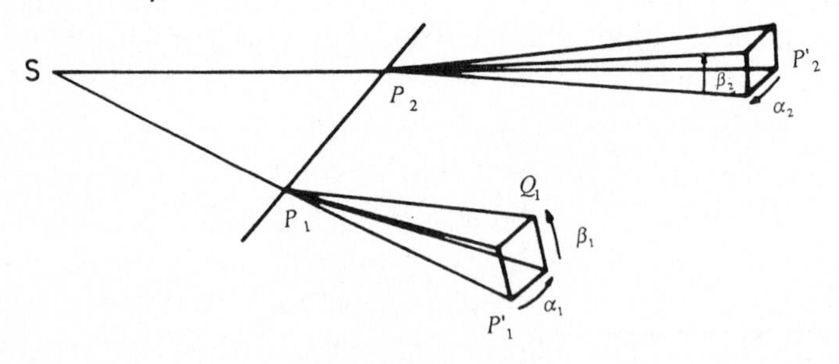

FIGURE 4.1

Schematic representation of fluctuations α in plane and β out-of-plane of a spherical wave at the emergence from a turbulent layer where apexes denote derivatives with respect to the distance ρ and $D_\theta(\rho)$ is the structure function of the phase fluctuations at the two points P_1, P_2. For this reason we call these angle-of-arrival fluctuations phase related effects.

In the simple geometrical optics approximation one has

4.5) $$D_\theta(\rho) = \int_0^L \int_0^L B_n(r) dx_1 dx_2$$

where L denotes the thickness of the turbulent layer. The integral is to be evaluated along the two unperturbed rays passing through P_1 and P_2, x_1 and x_2 denote points on the first and second ray respectively and r their distance.

The theory of Tatarskii, developed in terms of refractive index spectrum, in the case of a very long path, gives:

4.6) $$D_\theta(\rho) = 8\pi^2 L \int_0^\infty [1 - J_0(\kappa\rho)] \Phi_n(\kappa) \kappa \, d\kappa$$

where J_0 denotes the Bessel function of zero order. By using Eq.s 4.3 and 4.4 one obtains

4.7) $$B_\alpha(\rho) = 4\pi^2 L \int_0^\infty \left[J_0(\kappa\rho) - \frac{J_1(\kappa\rho)}{\kappa\rho} \right] \Phi_n(\kappa) \kappa^3 d\kappa$$

4.8) $$B_\beta(\rho) = 4\pi^2 L \int_0^\infty \frac{J_1(\kappa\rho)}{\kappa\rho} \Phi_n(\kappa) \kappa^3 d\kappa$$

The convergence of the integrals for $\rho = 0$ is guaranteed by the "cutoff" factor of $\Phi_n(\kappa)$ at high spatial frequencies so that B_α and B_β are sensitive to inner scale ℓ_0. As a consequence the correlation radius for angle of arrival fluctuations is of the order of the inner scale.

Tatarskii evaluated $B_\alpha(\rho)$ and $B_\beta(\rho)$ for a spectrum of the refractive index given by

4.9) $$\Phi_n(\kappa) = 0.033\, C_n^2 \kappa^{-11/3} e^{-\kappa^2/\kappa_m^2}$$

where

4.10) $$\kappa_m = 5.92 / \ell_0$$

This spectrum corresponds to Kolmogorov turbulence having a finite inner scale and infinite outer scale, see Eq.s 2.7 and 2.8. We refer to this model as the Kolmogorov-Tatarskii model.

In Fig. 4.2 from Tatarskii, quantities

$$b_\alpha(\rho) = B_\alpha(\rho) / \langle \alpha^2 \rangle \qquad \text{and} \qquad b_\beta(\rho) = B_\alpha(\rho) / \langle \beta^2 \rangle$$

where

4.11) $$<\alpha^2>= B_\alpha(0) =<\beta^2>= 3.3C_n^2 L\ell_0^{-1/3}$$

are plotted versus $\kappa_m \rho$

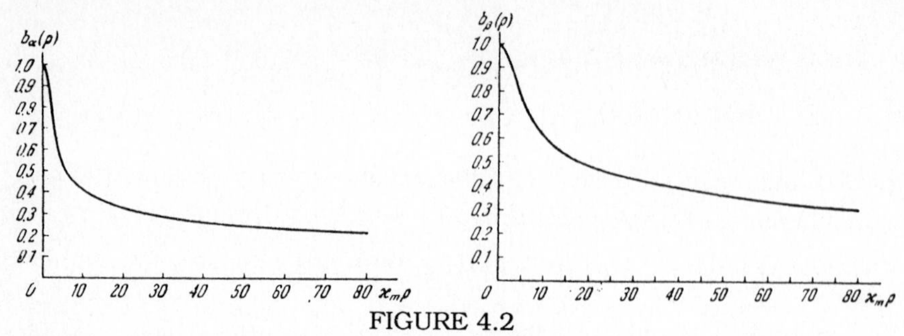

FIGURE 4.2
Correlation coefficients $b_\alpha(\rho)$ and $b_\beta(\rho)$ plotted versus $\kappa_m \rho$ from Tatarskii[2]

4.2 Thin beam lateral displacements (wandering)

By thin beams we mean beams whose transversal dimension can be considered small with respect to the inner scale of turbulence, but large enough to have negligible diffraction. Provided that the path is short enough these two requirements can be satisfied.
With reference to fig.4.3 let x denote the propagation direction of two parallel laser beams lying in the plane xy. For simplicity plane xy will be referred to as horizontal plane and z as vertical direction. Let d denote the distance between the beams and L the turbulent path, as before. In the absence of turbulence the two beams cross the output plane $x = L$ at two points P_1 and P_2. Due to turbulence the points of crossing P'_1 and P'_2 fluctuate randomly around P_1 and P_2 respectively.

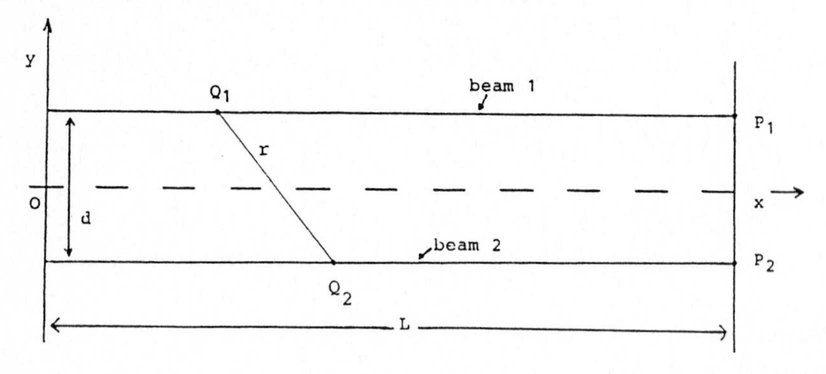

FIGURE 4.3
Geometry of parallel beam propagation

The fluctuating differences $\eta_1 = y'_1 - y_1$ and $\zeta_1 = z'_1 - z_1$ between the coordinates of P'_1 and P_1 describe the wandering of the first beam and the analogous quantities η_2 and ζ_2 describe that of the second beam. Quantities η will be referred to as horizontal fluctuations and ζ as vertical fluctuations. In the limits of geometrical optics one can write

$$\eta_1 = \int_0^L \alpha(Q_1)\,dx_1 \tag{4.12}$$

$$\zeta_1 = \int_0^L \beta(Q_1)\,dy_1 \tag{4.13}$$

where $\alpha(Q_1)$ and $\beta(Q_1)$ denote the in plane and out-of-plane angle-of-arrival fluctuations respectively at the point Q_1 along the first ray. Analogous relations hold for η_2 and ζ_2. By multiplying η_1 and η_2 and averaging one obtains

$$B_y(d) = <\eta_1\eta_2> = \int_0^L\int_0^L B_\alpha(r)\,dx_1 dx_2 \tag{4.14}$$

where

$$B_\alpha(r) = <\alpha(Q_1)\alpha(Q_2)>$$

is the correlation function of angles α at two points Q_1, Q_2 along the two rays respectively and $r = |Q_1Q_2|$. Here use has been made of the hypothesis of homogeneity. Analogously one has

$$B_z(d) = \int_0^L\int_0^L B_\beta(r)\,dx_1 dx_2 \tag{4.15}$$

with analogous meaning for $B_\beta(r)$. Although B_y and B_z are in the form of integrals one can try to make some useful considerations. Lateral wanderings η_1 and ζ_1 are obtained by integrating angle-of-arrival fluctuations along the path. Therefore B_y and B_z have a "higher" dependence on L than B_α and B_β . They are also expected to have the same correlation radius as the angles of arrival correlation functions. These two facts can enable these functions to be particularly suitable for measurements to derive inner scale information. This is confirmed by an evaluation made by us in the case of a simple model of turbulence . Although for this quantity care has to be taken in utilizing results which are valid for the infinite path, most of the conclusions of the following sections devoted to angle-of-arrival fluctuations should be valid also for lateral displacements.

5. FIRST VALUES OF THE INNER SCALE BASED ON ANGLE-OF-ARRIVAL FLUCTUATION MEASUREMENTS

As appears from figure 4.2, measurements of $B_\alpha(\rho)$ and $B_\beta(\rho)$ can permit an evaluation of the inner scale. Experimental values of the inner scale were initially obtained by us through measurements of angle-of-arrival fluctuations, both in laboratory generated turbulence and in the atmosphere [15]. For these measurements a spherical wave was used. After a path in the atmosphere it was made to impinge on a mask with small holes, 2.5 mm diameter, in both horizontal and vertical directions. This constituted a kind of Hartman test.The thin beams emerging from the mask were made to travel along a path free of turbulence to have a lever effect, then to impinge on a diffusing screen. Behind the screen a camera was placed and practically frozen-in photograms were obtained. Subsequently the positions of the spot centres, were manually read and the structure functions $D_\alpha(\rho)$ and $D_\beta(\rho)$ were computed. Fig.5.1 a) and b) represent a set of results obtained for horizontal points in the open atmosphere. Circles and crosses were obtained by two different readers of the same photograms. In fig.5.1 a) the measured $C_{\alpha\beta}$ is also represented as dots and triangles. The dashed curve is obtained from the Tatarskii theory in the case of a spherical wave, for $\ell_0 = 9mm$ and $C_n^2 = 5.4\ 10^{-13} m^{-2/3}$. The tendency to increase more rapidly than expected for large ρ can be explained as an outer scale effect or as an effect[15] due to the path between mask and screen.

From these measurements it appeared for the first time that the inner scale could be much larger than 1 mm as expected at that time. The result

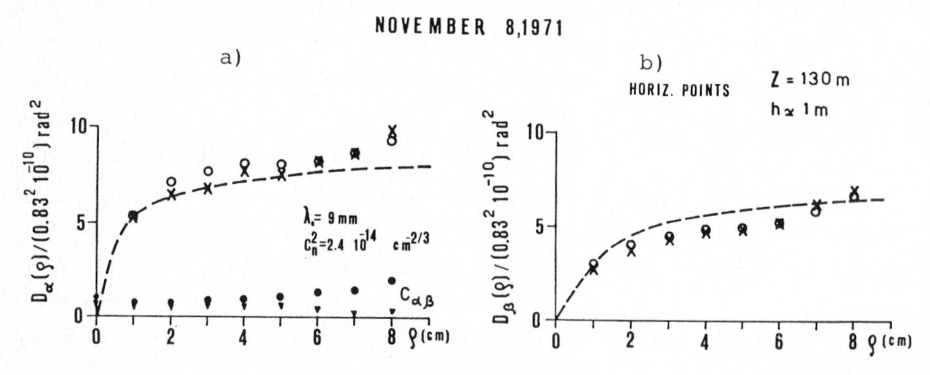

FIGURE 5.1
Measured structure functions of a spherical wave in the atmosphere. Dashed lines from Tatarskii theory

was confirmed by subsequent sets of measurements, done by us in the following year, and also by the results of Hill and Ochs[11].

However, due to the elaboration length, measurements of the inner scale with experiments of this kind appeared impractical, at that time. Recently however, the development of position sensors and the techniques for wavefront sensing required by adaptive optics have made this kind of measurement more practical. Moreover a property of the correlation functions of phase related quantities allows us to suggest a different method to measure inner scale that will be described in the next section, with reference to angle-of-arrival fluctuations.

6. A METHOD TO MEASURE INNER SCALE BASED ON A PROPERTY OF PHASE RELATED QUANTITIES

6.1- Property and definition of the "difference function".

Let us start by noting a property of the correlations functions of angle-of-arrival fluctuations, for a homogeneous medium.

As appears from Eq.s 4.7 and 4.8 , correlation $B_\alpha(\rho)$ and $B_\beta(\rho)$ are different functions of ρ. For a given ρ, they generally assume different values and are only equal in the two limiting cases $\rho=0$ and $\rho\to\infty$. For a given small ρ correlation $B_\alpha(\rho)$ is lower than correlation $B_\beta(\rho)$. This is due to physical reasons because ρ is the distance in the horizontal plane, so that the in-plane fluctuations start from a distance ρ, while the out-of-plane fluctuations start from a zero "vertical" distance. Therefore out-of-plane fluctuations are more correlated than in-plane-fluctuations. This property also suggests that the function (difference function)

6.1) $$\Delta(\rho)=B_\beta(\rho)-B_\alpha(\rho)$$

which vanishes for $\rho = 0$ and $\rho \to \infty$, be a positive function, firstly increasing with ρ then decreasing for large ρ and having a maximum value when ρ is equal to the inner scale.

This property seems very interesting because it can permit the measurement of the inner scale by finding experimentally the maximum of function $\Delta(\rho)$.

Starting from Eq.4.7 and 4.8 one immediately obtains

6.2) $$\Delta(\rho)=4\pi^2 L\int_0^\infty J_2(\kappa\rho)\Phi_n(\kappa)\kappa^3 d\kappa$$

Let us recall that this expression holds for very long paths $(L\to\infty)$.

Alternatively if the known quantity is the structure function of phase fluctuations $D_\varphi(\rho)$ one has

6.3) $$\Delta(\rho)=\frac{1}{2}\left[\frac{D'_{\theta}(\rho)}{\rho}-D''_{\theta}(\rho)\right]$$

We utilized Eq.6.2 or 6.3 to evaluate $\Delta(\rho)$ for a number of different models, as is described below and we checked it experimentally.

6.2 - Gaussian model.

The Gaussian model has simple matematical expressions for the correlation function of refractive index fluctuations $B_{nG}(r)$

6.4) $$B_{nG}(r)=<\delta^2\mu>e^{-r^2/r_0^2}$$

where $<\delta\mu^2>$ denotes refractive index variance. Quantity r_0 is called "scale" of the turbulence and can be considered equally inner or outer scale. Moreover being mathematically simple, this model can be utilized to describe turbulence with Reynolds number not too high, where the inertial range is small. Function $\Delta(\rho)$ for the Gaussian model, first defined by eq.3 of Ref.16, was there evaluated in the limits of the geometrical optics, by using Eq.6.3. One has

6.5) $$\Delta_G(\rho)=2B_{\alpha G}(0)\frac{\rho^2}{r_0^2}e^{-\rho^2/r_0^2}$$

where the subscript G in $\Delta_G(\rho)$ refers to the Gaussian model. [1]

Quantity $B_{\alpha G}(0)$ represents the variance of α. For a long-path $L>>r_0$ one has:

6.6) $$B_{\alpha G}(0)=\frac{2<\delta\mu^2>}{r_0}\sqrt{\pi}\,L$$

From Eq.6.6 it is evident that $\Delta_G(\rho)$ is a positive function having a maximum for

6.7) $$\rho=r_0$$

In fig.6.1 two examples of application are reported from ref.16 in the case of laboratory generated thermal turbulence. Measurements of angle-of-arrival structure functions were made by using the apparatus and photographic technique previously mentioned. Curves and parameters represent fitting values of the Gaussian model. Although there are negative values denoting an anisotropy, the presence of the maximum is evident.

6.3 - Kolmogorov - Tatarskii spectrum

As already stated, by Kolmogorov-Tatarskii spectrum we denote the model represented by Eq.s 2.7 and 2.8, where Eq.2.8 is assumed to be valid from ℓ_0 to infinity, which means an infinitely large outer scale. The corresponding spectrum is given by Eq.4.9. The difference function

[1] Eq.5 of ref 16 has a clear error of sign.

$\Delta(\rho)$ for this spectrum, here denoted by $\Delta_T(\rho)$, is given by

$$\Delta_T(\rho) = <\alpha^2> \frac{\kappa_m^2\rho^2}{24}\, {}_1F_1\left(\frac{7}{6}, 3, -\frac{\kappa_m^2\rho^2}{4}\right) \tag{6.8}$$

where ${}_1F_1$ denotes the hypergeometric confluent function defined as

$${}_1F_1(a,b,z) = \frac{\Gamma(b)}{\Gamma(a)} \sum_{n=0}^{\infty} \frac{\Gamma(n+a)}{n!\Gamma(n+b)} z^n \tag{6.9}$$

$\Gamma(x)$ is the Euler function, and $<\alpha^2>$ is given by Eq. 4.11.
In fig.6.2 a) the $\Delta(\rho)$ function of K-T model is represented, normalized to the corresponding angle-of-arrival variance plotted versus $\kappa_m\rho = 5.92\rho / \ell_0$. As appears from this figure a maximum is present, that allows one to obtain ℓ_0. As a comparison $\Delta_G(\rho)$ from Gaussian model is reported in Fig.6.2 b) plotted versus ρ / r_0.The different shape of the two curves can give additional information on the model.

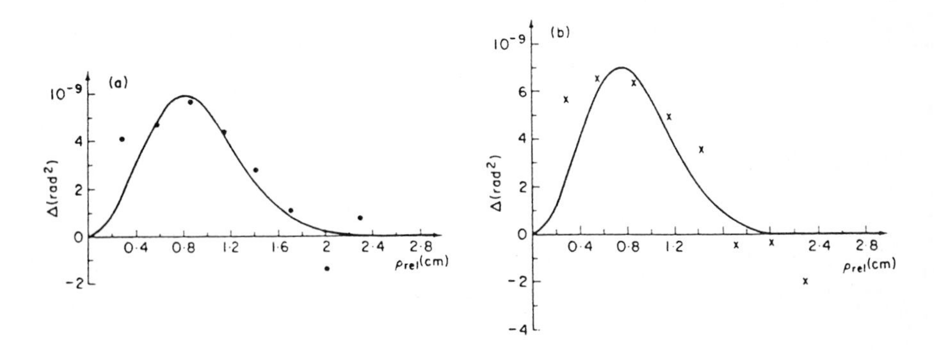

FIGURE 6.1
Two examples of experimental difference function, crosses and dots, plotted versus ρ. Curves are from Gaussian model with: a) r_0=8 mm, $<\alpha^2>$= 8 10^{-9} rad^2, and b) r_0= 7.5 mm, $<\alpha^2>$= 8 10^{-9} rad^2

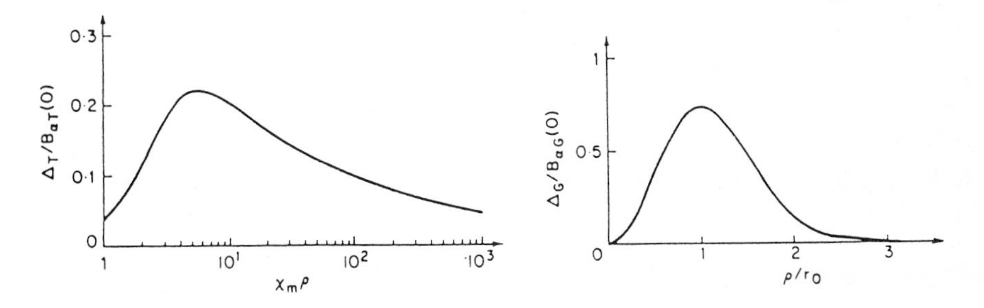

FIGURE 6.2
Normalized difference function from: a) Kolmogorov-Tatarskii spectrum ($L_0 = \infty$) plotted versus $\kappa_m\rho$, and b) Gaussian spectrum plotted versus ρ / r_0

6.4 - Karman modified model

The spectrum of this model, given by

6.10) $$\Phi_n(\kappa)=0.033C_n^2 e^{-\kappa^2/\kappa_m^2}\left[\kappa^2+\kappa_0^2\right]^{-11/6}$$

where $\kappa_0=1/L_0$ allows one to take into account outer scale L_0 of turbulence. This spectrum reduces to the previous one when $L_0\rightarrow\infty$. It is possible to give an expression of $\Delta_K(\rho)$ (K stands for Karman) in terms of hypergeometric functions, however for our purposes a numerical evaluation is more useful. In fig.6.3 the difference between B_β and B_α is plotted versus $\kappa_m\rho$ for a number of values of the ratio $p=\ell_0/L_0$ between inner and outer scale. As appears from the figure when $p\approx 10^{-5}$ the situation corresponding to an infinite outer scale is reached. The maximum lies in correspondence to the inner scale value. The effect of outer scale L_0 is felt essentially on the value of the maximum and on the shape for large value of ρ. Increasing L_0 gives a deeper decreasing of Δ_K as a function of ρ.

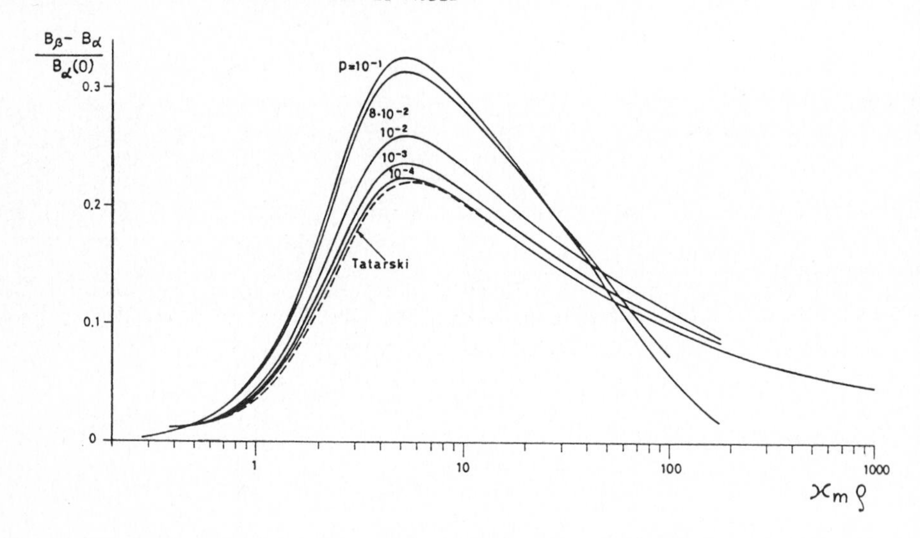

FIGURE 6.3
Normalized difference function from Karman modified model plotted versus $\kappa_m\rho$ for a number of values of $p=\ell_0/L_0$

6.5 - Moroder two-scale model

An analytical model for the structure function of the refractive index, suggested by Moroder some years ago[17], takes into account both inner and outer scale and describes Kolmogorov turbulence in both the inertial and dissipative regions. According to the model one has

$$D_n(r) = C_n R_0^{2/3}\left[1-e^{-r^2/\ell_0}\right]\left[1+\left(\frac{\ell_0}{R_0}\right)^{2/3} - e^{(-r/R_0)^{2/3}}\right] \tag{6.11}$$

where R_0 is of the order of the outer scale L_0. Moroder evaluated phase and angle-of-arrival structure functions. By utilizing Moroder results the function $\Delta_M(\rho)$ (where M stands for Moroder) of our interest is easily obtained. In Fig.6.4 function $\Delta_M(\rho)$ normalized to $B_\alpha(0)$ is plotted versus ρ/ℓ_0, for a number of values of the ratio $p_1=\ell_0/R_0$. Here the position of the maximum is slightly displaced towards values of ρ larger than the inner scale. As before the outer scale has influence on the decay region.

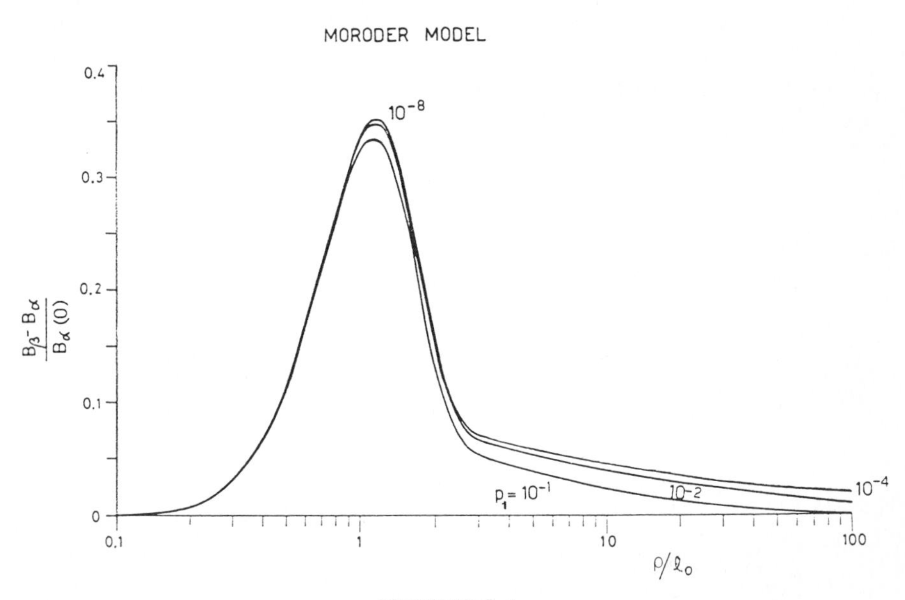

FIGURE 6.4

Normalized difference function from Moroder model plotted versus ρ/ℓ_0 for a number of values of the ratio $p_1=\ell_0/R_0$

6.6 - Hill model

A model suitable to describe the real spectrum in the inner scale region was given by Hill [18]. This model presents a "bump" in the spectrum as a function of κ in the inner scale region. We have numerically evaluated the difference function $\Delta_H(\rho)$ for Hill spectrum, by utilizing numerical values from Hill. In fig. 6.5 the difference function is plotted versus ρ/η where η is the Kolmogorov microscale. In air one has

6.12) $\ell_0 / \eta = 7.4$

From the figure it appears that the abscissa of the maximum is

6.13) $\rho \approx 9\eta = 1.2\ell_0$

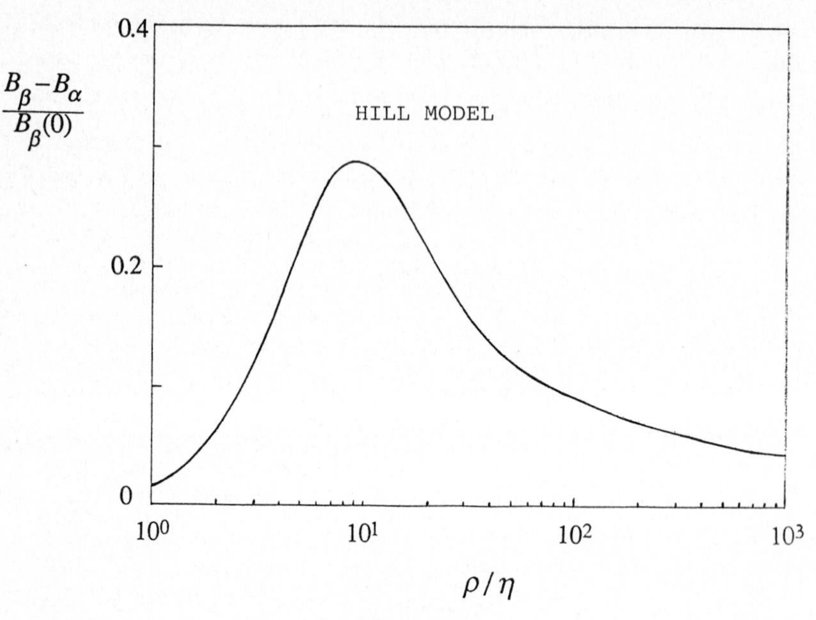

FIGURE 6.5
Normalized difference function from Hill model $(L_0 = \infty)$ plotted versus $\rho / \eta = 7.4\ \rho / \ell_0$

7. CONCLUSION

The effects of the inner scale of atmospheric turbulence on optical radiation propagated through the atmosphere have been reviewed. In particular the problem of the inner scale role in strong scintillation has been described. The influence of the inner scale on phase related quantities, such as (differential) angle-of-arrival fluctuations or wandering of thin beams, has been shown. A property of the correlation functions of phase related quantities has been presented. It allowed us to develop a method to measure inner scale. The method is based on the experimental determination of the maximum of a suitable correlation function "difference function". The method seems largely independent of the model of turbulence, as appears from the results of the following examined models: Gaussian, Kolmogorov-Tatarskii, Karman modified, Moroder, and

Hill. Practical application requires measurement of very small angle-of-arrival fluctuations even for non short paths. These measurement, already possible some years ago [19,15,20], are now easier due to the improved technology for wave front sensing. Lateral displacement (wandering) of thin beams is also a candidate for an analogous method [21], suitable for very short paths of the order of ten meters.

REFERENCES

1) A.G. Vinogradov, Yu.A. Kravtsov and V.I. Tatarskii, "Backscatter Amplification effect for bodies located in a medium with random inhomogeneities", Izv. Vyssh. Uchebn. Zaved Radiofiz. Vol.16, 1064-1070, 1973.
See also Yu.A. Kravtsov and A.I. Saichev.,"Effect of round-trip transmission of waves in randomly inhomogeneous media", Sov. Phys. Uspekhi Vol.25, 167, 1982.

2) V.I. Tatarskii, "The effects of the turbulent atmosphere on wave propagation", NTIS National Technical. Information Services US Dept. of Commerce, Springfield Va., (1971)

3) Akira Ishimaru, "Wave propagation and scattering in random media", Vol.1 and 2. Academic Press 1978.

4) R.F. Fante,"Inner-scale size effect on the scintillations of light in the turbulent atmosphere"J. Opt. Soc. Am. Vol. 73, 277 (1983)

5) B.J. Uscinski,"Intensity fluctuations in a multiple scattering medium. Solutions of the fourth momentum equation". Proc. R. Soc. Lond. A Vol. 380 137 (1982)

6) A.M. Whitman, M.J. Beran,"Two-scale solutions for atmospheric scintillations" J. Opt. Soc. Am. A Vol.12, 2133 (1985)

7) M. Spivack, and B.J. Uscinski," Accurate numerical solutions of the fourth moment equations for very large values of Γ" J. Mod. Opt. Vol.35, 1741 (1988)

8) A.M. Whitman and M.J. Beran,"Two-scale solutions for atmospheric scintillations from a point source" J. Opt. Soc. Am. A Vol.5, 735 (1988)

9) J.M. Martin and S.M. Flatté,"Intensity images and statistics from numerical simulation of wave propagation in 3-D random media" Appl. Opt. Vol.27, 2111 (1988)

10) R.J. Hill,"Comparison of scintillation methods for measuring the inner scale of turbulence" Appl. Opt. Vol.27, 2187 (1988)

11) R.J. Hill and G.R. Ochs,"Fine calibration of large-aperture optical scintillometers and an optical estimate of inner scale of turbulence",

Appl. Opt. Vol.17, 3608 (1978)

12) G.R. Ochs and R.J. Hill,"Optical-scintillation method of measuring turbulence inner scale", Appl. Opt. Vol.24, 2430 (1985)

13) Z. Azar E. Azoulay and M. Tur,"Optical bichromatic correlation method for remote sensing of inner scale" in CLEO Tech. Digest Series 1987 Vol.14 (Opt. Soc. Am., Washington, DC 1987), p.4.
See also: E. Azoulay, V. Thiermann, A. Jetter, A. Kohnle and Z. Azar, Optical measurement of the inner scale of turbulence, J. Phys D: Appl.Phys. ECOOSA/88, Vol.21, S41-S44 (1988); and also E. Goldner and A. Weitz, Two-color correlation of scintillations, Appl. Opt. Vol.25, 3486 (1986)

14) W.A. Coles and R.G. Frehlich,"Simultaneous measurements of angular scattering and intensity scintillation in the atmosphere, J.Opt. Soc. Am. Vol.72, 1042-48 (1982) sec also Appl. Opt. Vol.27, 2194 (1988)

15) A. Consortini,"Measurements of angle of arrival fluctuations of a laser beam due to turbulence. AGARD Conference Proceedings N°183 on Optical Propagation in the Atmosphere. ISBN 92-835-0164-0 p.26.1-26.7, May 1976.

16) A. Consortini, P. Pandolfini, C. Romanelli and R. Vanni,"Turbulence investigation at small scale by angle-of-arrival fluctuations of a laser beam" Optica Acta Vol.27, 1221-1228 (1980).

17) E. Moroder and F. Pasqualetti,"Study of a particular model of atmospheric turbulence" Atti della Fondazione Giorgio Ronchi, Vol.26, 335-348, (1971).

18) R.J. Hill and S.F. Clifford,"Modified spectrum of atmospheric temperature fluctuations and its application to optical propagation" J. Opt. Soc. Am. Vol.68, 892-899, (1978).

19) B.D. Borisov, V.M. Sazanovich and S.S. Khmelevtosov,"Investigation of the fluctuations in the angles of arrival of laser radiation in the surface layer of the atmosphere" Iz. V. U. k. Zav. Fiz. Vol.12, 1, 103-106, (1969).

20) J. Borgnino and F. Martin,"Correlation between angle-of-arrival fluctuations on the entrance pupil of a solar telescope" J. Opt. Soc. Am. Vol.67, 1065-1072, (1977)

21) A. Consortini, K.A. O'Donnell and G. Conforti,"Determination of the inner scale of atmospheric turbulence through laser beam wandering" Proceedings 14th Congress of the International Commission for Optics, Quebec, Canada (Aug. 1987). SPIE Vol.813, 117 (1987).

Scattering in Volumes and Surfaces
M. Nieto-Vesperinas and J.C. Dainty (Editors)

DIFFRACTION TOMOGRAPHY: POTENTIALS AND PROBLEMS

Leiv-J. Gelius and Jakob J. Stamnes

Norwave Development A.S, Forskningsveien 1, 0371 Oslo 3, Norway*

A tutorial description of diffraction tomography is first given. Then the validity of this method is examined using exact scattered data for a penetrable elliptical cylinder and reconstructions by means of the filtered backpropagation algorithm. Finally the potentials and problems of diffraction tomography are discussed.

1. INTRODUCTION

In recent years much effort has been expended at developing inversion algorithms for tomography with diffracting wavefields, the earliest developments[1-3] being based on Wolf's formulation of the inverse scattering problem.[4]

Diffraction tomography (DTG), as it is now called, is well suited for transmission data and has a lot in common with x-ray tomography.[5] But unlike the x-ray method, DTG is not based on straight ray paths and can account for diffraction in the reconstruction process.

The governing principle of DTG is the generalized projection-slice (GPS) theorem,[1] which (in two dimensions) relates the one-dimensional spatial Fourier transform of the measured data to the two-dimensional spatial Fourier transform of the object along a portion of a circle in the object's Fourier space. The GPS theorem is based on plane-wave expansions and normally requires a homogeneous background medium. In the past several different implementations of the GPS theorem have been proposed,[1,6,7] mostly for applications in medical diagnostics, but some authors [8,9] have discussed how to extend medical DTG to cover geophysical applications.

In this paper we first give a tutorial description of DTG and then examine its validity using exact scattered data for a penetrable elliptical cylinder and reconstructions by means of the filtered backpropagation algorithm.[5] Finally we discuss the potentials of DTG and also some of the problems that must be solved to make the method practicable.

*Research supported by the Royal Norwegian Council for Scientific and Industrial Research.

2. FUNDAMENTALS OF ACOUSTICAL DIFFRACTION TOMOGRAPHY

2.1. Physical model

For brevity we limit our discussion to line sources and 2D objects embedded in a homogeneous background medium. Also, we restrict our attention to weak scatterers so that we can apply Fourier methods based on the generalized projection-slice theorem to obtain a simple relationship between the scattered field data and the Fourier transform of the scattering object.

Thus we consider inverse methods that are based on the following assumptions:

1) Uniform (homogeneous) background
2) Line sources or plane-wave illumination and 2D objects
3) Detectors regularly sampled at the Nyquist rate
4) Weak scattering objects

Fig. 1 shows schematically the geometry of a typical crosshole aquisition in geophysics, where the incident wavefields are generated using line sources at different depths in the well to the left. Since we have a uniform background, it is more convenient to work with plane waves, which we may synthesize using conventional slant-stacking of the line-source fields. A typical set up for ultrasound tomography is shown in Fig. 2, where the incident fields are plane waves at different angles of incidence.

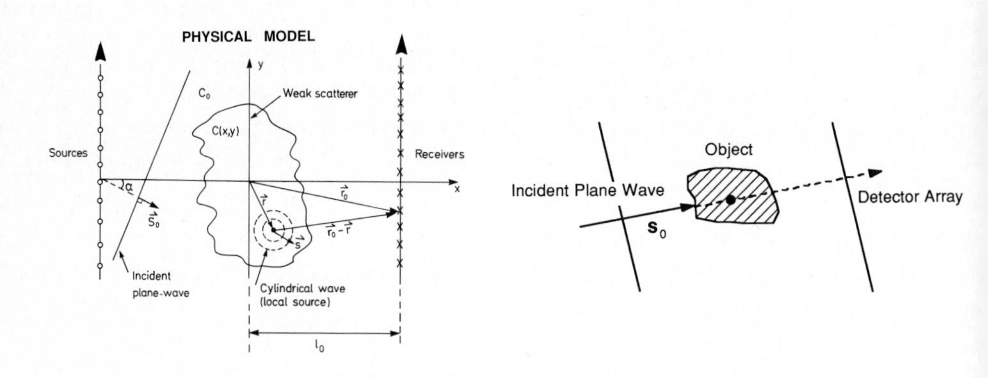

FIGURE 1
Scattering model and a crosshole aquisition geometry in geophysics.

FIGURE 2
Ultrasound tomography set up with plane-wave illumination.

Each incident plane wave hits the scatterer and generates secondary source fields. In general these secondary sources will interact, but in the weak-scattering approximation each infinitesimally small part of the scattering object acts as an <u>independent</u> secondary line source. The term weak scattering also implies that we neglect multiple scattering. The total wavefield then is the sum of the incident (or background) plane-wave field and the scattered field, where the latter is the sum of the fields generated by all the local, independent, secondary sources within the scatterer. This result is in accordance with Huygens' principle. Intuitively we expect the strength of each secondary source to be a function of the ratio c_0/c, where c_0 and c are the sound velocities of respectively the background and the scatterer.

2.2 Mathematical model

We now give a mathematical description of the physical model discussed above. For brevity we restrict our discussion to the geometry in Fig. 1 and refer the reader elsewhere[5] for a treatment of the geometry in Fig. 2. First we define the scattered field P_S and the object profile O as follows:

$$P_S(\vec{r}_0) = P(\vec{r}_0) - e^{ik\vec{s}_0 \cdot \vec{r}_0}, \tag{2.1a}$$

$$O(\vec{r}) = 1 - c_0^2/c^2(\vec{r}), \tag{2.1b}$$

where $\vec{r}$ and $\vec{r}_0$ are the position vectors of respectively a secondary source point and the observation point, i.e.

$$\vec{r} = x\,\vec{i} + y\,\vec{j} \quad ; \quad \vec{r}_0 = l_0\vec{i} + y_0\vec{j}, \tag{2.1c}$$

with the distance l_0 being as shown in Fig. 1. $\vec{i}$ and $\vec{j}$ are unit vectors along the x-axis and y-axis, respectively.

Equation (2.1a) shows that the scattered field $P_S(\vec{r}_0)$ is the difference between the total field $P(\vec{r}_0)$ and the incident plane wave $\exp(ik\vec{s}_0 \cdot \vec{r}_0)$, the latter being of unit amplitude and propagating in the direction of the unit vector $\vec{s}_0$.

<u>Forward problem</u>

In the forward problem the object profile and the incident field are known and our task is to determine the scattered field. To that end, we start with the acoustical wave equation

$$[\nabla^2 + k^2(\vec{r})]\, P(\vec{r}) = 0, \tag{2.2a}$$

where

$$k^2(\vec{r}) = \frac{\omega^2}{c^2(\vec{r})} = k^2\,[1-O(\vec{r})] \quad ; \quad k = \frac{\omega}{c_0}. \tag{2.2b}$$

Substituting Eq. (2.2b) in Eq. (2.2a), we can rewrite the latter as

$$(\nabla^2 + k^2)\, P(\vec{r}) = k^2\, O(\vec{r})\, P(\vec{r}). \tag{2.3}$$

The physical interpretation of Eq. (2.3) is as follows. If the scattering object is removed so that $O(\vec{r}) = 0$, then the solution of Eq. (2.3) is the incident field. When $O(\vec{r}) \neq 0$, the secondary sources on the right-hand side of Eq. (2.3) generates a scattered field $P_S(\vec{r}_0)$, whose <u>exact</u> solution is

$$P_S(\vec{r}_0) = k^2 \int d^2\vec{r} \ \ O(\vec{r})P(\vec{r}) \ G(\vec{r}-\vec{r}_0), \tag{2.4a}$$

where

$$G(\vec{r}-\vec{r}_0) = -\frac{i}{4} H_0^{(1)}(k|\vec{r}-\vec{r}_0|), \tag{2.4b}$$

is the Green function of a line source in a homogeneous medium. Here $H_0^{(1)}(k|\vec{r}-\vec{r}_0|)$ is the zeroth-order Hankel function of the first kind, and $k = \omega/c_0$ is the wavenumber of the homogeneous background.

Note that the integrand in Eq. (2.4a) contains the total field P, which is the sum of the incident field $P_0 = \exp(ik\vec{s}_0\cdot\vec{r})$ and the scattered field P_S. This makes the relationship between the scattered field P_S and the object profile O very complicated. To simplify the problem we now apply the weak-scattering approximation to Eq. (2.4a). Thus, inside the scatterer we replace the total field P by the unperturbed incident field, i.e. we set

$$P(\vec{r}) \simeq P_0(\vec{r}) = \exp(ik\vec{s}_0\cdot\vec{r}). \tag{2.5}$$

to obtain from Eqs. (2.4a,b)

$$P_S(\vec{r}_0) \simeq k^2 \int d^2\vec{r} \ O(\vec{r}) \ P_0(\vec{r}) \ G(\vec{r}-\vec{r}_0). \tag{2.6}$$

The interpretation of this result is that the scattered field is the sum of all the fields generated by the local, independent, secondary sources, each of strength $k^2 O(\vec{r})$. As noted earlier, $G(\vec{r}-\vec{r}_0)$ is the Green function associated with propagation from the local scatterer to the detector, and the incident plane wave $P_0(\vec{r}) = \exp(ik\vec{s}_0\cdot\vec{r})$ in Eq. (2.6) gives the correct relative phase of each independent, local source.

The weak-scattering approximation in Eq. (2.5), called the first Born approximation, is best suited for backscattered data.[10] In tomography, however, one deals with transmitted wavefields, for which the phase or travel time plays a more significant role than does the amplitude. To describe this properly, we now introduce the Rytov Transform F, which is an exact phase transform given by the following two equations:

$$P(\vec{r}) = e^{W(\vec{r})} = e^{ik\vec{s}_0\cdot\vec{r} + \delta W(\vec{r})}, \tag{2.7}$$

$$ikF(\vec{r}) = \delta W(\vec{r})e^{ik\vec{s}_0\cdot\vec{r}}, \tag{2.8}$$

where the phase perturbation $\delta W(\vec{r})$ is purely imaginary in non-attenuating media.

Combining Eqs. (2.3), (2.7) and (2.8) gives the following inhomogeneous wave equation

$$(\nabla^2 + k^2)\, F(\vec{r}) = \frac{1}{ik}\{k^2 O(\vec{r}) - [\nabla(\delta W)]^2\}P_0(\vec{r}), \tag{2.9a}$$

and, hence, (cf. (2.3) and (2.4)) the phase perturbation is given by

$$\delta W(\vec{r}_0) = k^2\, e^{-ik\vec{s}_0\cdot\vec{r}_0} \int d^2\vec{r}\{O(\vec{r}) - \frac{1}{k^2}[\nabla(\delta W)]^2\}P_0(\vec{r})\; G(\vec{r}-\vec{r}_0), \tag{2.9b}$$

which on making the weak-scattering assumption

$$\frac{1}{k^2}\,|[\nabla(\delta W)]^2| << |O(\vec{r})|, \tag{2.9c}$$

becomes

$$\delta W(\vec{r}_0) \simeq k^2\, e^{-ik\vec{s}_0\cdot\vec{r}_0} \int d^2\vec{r}\; P_0(\vec{r})\; O(\vec{r})\; G(\vec{r}-\vec{r}_0). \tag{2.9d}$$

Thus, using the assumption in Eq. (2.9c), called the first Rytov approximation, we obtain the approximate result in Eq. (2.9d), according to which $\exp(ik\vec{s}_0\cdot\vec{r}_0)\delta W(\vec{r}_0)$ is given by precisely the same expression as $P_S(\vec{r}_0)$ in Eq. (2.6).

The difference between Eqs. (2.6) and (2.9d) lies in the way in which we represent the data. In the Born approximation we work directly with the scattered pressure field $P_S(\vec{r}_0)$, whereas we use the phase perturbation $\delta W(\vec{r}_0)$ in the Rytov approximation. To enable us to combine Eqs. (2.6) and (2.9d), we define our measurement data as follows:[8]

$$D(\vec{r}_0) = \begin{cases} \frac{2i}{k^2}\,[P(\vec{r}_0) - e^{ik\vec{s}_0\cdot\vec{r}_0}] & \text{Born data} \\[2ex] \frac{2i}{k^2}\, e^{ik\vec{s}_0\cdot\vec{r}_0}\, \ln[P(\vec{r}_0)e^{-ik\vec{s}_0\cdot\vec{r}_0}] & \text{Rytov data} \end{cases} \tag{2.10}$$

With this definition we can replace Eqs. (2.6) and (2.9d) by one equation covering both the Born and Rytov approximation, i.e.

$$D(\vec{r}_0) = \frac{1}{2} \int d^2\vec{r}\; e^{ik\vec{s}_0\cdot\vec{r}}\; O(\vec{r})\; H_0^{(1)}(k|\vec{r}-\vec{r}_0|). \tag{2.11}$$

Inverse problem

Equation (2.11) constitutes the forward model of our scattering problem. To invert this equation we proceed as follows. First we introduce the angular spectrum representation of the Green function (Weyl's expansion):[11-12]

$$H_0^{(1)}(k|\vec{r}-\vec{r}_0|) = \frac{1}{\pi} \int_{-\infty}^{\infty} \frac{d\kappa}{\gamma}\; e^{-ik\vec{s}\cdot(\vec{r}-\vec{r}_0)}. \tag{2.12a}$$

Here γ and κ are the x and y components of the wave vector $k\vec{s}$, i.e.

$$k\vec{s} = \gamma\vec{i} + \kappa\vec{j} \quad ; \; \gamma = (k^2 - \kappa^2)^{1/2}, \tag{2.12b}$$

where the latter equation is the dispersion relation. Next, we substitute Eq. (2.12a) in Eq. (2.11) to obtain

$$D(\vec{r}_0) = \frac{1}{2\pi} \int_{-\infty}^{\infty} \frac{d\kappa}{\gamma}\; \tilde{O}[k(\vec{s}-\vec{s}_0)]e^{ik\vec{s}\cdot\vec{r}_0}, \tag{2.13a}$$

where

$$\tilde{O}[k(\vec{s}-\vec{s}_0)] = \int d^2\vec{r}\; O(\vec{r})\; e^{-ik(\vec{s}-\vec{s}_0)\cdot\vec{r}}. \tag{2.13b}$$

But since also

$$D(\vec{r}_0) = \frac{1}{2\pi} \int_{-\infty}^{\infty} d\kappa\; \tilde{D}(l_0,\kappa)e^{i\kappa y} = e^{-i\gamma l_0} \frac{1}{2\pi} \int_{-\infty}^{\infty} d\kappa\; \tilde{D}(l_0,\kappa)e^{-ik\vec{s}\cdot\vec{r}_0}, \tag{2.14}$$

we see on comparison with Eq. (2.13a) that

$$\tilde{O}[k(\vec{s}-\vec{s}_0)] = \gamma\; e^{-i\gamma l_0}\; \tilde{D}(l_0,\kappa;\vec{s}_0), \tag{2.15}$$

where the Fourier vector $\vec{K}$ of the object's Fourier transform $\tilde{O}(K_x,K_y)$ is given by

$$\vec{K} = K_x\,\vec{i} + K_y\,\vec{j} = k(\vec{s}-\vec{s}_0). \tag{2.16}$$

Equation (2.15) constitutes the so-called generalized projection-slice theorem, which relates the one-dimensional Fourier transform of the data $\tilde{D}$ to the two-dimensional Fourier transform of the object profile $\tilde{O}$ along the semi-circle $k(\vec{s}-\vec{s}_0)$, where k is the wavenumber of the incident plane wave. As hown in Fig. 3, this semi-circle of radius $k = \omega/c_0$ is centered at $-k\vec{s}_0$, $\vec{s}_0$ eing the direction of the incident plane wave, and it passes through the origin.

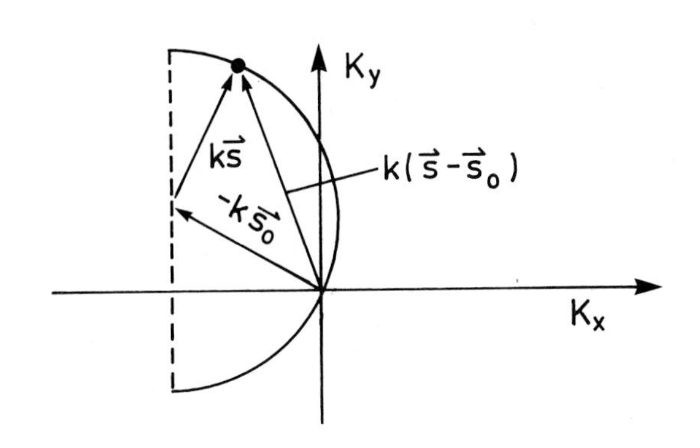

FIGURE 3
Generalized projection-slice theorem

By carrying out experiments with plane waves incident at many different directions $\vec{s}_0$ or angles α (cf. Fig. 1) and measuring the data D in each experiment, one can build up information about the Fourier components of the object.

2.3 Numerical reconstruction

The most straightforward numerical reconstruction is obtained directly from the generalized projection-slice theorem given in Eq. (2.15). The first step is then to interpolate between the measured values of the object's Fourier components, which as noted earlier, are known on sections of semi-circles. In this manner one can obtain Fourier components on a regular sampling grid. An inverse Fourier transform then provides us with a reconstruction of the object profile.[1,2,6] Such interpolation methods, however, have a number of drawbacks, the most severe being the "image artefacts" caused by aliasing due to undersampling in the spatial frequency domain.

An alternative reconstruction algorithm which does not involve interpolation, has been developed by Devaney. It is called the filtered backpropagation algorithm.[5] To explain it we start with the Fourier representation of the object profile, i.e.

$$O(\vec{r}) = \left(\frac{1}{2\pi}\right)^2 \int d^2\vec{K}\, \tilde{O}(\vec{K})\, e^{i\vec{K}\cdot\vec{r}}, \qquad (2.17a)$$

where the 2D Fourier transform $\tilde{O}(\vec{K})$ of the object profile is related to the 1D Fourier transform $\tilde{D}(l_0, \kappa ; \vec{s}_0)$ of the measured data via the projection-slice theorem in Eq. (2.15).

Next, we change integration variables from spatial frequencies (K_x, K_y) to (κ, α), where κ is the spatial frequency of the data and α is the view angle, i.e. (cf. Fig. 1)

$$\vec{s}_0 = \cos\alpha \, \vec{i} + \sin\alpha \, \vec{j}. \tag{2.17b}$$

Using Eqs. (2.15), we can rewrite Eq. (2.17a) as follows:

$$O(\vec{r}) = \frac{1}{2\pi} \int O_\alpha(\vec{r}) \, d\alpha, \tag{2.18a}$$

$$O_\alpha(\vec{r}) = \int \tilde{F}(\kappa,\alpha) \, \tilde{D}(\kappa,\alpha) \, \tilde{B}(\kappa,\alpha) d\kappa, \tag{2.18b}$$

where

$$\tilde{F}(\kappa,\alpha) = \frac{1}{2\pi} \gamma |J|, \tag{2.19a}$$

$$\tilde{B}(\kappa,\alpha) = e^{i\vec{K}\cdot\vec{r} - i\gamma l_0}, \tag{2.19b}$$

with J being the Jacobian associated with the transformation from (K_x, K_y) to (κ,α) (cf. Eqs. (2.12b), (2.16), and (2.17b).

Thus the reconstruction of the object profile includes two steps. For each angle of incidence α one first computes the partial reconstruction $O_\alpha(\vec{r})$ as shown in Eq. (2.18b). Next one adds all these partial reconstructions as indicated in Eq. (2.18a).

The Jacobian contained in the spatial filter $\tilde{F}$ in Eq. (2.19a) corrects for the non-uniform sampling of the object's Fourier spectrum. In addition this filter should include a rejection factor to avoid multiple coverage in the Fourier spectrum of the object.

The role of the backpropagation operator $\tilde{B}$ in Eq. (2.19b) is to remove the wave propagation effects. It plays a similar role here as do the inverse wave-extrapolators in migration.[13]

3. GENERATING SCATTERED DATA

To determine under what circumstances a given algorithm provides reliable reconstructions one can first compute the scattered field for a known object and then run these data through the inversion algorithm to see how close the reconstruction comes to the original object.

In previous studies scattered data have usually been obtained either from approximate solutions based on the weak-scattering assumption or from exact solutions for perfectly reflecting objects with soft or hard boundary conditions.

In a quantitative study it is important to have reliable scattered data for objects of variable contrast and shape. For this reason we have developed exact solutions for acoustic scattering by a penetrable elliptical cylinder.[14]

We now show an example of scattered data computed by the exact method and compare them with corresponding data obtained from the weak-scattering approximation. The scattering object is the penetrable elliptical cylinder shown in Fig. 4. Figs. 5 and 6 show the real part and the imaginary part of

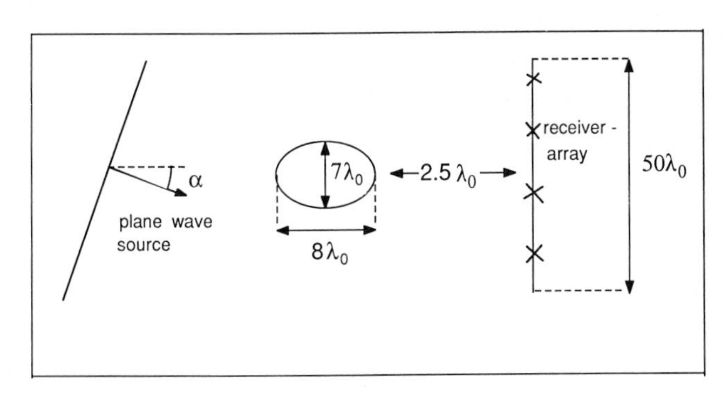

FIGURE 4
Aquisition geometry

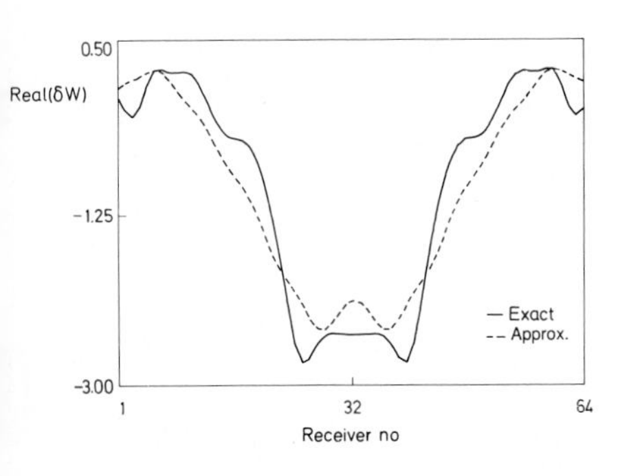

FIGURE 5
Real part of δW ($\alpha=0^0$). Exact and approximate solution.

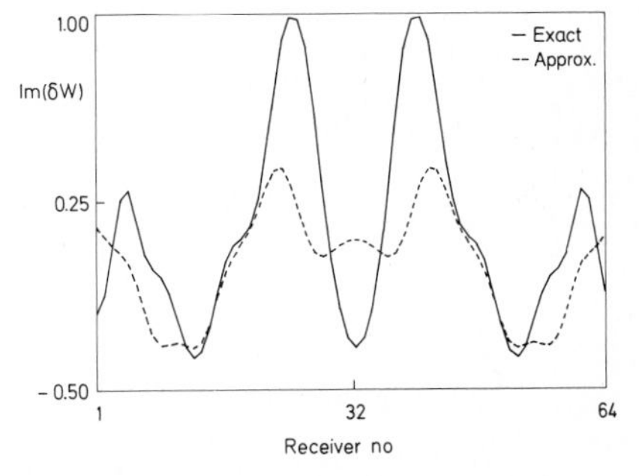

FIGURE 6
Imaginary part of δW ($\alpha=0^0$). Exact and approximate solution.

the phase perturbation δW (cf. Eq. (2.9d)), for both exact data and approximate data. The differences between the data are significant, even in this case where the scatterer is rather weak, the object velocity and background velocity being 2100 m/sec and 2000 m/sec, respectively. Of course, these differences will be reflected in the reconstructed images.

4. BORN VERSUS RYTOV

4.1 Interpretation of data

According to Eq. (2.7) the phase perturbation δW of the recorded wavefield is given exactly by

$$\delta W(\vec{r}_0) = \ln[P(\vec{r}_0)\, e^{-ik\vec{s}_0\cdot\vec{r}_0}], \tag{4.1}$$

so that for Rytov data Eq. (2.10) gives

$$D_R(\vec{r}_0) = \frac{2i}{k^2}\, e^{ik\vec{s}_0\cdot\vec{r}_0} \cdot \delta W(\vec{r}_0). \tag{4.2}$$

If we write the Born data in Eq. (2.10) in the same form, i.e.

$$D_B(\vec{r}_0) = \frac{2i}{k^2}\, e^{ik\vec{s}_0\cdot\vec{r}_0} \cdot \delta W_B(\vec{r}_0), \tag{4.3}$$

we find that

$$\delta W_B(\vec{r}_0) = e^{-ik\vec{s}_0\cdot\vec{r}_0} \cdot P_S(\vec{r}_0) = [e^{\delta W(\vec{r}_0)} - 1] \simeq \delta W(\vec{r}_0). \tag{4.4}$$

Equation (4.4) shows that the Born approximation provides an accurate representation of the phase perturbation δW, only if it is small. Since δW often is larger in transmission than in backscattering, Eq. (4.4) implies that the Born approximation is better suited for backscattered data, while the Rytov approximation is better suited for transmission data.

4.2 Scattering potential

In order to discuss the Born and Rytov approximations from a different point of view, we now return to the exact relationship between the object and its scattered field, given in Eq. (2.4a), and rewrite it as follows

$$P_S(\vec{r}_0) = k^2 \int d^2\vec{r}\; S_B(\vec{r})\, P_0(\vec{r})\, G(\vec{r}-\vec{r}_0), \tag{4.5}$$

where the secondary-source function or scattering potential $S_B(\vec{r})$ for <u>one</u>

single incident plane wave is given by

$$S_B(\vec{r}) = [1 + \frac{P_S(\vec{r})}{P_0(\vec{r})}]O(\vec{r}). \tag{4.6}$$

A comparison between Eqs. (2.6) and (4.5) shows that what we reconstruct using the Born-approximation scheme, is not the object profile O, but rather the scattering potential S_B, which is related to O by Eq. (4.6). But if

$$|P_S(\vec{r})/P_0(\vec{r})| << 1, \tag{4.7}$$

as is assumed in the Born approximation, then what we reconstruct is approximately equal to the object profile, i.e.

$$S_B(\vec{r}) \simeq O(\vec{r}). \tag{4.8}$$

It follows from Eq. (4.6) that $S_B(\vec{r})$ and $O(\vec{r})$ have the same zeros. This implies that an inversion algorithm based on Born data will reconstruct the shape or outer boundary of the object well. Outside the scattering object both $S_B(\vec{r})$ and $O(\vec{r})$ vanish.

Next consider the Rytov transformation, which gives the following exact solution for the phase perturbation δW (cf. Eq. (2.9b))

$$P_0(\vec{r}_0)\delta W(\vec{r}_0) = k^2\int d^2\vec{r}\ S_R(\vec{r})\ P_0(\vec{r})\ G(\vec{r}-\vec{r}_0), \tag{4.9}$$

where

$$S_R(\vec{r}) = [O(\vec{r}) - (\nabla\delta W)^2/k^2]. \tag{4.10}$$

Comparing Eqs. (4.9) and (2.9d) we see that what we reconstruct using Rytov data, is not the object profile O, but the scattering potential S_R, which is related to O by Eq. (4.10). But if the Rytov approximation is fulfilled, i.e. if (cf. Eq. (2.9c))

$$|(\nabla\delta W)|^2 <<|k^2 O(\vec{r})|, \tag{4.11}$$

then

$$S_R(\vec{r}) \simeq O(\vec{r}). \tag{4.12}$$

It follows from Eq. (4.11) that the Rytov approximation is violated if the scattering object has sharp corners or other features that would give rise to a rapidly varying phase, and, hence, to a large phase gradient.

Note, however, that contrary to the scattering potential S_B obtained from Born data, the scattering potential S_R obtained from Rytov data, may be non-zero outside the object O. Therefore, a reconstruction from Rytov data does not outline the voids and shape of an object as well as a reconstruction from Born data.

4.3 Scattering example

To illustrate some of the features discussed above, we consider the scattering geometry in Fig. 4, where the "unknown" object is an elliptical cylinder. The background velocity is 2000 m/sec and the incident plane waves have a frequency of 100 Hz, so that the background wavelength λ_0 is 20 meters. The incident plane waves cover all angles between -90^0 and $+90^0$, but the receiver array, which is 1000 m, limits the effective view angles to lie within $\pm 70^0$. The scatterer is weak, the object velocity being 2100 m/sec. Note that the receiver array is only 2.5 wavelengths away from the object, so the data represent near-field measurements.

Fig. 7 shows a velocity profile through the minor (vertical) axis of the scatterer, with Born data as input. This profile, shows that the outer boundary of the original object is well recovered in accordance with the predictions of Eq. (4.6). But the estimate of the object velocity is rather poor.

The next figure shows the corresponding result using the Rytov inversion scheme. In contrast to Fig. 7, the velocity profile in Fig. 8 provides a good

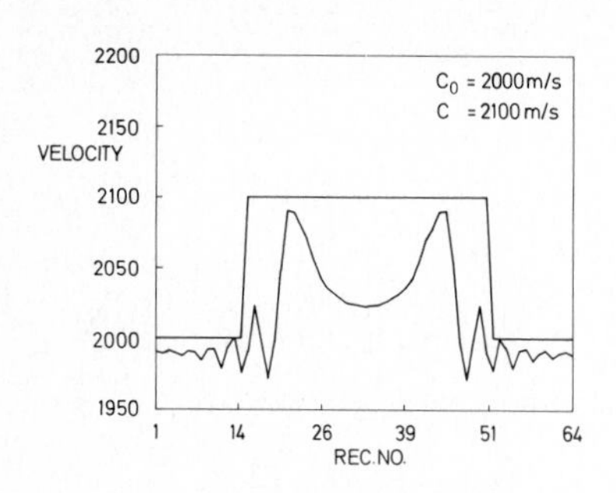

FIGURE 7
Velocity profile, Born data.

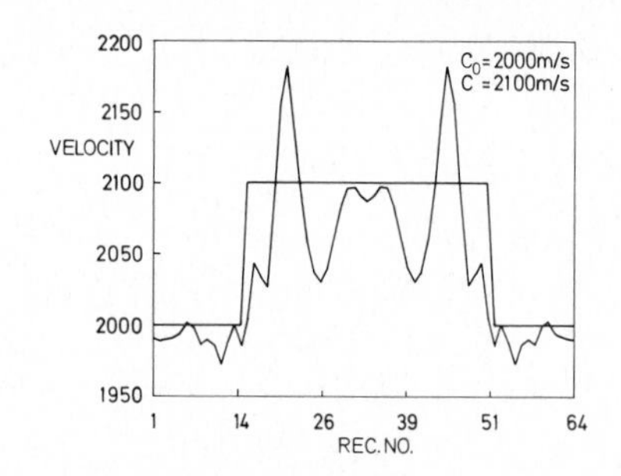

FIGURE 8
Velocity profile, Rytov data.

estimate of the inner part of the object velocity, but shows large over-shoots near the edges.

In both examples the reconstructed object is somewhat smaller than the original object. This may be due to the fact that the data also contain vanescent energy since the receiver is only 2.5 wavelengths away from the catterer. The reconstruction scheme does not take into account evanescent ave modes.

. SYNTHETIC DATA COMPARISONS

We now present reconstructions based on synthetic data obtained from the xact solutions mentioned earlier. If not stated otherwise, we use Rytov ata, since such data are better when dealing with transmitted fields.

Fig. 9 shows the scattering geometry. Incident plane waves are generated y sources on the surface (VSP (vertical seismic profiling) aquisition) or in neighbouring borehole (crosshole aquisition). The detector array has length and is symmetrically positioned about the horizontal axis of the elliptical catterer. The frequency of each incident plane wave is 100 Hz and the ackground velocity is 2000 m/sec.

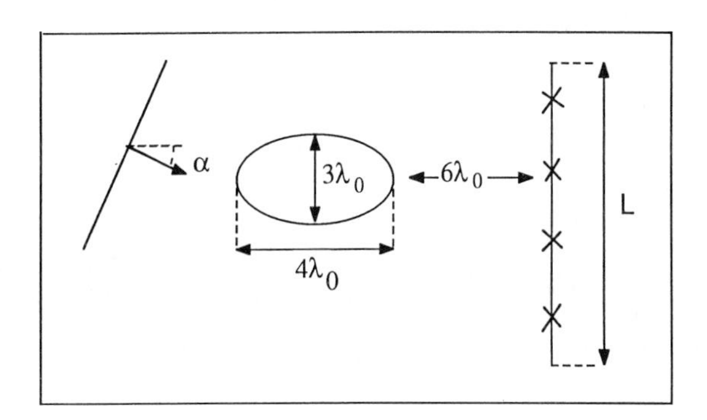

FIGURE 9
Scattering model

The weak-scattering assumption

We first examine the weak-scattering assumption by considering a crosshole xperiment with a strongly scattering object having a velocity of 2500 m/sec, 5% above that of the background. The distance between the boreholes is 250 eters. The velocity reconstructions in Figs. 10 and 11 show respectively a ontour plot and a profile through the minor axis of the scatterer. We see hat the velocity profile is severely distorted, but nevertheless the shape of he object is preserved. From this and many similar experiments we draw the ollowing conclusions:

) For a scatterer with a contrast less than 5 - 10% the weak-scattering assumption works well.

2) We can determine the geometry of a strong scatterer by applying a non-iterative diffraction-tomography algorithm. If we also want to determine its contrast, we may apply a two-step or iterative algorithm, in which we first determine the geometry and then the velocity.

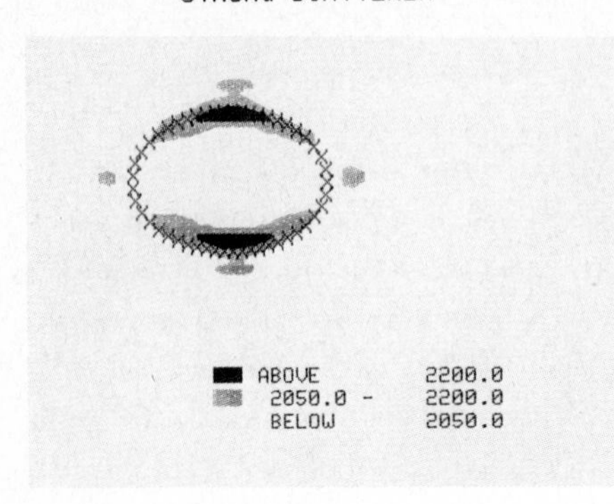

FIGURE 10
Velocity contour plot. Strong scatterer.

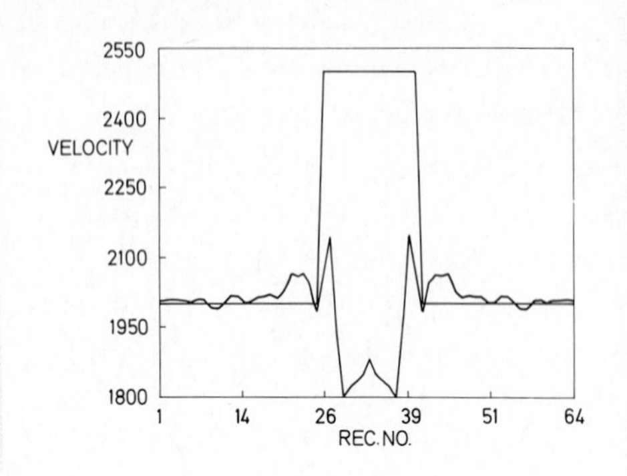

FIGURE 11
Velocity profile. Strong scatterer.

Limited view

When evaluating the diffraction-tomography method it is important to employ realistic aquisition conditions. We must account for the finite length of each borehole in a crosshole experiment, and also for the finite maximum offset in a VSP experiment. As mentioned earlier, the finite extent of the aquisition geometry yields incomplete coverage of the Fourier components of the object, and this in turn gives image artefacts. We now demonstrate how to improve the limited-view-angle image. First we consider reconstructions based on crosshole data and then we consider multi-data imaging.

Crosshole data

Consider a crosshole experiment in which the source and receiver array both have a length of 250 meters, which corresponds to incident angles between -40^0 and $+40^0$. The velocity of the scattering object is 2100 m/sec. The loss of resolution in the reconstruction is quite noticeable in the velocity contour plot in Fig. 12. The modulus of the Fourier transform of the reconstructed object is shown in Fig. 13, from which we see that the different quadrants of the object's Fourier space have received different coverage in the experiment due to aperture limitations.

Fig. 14a shows the full area in the object's Fourier space that can possibly be covered in a crosshole experiment, full coverage implying infinite source and receiver arrays. In practice one can only get a partial coverage, s indicated by the hatched area in Fig. 14b. This area, however, is <u>not</u> the overage of a real object, which must have the symmetry property illustrated n Fig. 15. But by invoking the reciprocity principle one can extend the artial coverage in Fig. 14b to include also the area circumferenced by the roken curves, and thus obtain the Fourier spectrum of a real object.

CROSSHOLE (NO RECIPROCITY)

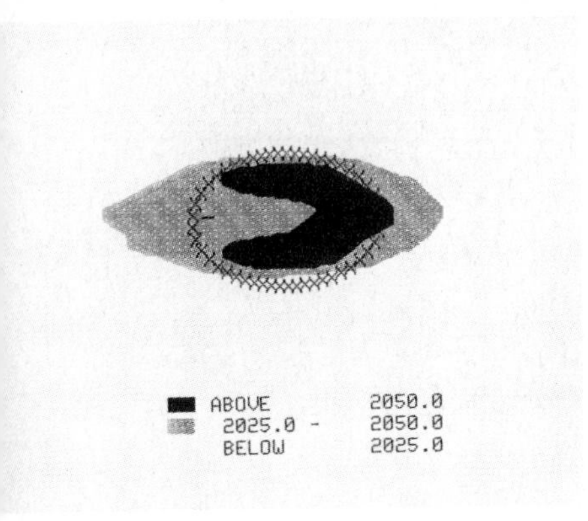

FIGURE 12
Velocity contour plot.
$-40^0 < \alpha < 40^0$. L=250m.

CROSSHOLE SPECTRUM (NO RECIPROCITY)

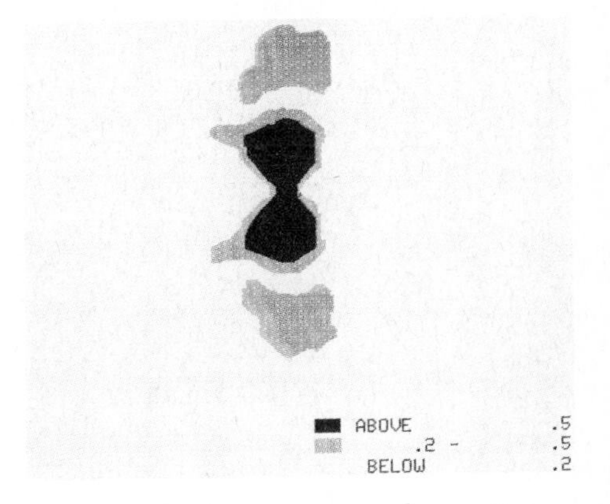

FIGURE 13
Fourier spectrum of
the object in Fig. 12

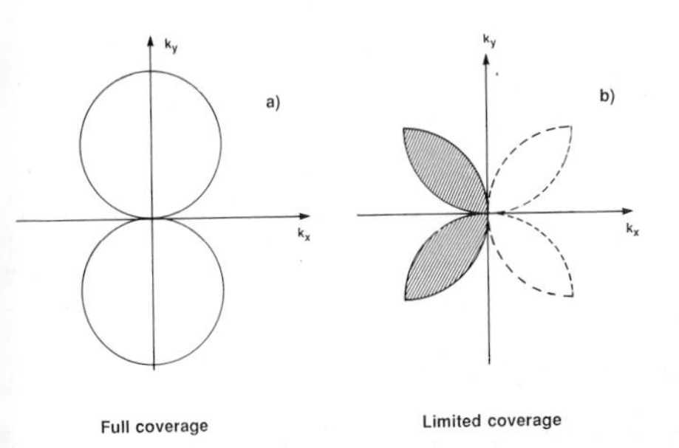

FIGURE 14
Theoretical sampling of the object's Fourier space in a crosshole experiment. Full and limited coverage.

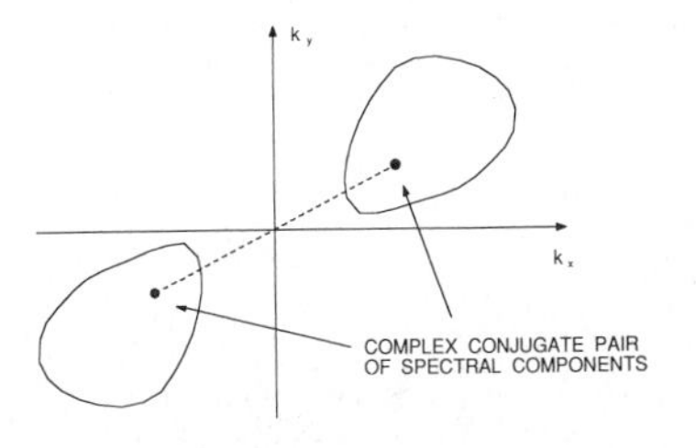

FIGURE 15
Spectrum of a real object.

To apply the reciprocity principle to our crosshole experiment we simulate that the sources and receivers are interchanged and add the new Fourier components to those aquired previously. Then the Fourier spectrum in Fig. 13 changes to that in Fig. 16, which is seen to have the symmetry property in Fig. 15.

Note also that this spectrum is symmetric about the K_x axis, as it ought to be, since both the object and the aquisition geometry are symmetric about the x axis. An inverse Fourier transform of the spectrum in Fig. 16 yields the reconstruction in Fig. 17, which on comparison with that in Fig. 12 reveals the improvement gained by invoking the reciprocity principle in a crosshole experiment.

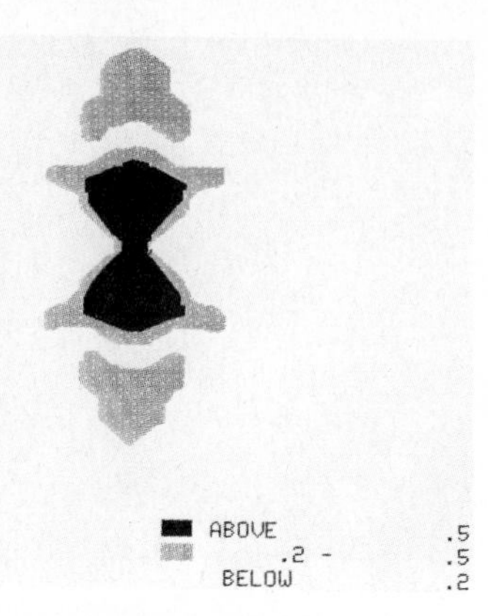

FIGURE 16
Fourier spectrum of the object in Fig. 12 obtained by invoking the reciprocity principle.

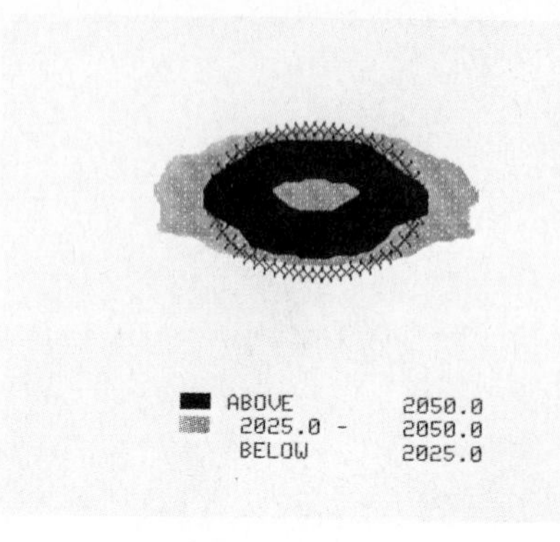

FIGURE 17
Velocity contour plot of the object, whose spectrum is shown in Fig. 16.

<u>Multi-data imaging</u>

We now consider what can be gained by combining VSP and crosshole data in the same reconstruction. Obviously, the degree of improvement will depend somewhat on the geological model and the aquisition geometry, but in general the information content will be improved when different types of data are combined.

Fig. 18 shows a scattering object corresponding to a shallow gas-filled

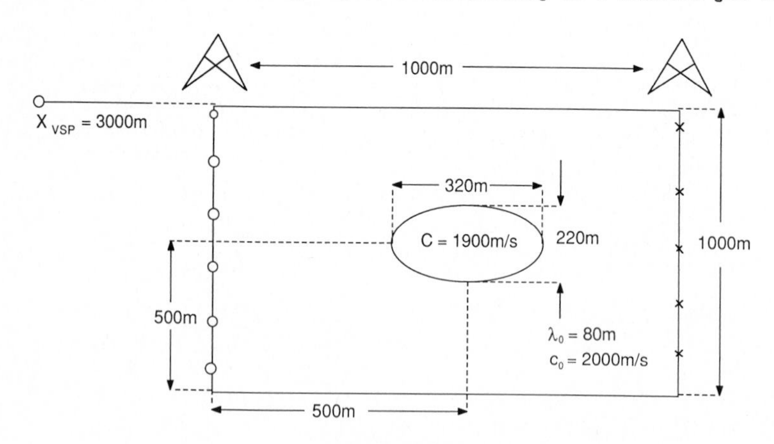

FIGURE 18
Aquisition geometry for combined offset VSP and crosshole experiment.

and body. The object velocity is 1900 m/sec and the source frequency is 25 z, corresponding to a wavelength of 80 meters. The maximum source-to-bore-ole offset in the VSP aquisition is 3000 meters, and the distance between the wo wells is 1000 meters. For this particular geometry the view angles lie etween -55^0 and $+55^0$ in the crosshole experiment, and between 10^0 and 145^0 in he VSP experiment.

Consider first the reconstructions based on crosshole data only. Fig. 19 hows the result obtained when reciprocity has been taken into account. Due to he narrow band of view angles the reconstructed object is smeared out in the ateral direction and the contrast is too low. Next we consider the image omputed from VSP data only. Fig. 20 shows the reconstruction based on a sum

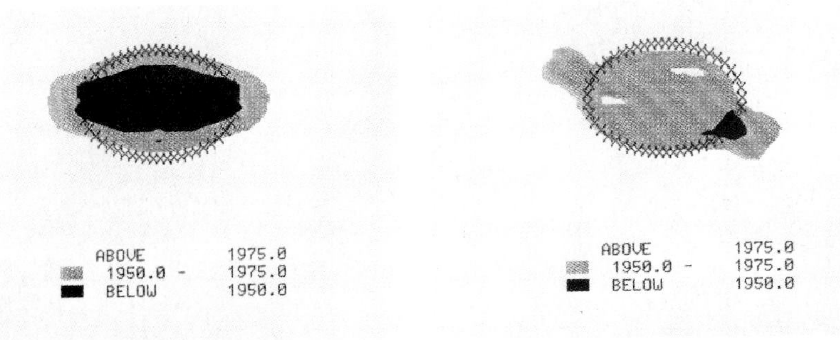

FIGURE 19
Crosshole image with reciprocity included.

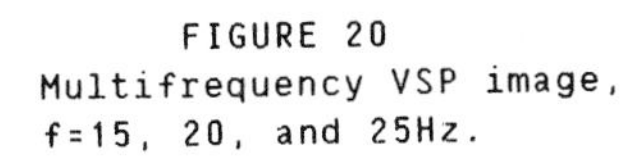

FIGURE 20
Multifrequency VSP image, f=15, 20, and 25Hz.

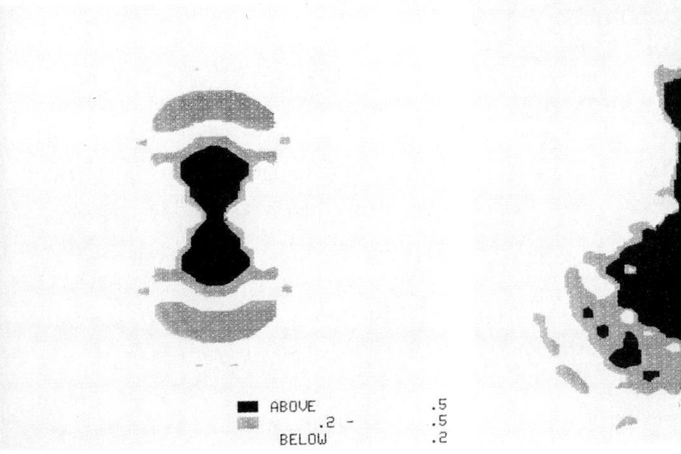

FIGURE 21
Fourier spectrum of the crosshole reconstruction in Fig. 19.

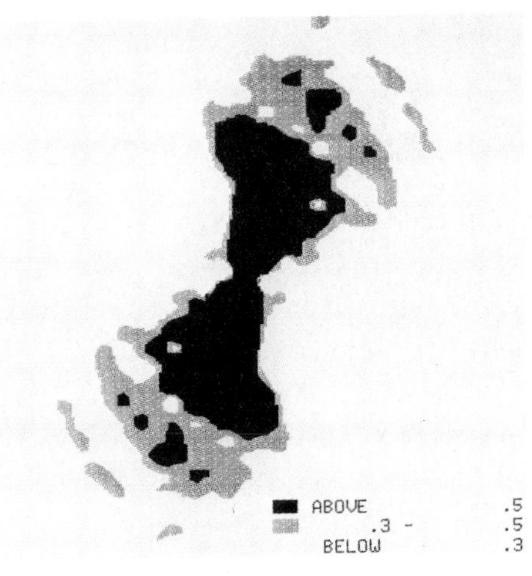

FIGURE 22
Fourier spectrum of the VSP reconstruction in Fig. 20.

of three frequencies (15, 20 and 25 Hz). Although the VSP image in Fig. 20 suffers from distortions and loss in contrast, it gives a better definition of the object's shape than the crosshole image in Fig. 19.

Figs. 21 and 22 show the Fourier spectrum of the crosshole and the VSP image, respectively. We now combine these two Fourier spectra using a simple exclusive-or sum to obtain the combined spectrum in Fig. 23, based on both VSP and crosshole data. An inverse Fourier transform of this spectrum gives the reconstruction in Fig. 24, which is seen to be far better than that based on either crosshole or VSP data alone.

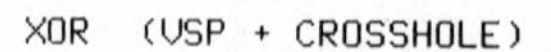

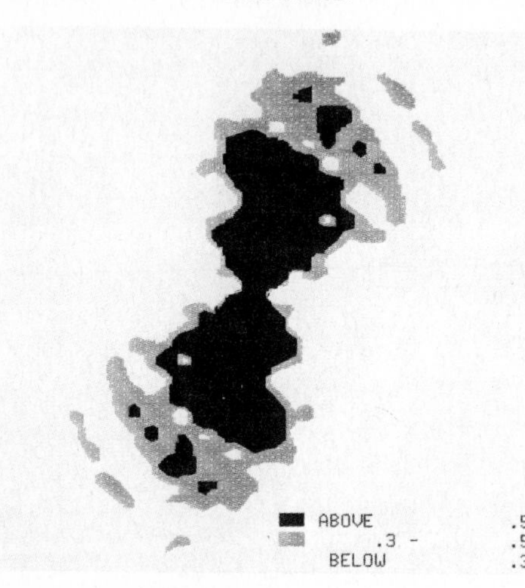

FIGURE 23
Combined spectrum (X-or sum).

MULTI DATA IMAGE

FIGURE 24
Multi-data image.

6. POTENTIALS AND PROBLEMS

The possible applications of diffraction tomography are many: medical diagnostics, ground investigations (road tunnels, sites for nuclear waste, etc.), exploration (oil and gas, minerals, etc.), and non-destructive testing of materials, to mention a few. But in most of these applications we are not looking for weak scatterers embedded in a uniform, non-attenuating background.

For medical diagnostics we need to make sure that the inversion algorithm works for weakly scattering objects in a strongly attenuating background, and in most other applications we need an inversion method that works for objects embedded in a non-uniform background. Also, the algorithms must be able to handle other, more general, aquisition geometries than those shown in Figs. 1 and 2. Finally, for many of the applications mentioned above, the vectorial character of the problem need to be taken into account.

In conclusion, there is still a lot of research and development to be done the field of diffraction tomography, but the results obtained so far are omising enough and the practical applications are important enough to stify further work.

FERENCES

R.K. Mueller, M. Kaveh, and G. Wade, Proc. IEEE 67 (1979) 567.

R.K. Mueller, M. Kaveh, and R.D. Inverson, A new approach to acoustic tomography using diffraction techniques, in: Acoustical Imaging, Vol. 8, ed. A.F. Metherell (Plenum, New York, 1980) pp. 615-628.

K. Iwata and R. Nagata, Japan J. Appl. Phys. 14 (1974) 379.

E. Wolf, Optics Commun. 1 (1969) 153.

A.J. Devaney, Diffraction tomography, in: Inverse Methods in Electromagnetic Imaging, Part 2, eds. W.-M. Boerner et al. (D. Reidel Publishing Company, 1985) pp. 1107-1135.

A.C. Kak, Proc. IEEE 67 (1979) 1245.

A.J. Devaney, Ultrasonic Imaging 4 (1982) 336.

A.J. Devaney, IEEE Trans. Geosci. Rem. Sensing GE-22 (1984) 3.

R.S. Wu and M.N. Toksøz, Geophysics 52 (1987) 11.

0) M.L. Oristaglio, J.Opt. Soc. Am. A 2 (1985) 1987.

1) H. Weyl, An. Phys. Lpz. 60 (1919) 481.

2) J.J. Stamnes, Waves in Focal Regions (Adam Hilger, Bristol, 1986) Section 5.2.

3) J.F. Claerbout, Imaging the Earth's Interior (Blackwell Scientific Publications, 1985) 144.

4) J.J. Stamnes and B. Spjelkavik, Acoustic and Electromagnetic scattering by elliptical cylinders, to be published.

Scattering in Volumes and Surfaces
M. Nieto-Vesperinas and J.C. Dainty (Editors)

THE PHYSICAL OPTICS OF ENHANCED BACKSCATTERING

Eric Jakeman

Royal Signals and Radar Establishment, Malvern, Worcestershire, UK

1 INTRODUCTION

Several observations of enhanced backscattering of coherent light from volume scatterers such as particle suspensions have been reported over the last few years[1-5]. These can be explained by the coherent addition of multiply scattered waves which add in phase only in the backward direction. They have aroused a great deal of interest in the scientific community partly because of the analogy which can be drawn with the phenomenon of "weak localisation" familiar to solid state physicists[6]. However, strong enhancement in the backward direction has also recently been observed in light scattered by smoothly varying random surfaces with inhomogeneities whose dimensions are significantly larger than the incident wavelength (this volume)[7-8]. In this paper the implications of a simple tutorial enhanced backscattering system which is, strictly speaking, neither simply a volume nor a surface scatterer will be discussed. The scattering geometry is illustrated in figure 1 and consists of back reflection from a plane mirror through a deeply modulated random phase changing screen. This is, by definition, a physical optics scattering system governed by the equations of scalar diffraction theory and polarisation effects are therefore excluded from the problem. The geometry is equivalent to squential scattering by two identical phase screens as illustrated in figure 1b. It is worth noting here that there is an extensive Soviet literature on back reflection problems of this type though with the emphasis on extended media with power law refractive index spectra rather than the far field geometry with smoothly varying refractive model which will be considered here. An excellent review of the early Soviet literature on the subject is given by Kravtsov and Saichev[9] with Lidar applications being discussed in a recently translated book by Banakh and Mironov[10].

In the next section a theoretical treatment of the problem will be outlined which is based on analytical techniques developed over the last decade or so to deal with single phase-screen scattering. In section 3 a review will be given of experimental data obtained by my colleagues on two types of scattering system which can be modelled broadly by the configuration illustrated in figure 1a. The final section gives a summary and indicates future directions for the work.

2 THEORY

A brief summary is given here of the theoretical treatment of the scattering geometry shown in figure 1[11]. It will be assumed that the screen/mirror combination is in the far field with respect to the source/receiver plane but that the mirror is in the Fresnel region with respect to the phase screen. The field incident on the screen at the second pass is then given in a Huyghens-Fresnel approximation by[12].

$$\mathcal{E}_i(r'') = \frac{A}{\lambda d} \int \exp\left\{ -i\underset{\sim}{k}.\underset{\sim}{r}' \sin\theta_i - \frac{r'^2}{W^2} + i\phi(r') + \frac{ik}{4d}(r'-r'')^2 \right\} d^2r' \quad (1)$$

where d is the mirror/screen separation, W is the width of a Gaussian beam (wavelength $\lambda = 2\pi/k$) incident at an angle θ_i with its waist at the screen, $\phi(r)$ is the distortion introduced by the screen and A is a constant. The field rescattered to the detector in the direction θ_i is thus

$$\mathcal{E}_s(\theta_s) = \int \mathcal{E}_i(r'') \exp\left\{ i\phi(r'') - i\underset{\sim}{k}.\underset{\sim}{r}'' \sin\theta_s \right\} d^2 r'' \quad (2)$$

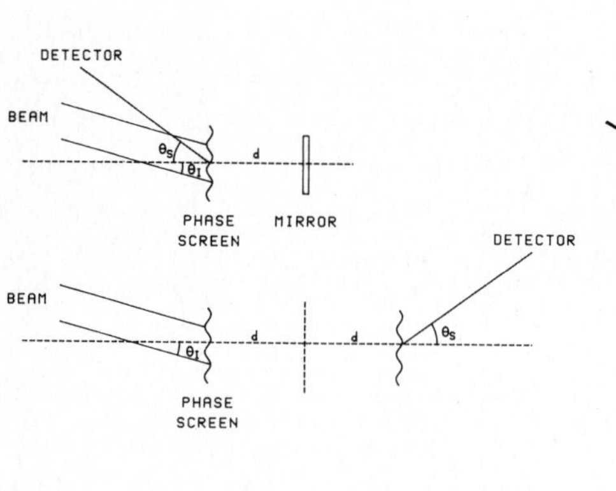

FIGURE 1
Scattering Geometry: Top (a) back reflection, bottom (b) identical screens.

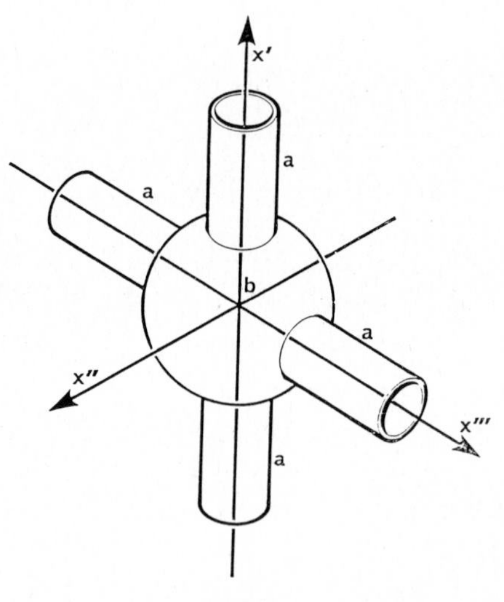

FIGURE 2
Region of integration (equation 3) showing (a) coherent and (b) incoherent contributions.

It is clear from (1) and (2) that the calculation of the mean scattered intensity, $\langle|\mathcal{E}_s|^2\rangle$, is similar to a fourth moment calculation for scattering by a single screen. This type of problem is well known and much studied but needs analysing with care. Direct numerical analysis is not straightforward and a combination of steepest descents and function modelling will be used to reveal the structure of the problem[11].

Note that polarisation effects are not included in the model but obliquity factors may be included to strengthen the analogy with surface scattering if desired.

In order to proceed it will be assumed that ϕ is a stationary, corrugated Gaussian process. This allows the mean intensity ($I = |\mathcal{E}|^2$) to be written in the form

$$\langle I(\theta_s)\rangle \propto \int_{-\infty}^{\infty} dx'dx''dx'''\ A(\theta_i, \theta_s, x', x'', x''')\ \exp\left\{-\phi_o^2\ G\ (x', x'', x''')\right\} \tag{3}$$

where

$$G(x',x'',x''') = 2 - \rho(x''-x') - \rho(x''+x') - \rho(x'''-x'') - \rho(x'''+x''') + \rho(x'-x''') + \rho(x'+x'''),$$

$$\rho(x) = \langle\phi(o)\ \phi(x)\rangle/\phi_o^2$$

and the remaining exponential factors are subsumed in A. A single-scale, smoothly varying model will be adopted for the phase correlation function:

$$\rho(x) = \exp\ (-x^2/\xi^2) \tag{4}$$

This corresponds to a phase distortion which is infinitely differentiable and characterised by inhomogeneities of lateral scale ξ. It is known that such a model leads to the generation of caustics and focusing effects in the scattered intensity pattern[13].

When the phase screen is deeply modulated, $\phi_o^2 >> 1$, the integral (3) is dominated by contributions from regions near the x', x''' axes as shown in figure 2. An excellent approximation for $\phi_o^2 > 3$ is obtained by using the method of steepest descents to perform the x'' integral. This is accomplished by expanding G to second order in x'' and yields the formula [11]

$$\langle I(\theta_s)\rangle \propto \int_0^{\infty} dx' \int_0^{x'} dx'''\ B(\theta_i, \theta_s, x', x''')\ \exp\left\{-\ \phi_o^2\ F(x', x''')\right\} \tag{5}$$

where $F(x',x''') = 2 - 2\rho(x') - 2\rho(x''') + \rho(x'-x''') + \rho(x'+x''')$

and B again includes various slowly varying factors. It is not difficult to compute (5) but it is useful to pursue the problem further analytically to clarify the various scattering mechanisms involved. This is done by function modelling. The x' - range is divided at $\xi/\sqrt{3}$ and F is approximated as follows:

$$F(x', x''') = \frac{6\,x'^2\,x'''^2}{\xi} \quad \text{for } x' < \xi/\sqrt{3}$$
$$= -\frac{x'''^2}{\xi} \quad \text{for } x' > \xi/\sqrt{3} \tag{6}$$

The integrals (5) can then be reduced further before computation, but, in the case where the number of correlation areas, N, of the screen contributing to the field at any point in the second pass is large, a simple analytical formula can be obtained [11].

$$\langle I(\theta_s)\rangle \propto \exp(-2\bar{\theta}^2/m_o^2)\left\{1 + \frac{\ell^2}{\ell_s^2}\exp(-2k^2\ell^2\,\Delta\theta^2) - f(\Delta\theta)/N\right\} +$$
$$+ (1/N)\exp(-\bar{\theta}^2/m_o^2)\left\{E_i(2\Delta\theta^2/m_o^2) + 2Ci\left(\frac{2k}{\sqrt{3}}\,\xi\Delta\theta\right)\right\} \tag{7}$$

Here $\bar{\theta} = (\sin\theta_1 + \sin\theta_s)/2$, $\Delta\theta = (\sin\theta_i - \sin\theta_s)/2$, $m_o = 2\phi_o/k\xi$ is the rms slope of the phase distortion introduced by the screen, $\ell_s = dm_o$, $\ell^{-2} = W^{-2} + \ell^{-2}$ and $N \sim m_o d/\xi$. When N is large the second term in the first curly brackets on the right hand side may be neglected but the second set of terms multiplied by N^{-1} may be large if the wave vector is sufficiently large because the exponential integrals contain a logarithmic divergence with k.

As an aside, it is worth noting that the structure of equation (7) is similar to that obtained for the second intensity moment of radiation singly scattered by a particulate. This can be represented as the two dimensional random walk

$$\mathcal{E} = \sum_{j=1}^{N} a_j e^{i\phi_j}$$

If the $\{a_n\}$ and the $\{\phi_n\}$ are statistically independent random variables and the $\{\phi_n\}$ are uniformly distributed, then the second intensity moment

$$\langle I^2\rangle = \sum \langle a_j\, a_k^*\, a_l\, a_m^*\rangle\ \langle \exp i(\phi_j - \phi_k + \phi_l - \phi_m)\rangle$$

is composed of only two classes of significant terms: Gaussian "interference" contributions when $j = k$, $l = m$ or $j = m$, $k = l$ corresponding to the axial regions where the integrand of equation (3) is significant (see figure 2), and single scatterer terms where $j = k = l = m$ corresponding to the contribution of the central region of figure 2. This contribution is model dependent ie dependant on the $\{a_n\}$. When only these two groups of terms are retained the normalised second moment may be written in the form

$$\frac{\langle I^2 \rangle}{\langle I \rangle^2} = 2\left(1 - \frac{1}{N}\right) + \frac{1}{N}\,\frac{\langle a^4 \rangle}{\langle a^2 \rangle^2} \tag{8}$$

This is similar to equation (7) with $\Delta\theta = 0$ and $\ell = \ell_s$ and suggests that the first set of terms in curly brackets on the right hand side of equation (7) arise from interference between reversible paths to the source which have traversed different scattering inhomogeneities, as illustrated in figure 3a, whilst the second set of terms multiplied by N^{-1} arise as a result of random geometrical focusing by single lens-like inhomogeneities in the screen, as illustrated in figure 3b.

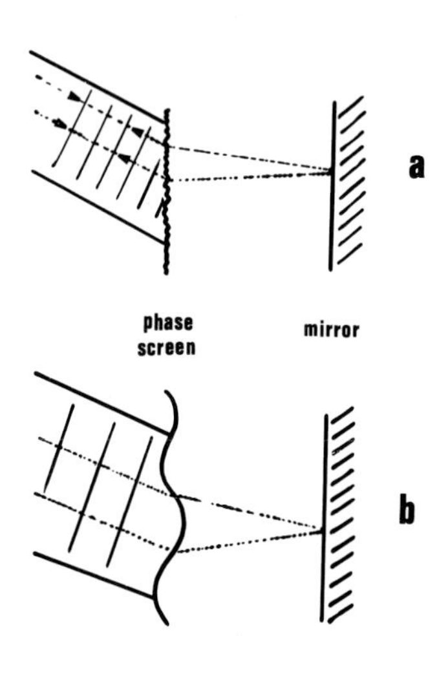

FIGURE 3
Enhancement mechanisms (a) coherent effect (b) Geometrical effect.

When N is sufficiently large only the interference terms in equation (7) survive and it is clear that a maximum enhancement factor of two is predicted in the backward direction $\theta_i = \theta_s$ by comparison with the intensity expected for two independent screens in that direction (given by the multiplicative exponential term). This backscattering enhancement is precisely that of a doubly scattering particulate. It is a coherent effect, visible when the screen is illuminated by a Gaussian speckle pattern at the second pass, but it is only weakly wavelength dependent and therefore visible in white light illumination. ℓ_s is the characteristic lateral separation of inhomogeneities involved in the double scattering determined by the geometrical spread of rays at the first pass whilst ℓ is the lateral scattering length for reversible paths, bearing in mind that an increased fraction of paths are returned outside the beam (and are therefore not reversible) as ℓ_s is increased or W reduced. The width of the enhancement peak $(k\ell)^{-1}$ narrows as ℓ_s is increased (ie with increased mirror/screen separation) whilst its magnitude diminishes.

The second set of terms multiplied by N^{-1} in equation (7) are largest near $N \sim 1$ when the mirror lies in the focusing region of the phase screen. A random "cats eye" or corner cube effect then occurs, as illustrated in figure 3b, giving an incoherent backscattering enhancement of geometrical origin. This is a broad feature by comparison with the coherent effect, being comparable to the geometrical spread of rays scattered by the phase screen. The magnitude of the effect can be large due to the logarithmic divergence with k present in the exponential integral Ci which arises from averages over the squares of caustics and is a well known feature in the theory of phase screen scattering[14].

Qualitatively similar results to the above are obtained for an isotropic random phase screen with a smoothly varying single scale profile[11]. Coherent backscattering enhancement will also be generated by fractal and subfractal screens. However, since refractive index fluctuations with unmodified power law spectra do not generate geometrical optics effects[15] it is unlikely that back-reflection through a fractal phase screen will give rise to significant incoherent enhancement.

Investigation of the spatial coherence properties of the field scattered in the geometry illustrated in figure 1 indicates that a normal Gaussian speckle pattern is to be expected provided the illuminated area is much larger than the characteristic phase inhomogeneities introduced by the screen (ie $W \gg \xi$) which was assumed throughout the above calculations. It is known, however, that when back-reflection takes place from a mirror which is sufficiently small, enhanced non Gaussian fluctuations are predicted by the theory. Indeed, the scattered field turns out to be proportional to the square of the field incident on the screen at the second pass. If this is a Gaussian (speckle) field then the scattered intensity is K-distributed[16] with a second moment of six. At the time of writing we have not investigated phenomena in this regime experimentally. However, the plane mirror geometry is not difficult to arrange in the laboratory and has now been studied with a number of model diffusers.

3 EXPERIMENT

A simple laboratory experimental arrangement corresponding to figure 1 is shown in figure 4 and is self explanatory. Detailed measurements of the statistics were made from the translation stage whilst the scattered intensity pattern was viewed and video recordings made at the viewing screen[17]. Experiments were carried out using two 10 cm glass plates for the phase screen, one with fine scale roughness induced by sand blasting with 50 μm particles and one being standard obscure bathroom glass with mm scale smooth variations.

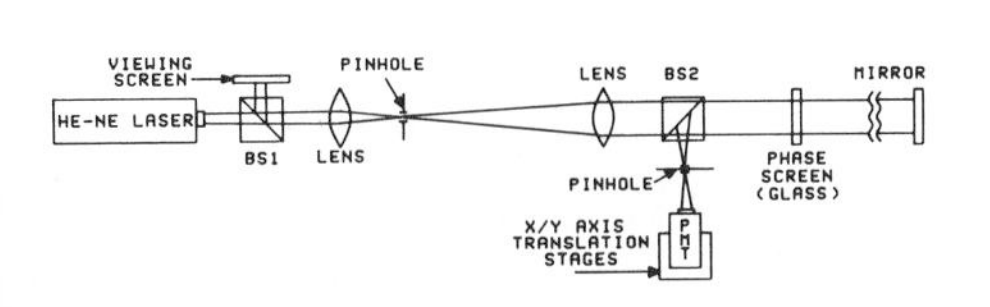

FIGURE 4
Experiments with roughened glass plates: diagram of apparatus.

The coherent enhanced back-scattering effect was generated when the fine scale glass was placed about 1 mm in front of the plane mirror. Fig 5(a) shows the appearance of the observed stationary speckle pattern and fig 5(b) shows the enhancement effect revealed when the glass sample was rotated. Figure 6(a) shows the enhancement on axis compared with theory. The predicted factor of two is almost achieved, falling short probably due to aperture averaging of the pattern. Figure 6(b) shows the same effect with non-normal incidence and demonstrates that the phenomenon occurs in the backscattering direction. Figure 6(c) and 6(d) demonstrate the effect of increasing the mirror/screen separation so that the proportion of reversible paths decreases. The magnitude of the enhancement is reduced as predicted by the theory.

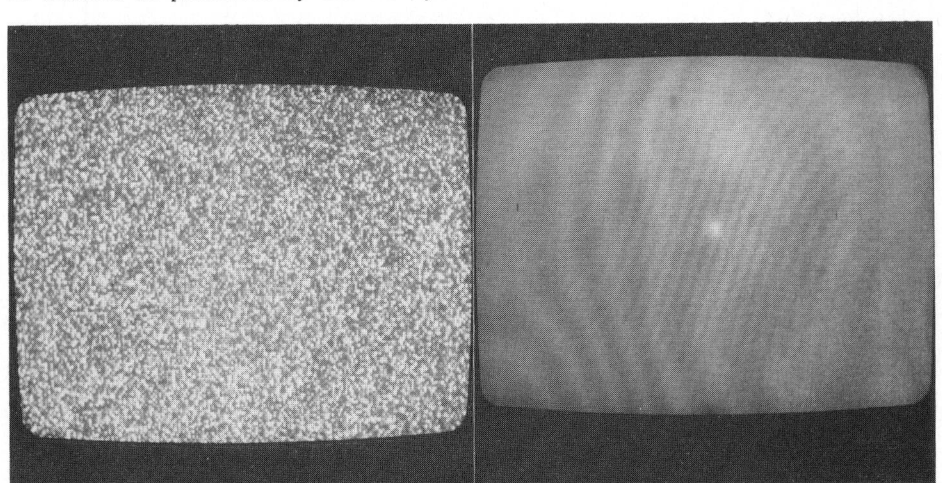

FIGURE 5a FIGURE 5b
Intensity pattern generated by back reflection of laser light through a fine scale diffuser (a) diffuser stationary (b) diffuser rotating and showing enhancement in backward direction.

The incoherent effect required too close a spacing of the diffuser and mirror for the sample with fine scale roughness and the standard bathroom glass was used instead. Fig 7(a) shows the observed stationary speckle pattern and fig (7b) the enhancement revealed by rotating the sample. Figure 8 shows a comparison of theory with experiment when the mirror was placed in the focusing region about 2 mm from the glass. The general shape of the data curve is qualitatively similar to that predicted theoretically, but quantitative agreement is poor. This could be because (i) the phase fluctuations introduced

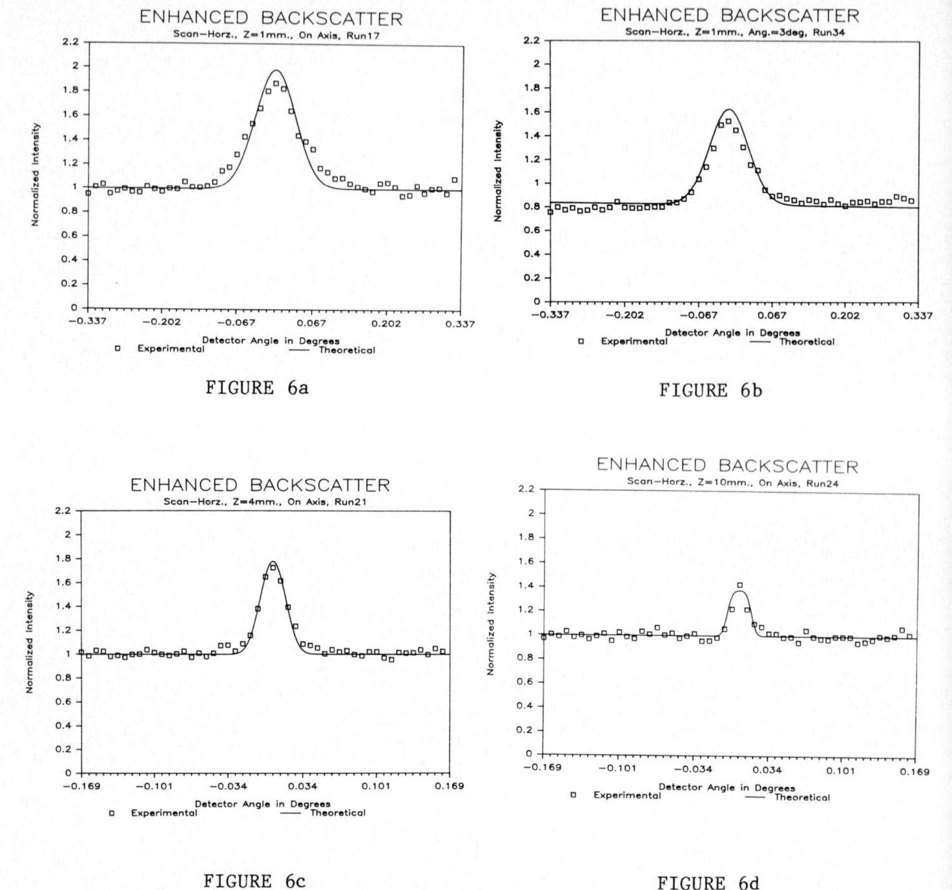

FIGURE 6a

FIGURE 6b

FIGURE 6c

FIGURE 6d

Comparison of measured data with the coherent enhancement effect generated by fine scale diffuser (a) normal incidence (b) 3° off-axis (c) and (d) reduction of effect due to increased mirror/screen separation.

by the glass were not isotropic and Gaussian (as assumed by the theory), (ii) the beam was too narrow (evidenced by the observation of a non-Gaussian speckle pattern fig 7(a) or (iii) the analytical result plotted for comparison was in fact inaccurate. In any case it may be significant that the shape of the curve is similar to that obtained by Mendez and O'Donnell[7] (fig 9) for enhanced backscattering from a smoothly varying, Gaussian reflecting surface, suggesting the importance of a geometrical optics enhancement in their measurements.

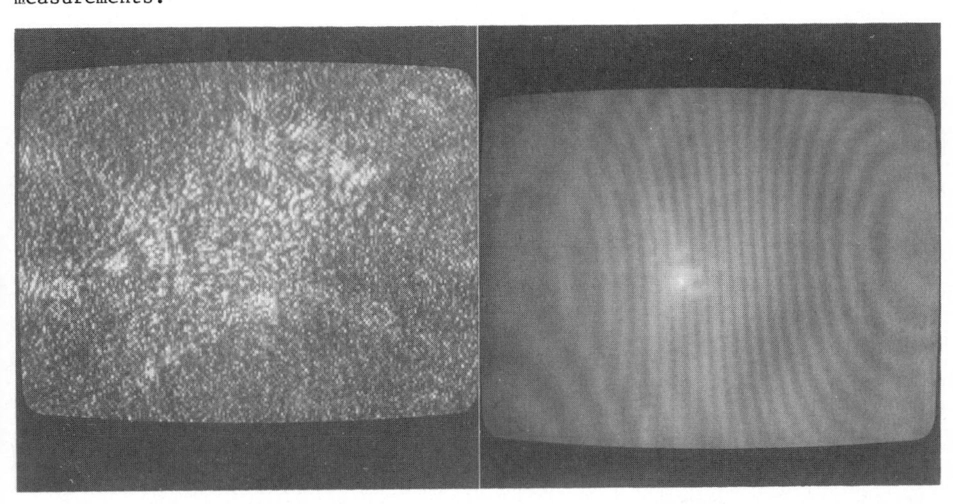

FIGURE 7(a) FIGURE7(b)
Intensity pattern generated by back reflection of laser light through a standard "obscure" glass (a) diffuser stationary; note that the pattern is not Gaussian speckle (b) diffuser rotating showing broader enhancement peak than coherent case.

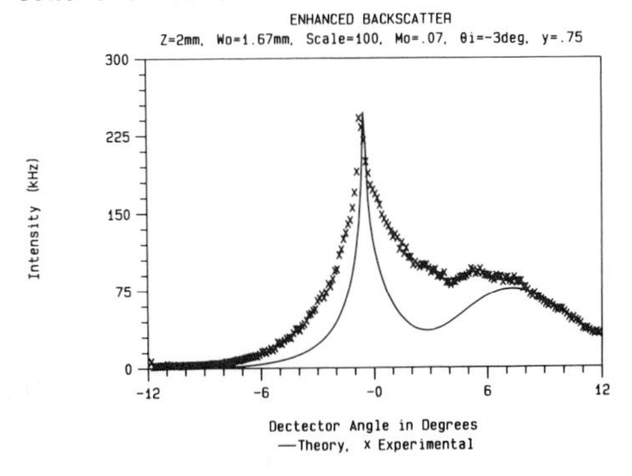

FIGURE 8
Comparison of measured data and theory with the mirror near the focussing region of the standard obscure glass

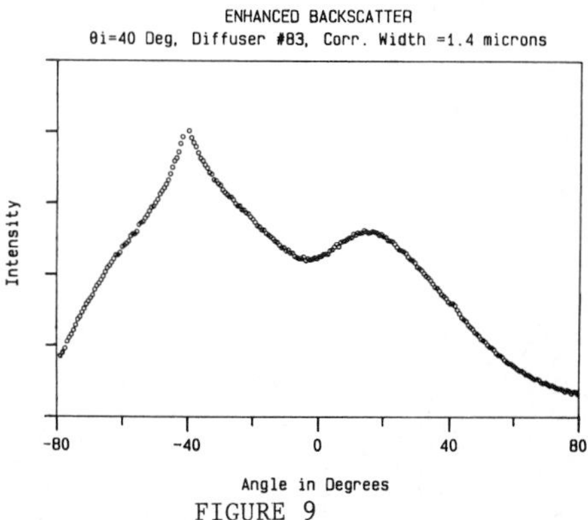

FIGURE 9
An experimental result obtained by Mendez and O'Donnell[7] for a steeply sloped reflecting surface.

We have carried out a second series of experiements on a much larger scale system. The phase screen in this case was a turbulent layer generated by a fan assisted industrial propane gas burner. This was illuminated by a 10 cm collimated laser beam back reflected from a 15 cm plane mirror over a 100m (maximum) indoor range. A diagram of the experimental arrangement is shown in figure 10. Figure 11(a) shows the returned beam in the absence of the turbulence. The disortion is due to astigmatism of the collimating mirror. Figure 11(b) shows the effect of introducing turbulence. Integration by the camera reveals an enhanced backscattering peak which is only a little wider than the expected diffraction limited return beam. Thus aberrations in the transmission optics are, on average, compensated by a "partial phase conjugation" phenomenon. Note that this can only occur when $\ell_s \gtrsim W$ so that the magnitude of the enhancement effect is reduced, according to theory, as the diffraction limit is approached. The quantitative results presented in figures 12(a), (b) confirm this behaviour. Figure 12(a) shows a data trace across the entire return intensity profile with a single point marking the presence of the back-scattering enhancement. Figure 12(b) shows a "close up" of the enhancement peak and confirms that the coherent enhancement factor of 2 is not achieved although the turbulent layer lies in the Gaussian speckle region with respect to the first pass. A detailed description of these experiments has recently appeared[18].

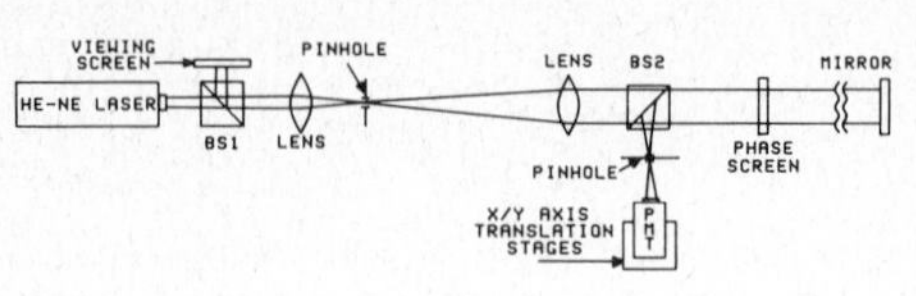

FIGURE 10
Experiments with a turbulent layer: diagram of apparatus.

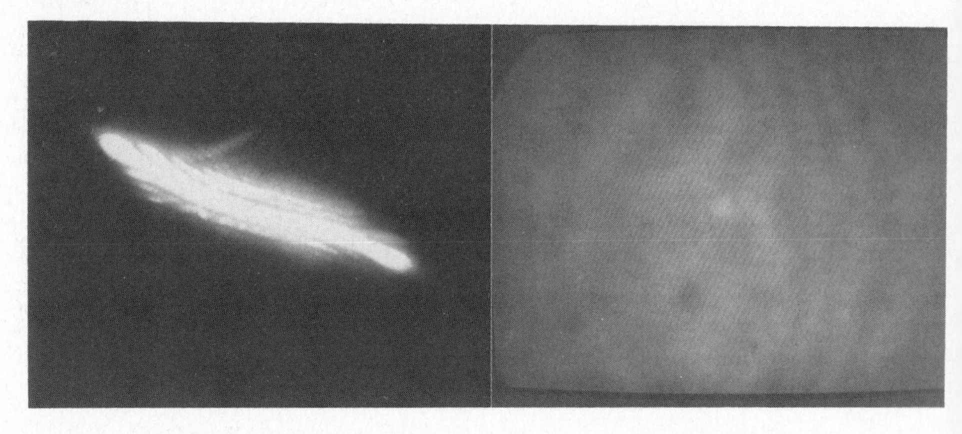

FIGURE 11a FIGURE 11b
Returned beam profile (a) in the absence of turbulence and (b) when the fan assisted gas burner was operating.

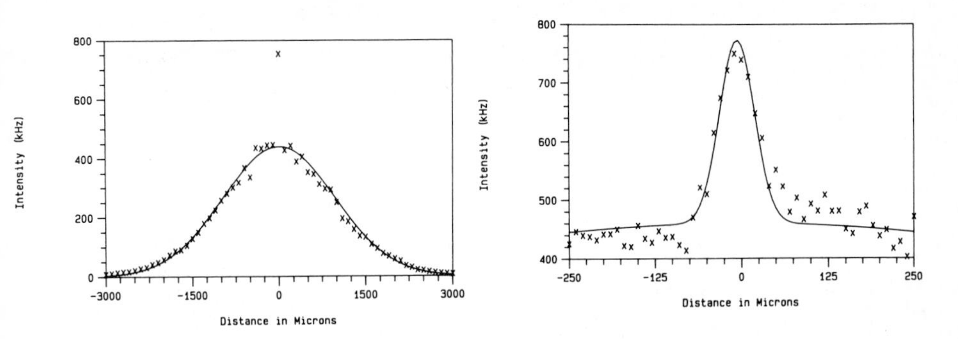

FIGURE 12(a) FIGURE 12(b)
Comparison of theory with experiment (a) scan across the returned intensity profile (b) close up of the enhancement peak.

Finally, preliminary experiments have been made on a third, even larger scale system: back reflection of laser light over a 1.2 Km atmospheric path. At the time of writing the quantitative data from these experiments has not been fully analysed. However, enhanced backscattering was observed visually though on an intermittent basis according to the prevailing turbulence conditions. Some observations of this effect have also been reported in Soviet literature[10].

4 FUTURE DIRECTIONS

In this paper a tutorial model for enhanced backscattering has been analysed theoretically and experiments to test the theoretical predictions have been reviewed. The model can generate those phenomena encompassed within physical optics where scalar diffraction theory is a valid approach and cannot be used, as it stands, to describe polarisation effects which often accompany multiple scattering. It does, however, reveal that in addition to the coherent enhancement associated with multiple scattering in many-scatterer systems, smoothly varying random media can generate incoherent enhancement in the backscattering direction due to a "cats eye" or random corner cube phenomenon. Double scattering is achieved through back reflection in the present model but the results suggest that steeply sloped smoothly varying reflective surfaces might well give rise to the same geometrical optics effect when the tangent plane approximation[19] breaks down.

There are clearly many other back reflection geometries requiring investigation although several have been studied already in some detail[9,10]. A variety of other statistical models for the scattering media are also of interest[15] as is the effect of broad band illumination, an extended source and detector integration. The work described here is restricted to the static medium case, whereas in many situations of interest the scatterer will move bodily and/or evolve with time. This presents another potential area of interest for noise modelling and remote sensing. Potential applications of the back reflection work worthy of investigation include imaging through turbulence, phase conjugation and noise modelling in multiple scattering and multipath system configurations[20-24].

ACKNOWLEDGEMENTS

I would like to thank my colleagues, Dr P R Tapster Dr D L Jordon and our visiting Research Fellow Dr A R Weeks for their invaluable experimental support of this work.

REFERENCES

1) Y Kuga and A Ishimaru, J Opt Soc Am A 8, (1984) 831-839.

2) M P Albada and A Lagendijk, Phys Rev Letts 55, (1985) 2692-2695.

3) P E Wolf and G Morey. Phys Rev Letts 55, (1985) 2696-2699.

4) S Etemad, R Thomson, M J Andrejco, S John and F C Mackintosh, Phys Rev Lett 59 (1987) 1420-1423.

5) M P Van Albada, M B Van der Mark and A Lagendijk, Phys Rev Lett, 58, (1987) 361-364.

6) G Bergmann, Phys Rev, 107, (1984) 1-58.

7) E R Mendez and K A O'Donnell, Opt Commun, 61, (1987) 91-95.

8) K A O'Donnell and E R Mendez, J Opt Soc Am 4, (1987) 1194-8.

9) Yu A Kravtsov and A I Saichev, Sov Phys Usp, 25, (1983) 494-508.

10) V A Banakh and V L Mironov: Lidar in a Turbulent Atmosphere (Artech House, New York 1988).

11) E Jakeman, J Opt Soc Am A 5, (1988) 1638-1648.

12) E Jakeman and J G McWhirter J Phys A 10 (1977) 1599-1643.

13) M V Berry J Phys A 10 (1977) 2061-2081.

14) E E Salpeter, Astrophys J 147, (1967) 433-448.

15) E Jakeman Wave Propagation and Scattering Ed B J Uscinski, (Clarendon Press, Oxford 1986) 49-63.

16) E Jakeman J Phys A 13, (1980) 31-48.

17) E Jakeman, P R Tapster and A R Weeks, J Phys D, 21 (1988) 532-536.

18) P R Tapster, A R Weeks and E Jakeman, J Opt Soc Am. A6, (1989) 517-522.

19) P Beckman and A Spizzichino: The Scattering of Electromagnetic Waves from Rough Surfaces (Pergamon Press, Oxford 1963).

20) I B Esipov and V V Zosimov, Opt Spectrosc, 60, (1986) 234-236.

21) V A Banakh, V M Buldakov and V L Mironov, Opt Spectrosc, 58, (1985)

22) V P Aksenov, V A Banakh and B N Chen, Opt Spectrosc, 57, (1985) 445-446.

23) C J Baker, R J A Tough and J M Pink, IEEE Radar 88 Conference Proceedings, pp241-243.

24) Ibid submitted to IEE Proceeding F.

Scattering in Volumes and Surfaces
M. Nieto-Vesperinas and J.C. Dainty (Editors)

SCATTERING EXPERIMENTS WITH SMOOTHLY VARYING RANDOM ROUGH SURFACES AND THEIR INTERPRETATION

Eugenio R. MENDEZ

División de Física Aplicada
C.I.C.E.S.E., Ensenada, Baja California, México

Kevin A. O'DONNELL

School of Physics
Georgia Institute of Technology, Atlanta, Georgia, U.S.A.

1. INTRODUCTION

There have been many theoretical and experimental studies of scattering from random rough surfaces. From a theoretical standpoint, the problem of determining the mean scattered intensity from a surface of given statistical properties has a considerable degree of complexity. This has led to various approximations in the analysis, and to a vast and diverse literature.

Simplifications of great practical significance due to their mathematical tractability and their applicability in many situations are models based on the Kirchhoff approximation[1]. Of these theories, perhaps the most successful has been the formulation discussed by Beckmann[2], which has been widely used in many regions of the spectrum. In this method, the field and its normal derivative along the rough surface are approximated by the values that would exist on an infinite plane tangent to each point. While there are situations in which this approximation would undoubtedly be inappropriate, its region of validity has not been clearly established. Recently, experiments with characterized surfaces[3] and also computer simulations[4,5] have found significant departures from the Kirchhoff theory.

From a fundamental point of view, the Kirchhoff approximation should fail when the surface contains features smaller than or comparable to the wavelength (resonant effects), or else when multiple scattering takes place. It is then not surprising that the most recent theoretical approaches to the problem have attempted to account for such effects[1]. Most of these theories assume that the random height variations on the surface constitute a Gaussian random process and involve approximations of one sort or another. These approximations impose restrictions that, very often, do not have clear implications. It is then important to establish the relative merits of these different treatments and compare their predictions with the results of computer simulations and experiments.

Although there has been a great deal of experimental work concerning wave scattering from surfaces, much of this suffers from various shortcomings. One of the most serious of these is that in many cases the statistical properties of the surfaces used are poorly known, so that direct comparisons of data with theory are either not possible or else involve rather arbitrary assumptions. This state of affairs has led to a deficiency of rigorous tests of the developing theoretical techniques, and in an important sense the surface scattering problem has thus remained poorly understood.

Our recent experiments have tried to alleviate this situation[3,6]. Using a method due to Gray[7], we have produced random rough surfaces that have height variations which approximate, at least in principle, a Gaussian random process with a Gaussian correlation function. In the cases in which it was possible, the statistical properties of the surface were estimated with a stylus instrument and agreed well with the expected values. For the surfaces with lateral correlation lengths a comparable to the standard deviation of heights σ, a significant part of the scattered intensity is in the cross-polarized component. These surfaces also exhibit backscattering enhancement. This phenomenon has recently attracted much attention, in part due to its connection with the localization of electrons in a random potential[8,9]. These observations cannot be explained by theories based on the single-scatter Kirchhoff approximation, and instead we believe that these effects are due to multiple scattering. In the following, we present selected results of this experimental study, and discuss our interpretation of these results.

2. EXPERIMENTAL RESULTS

The surfaces were made by exposing photoresist-coated plates to laser speckle patterns. Due to the exposing geometry and the number of superimposed patterns, the height variations on the surface should constitute a good approximation to a Gaussian random process with a Gaussian correlation function[7], as many of the rigorous theories assume. To produce highly reflective surfaces, the photoresist plates were vacuum-coated with gold, and the experiments were done in reflection.

Surfaces were fabricated which, according to their statistical properties, may be separated into three groups. The first group is represented here by diffuser #80, with an estimated correlation length a of 20.9 μm (hereafter defined as the $1/e$ radius of the correlation function) and a standard deviation of heights σ of 2.27 μm. For the second group (diffuser #45) we estimate $a \cong 4.5$ μm and $\sigma \cong 0.9$ μm. Due to the relatively large slopes present, perhaps the most interesting surfaces are those of the third group, represented by diffuser #83, with a in the range 1.4 - 1.8 μm and $\sigma \cong$ 1 μm. The first group of surfaces were characterized by means of a stylus instrument, giving reliable results. In particular, both the histogram of height fluctuations and the lateral correlation function of surface #80 were found to be of nearly Gaussian form. The surfaces of the second group contain features very close to the resolution limit of the profiler and, although the measured statistics were close to the expected ones, the validity of the stylus data is uncertain. The diffusers of the third group could not be characterized in this way, and we only quote the statistical parameters expected from the exposing geometry and the exposure time.

In a typical experiment, a laser beam of P or S polarization was incident on the sample. Data were taken at a variety of wavelengths (0.514, 0.633, and 10.6 μm). In each scan the angle of incidence was held fixed, and a detector mounted on an arm was moved along an arc centered on the surface. A field lens in front of the detector integrated over a small solid angle (providing about

1 degree of angular resolution) to reduce speckle noise; hence the measurements represent the diffuse scatter of the surface as a function of scattering angle. A polarizer in front of the detector allowed the co- and cross-polarized scatter to be measured; in the notation used here *SP* data, for example, implies that *S* polarization was incident and *P* polarization was detected. A more detailed discussion of surface fabrication and experimental procedure may be found in Ref. 3.

Surfaces of the first group have mild slopes and a scale parameter a large compared with the illumination wavelength. Figure 1 shows experimental data for diffuser #80 with 10.6 μm wavelength illumination, together with a comparison of calculations from the Beckmann theory. The angles plotted here and in other figures are with respect to the surface normal. It can be seen that both data and theory show a Gaussian-like distribution of light centered on the direction of specular reflection. It is notable that this is not a best-fit plot, as the parameters required by the theory are taken from the profilometer data. Thus even with $a \cong 2\lambda$, the Beckmann theory provides an excellent fit. This in itself is significant, as to our knowledge this represents the first test of the Beckmann theory with a surface that is well-characterized. Further data taken with this surface at infrared and visible wavelengths also agree well with this theory, as long as the angle of incidence is not too large[3].

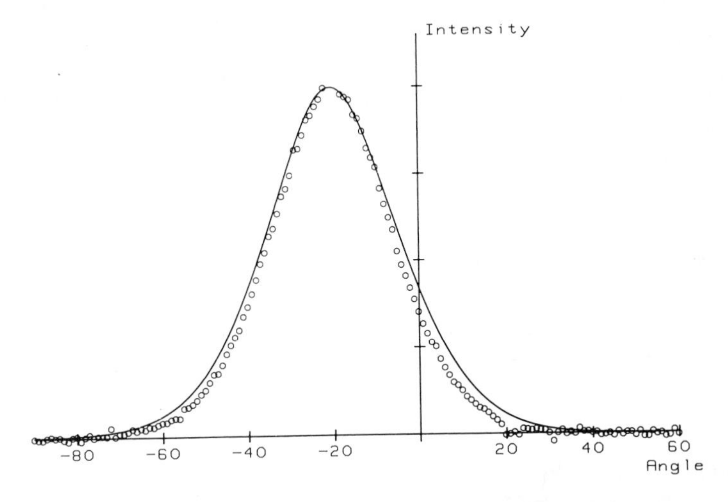

FIGURE 1. Diffuse scatter from diffuser #80 (a = 20.9 μm and σ = 2.27 μm from profilometer data) as a function of detector angle. The angle of incidence is +20 degrees, illumination wavelength 10.6 microns, and only co-polarized scatter was observed. A specular component was present but is not shown. The solid line is the theory due to Beckmann [2].

The fine-scale surfaces in the second and third groups showed quite different and unusual scattering behavior as compared with those of the first group, but the unusual behavior is more marked for the third group; for brevity we will then only discuss results obtained with diffuser #83. For this surface, $a \cong 2\lambda$ at the 0.633 μm wavelength. However, this does not imply that scattering results similar to Fig. 1 should be expected, as the slopes present on surface #83 are much stronger than those of #80. It is well known that for a Gaussian process the slope is also Gaussian-distributed with standard deviation equal to $\sqrt{2}\sigma/a$. Hence surfaces in this group contain both very high slopes (of the order of unity) and features only slightly larger than the wavelength of the incident radiation. Even in the geometrical optics limit, the Kirchhoff approximation should then break down due to multiple scattering and shadowing effects.

Figure 2 shows, for $\lambda = 0.633$ μm, the mean scattered intensity as a function of scattering angle for S polarized waves incident normally on the surface. There are several remarkable aspects of this data. First, there is a large amount of cross-polarized scatter; at a typical angle it is roughly 50% of the height of the co-polarized scatter. Perhaps most remarkable are the large peaks in both polarization components at 0 degrees, along with significant secondary minima and maxima. It must be stressed that these peaks are not associated with a coherent component, as this surface was quite rough as compared with the incident wavelength. Figure 3 shows data taken with the same experimental conditions, except the angle of incidence has been increased to 10 degrees. It can be seen that there are still very strong peaks present in the *backscattering* direction in both polarization components, and the secondary maxima have become rather skewed. In the Beckmann theory, the incoherent angular scatter is related to the distribution of surface slopes. However, for diffuser #83 the scattering pattern is very narrow as compared with the surface slope distribution, which, along with the backscattering structure, strongly suggests that the Kirchhoff theory is quite inappropriate for this surface. Further measurements have shown that the backscattering enhancement slowly weakens as the angle of incidence increases, and is still quite observable at angles of incidence greater than 60 degrees.

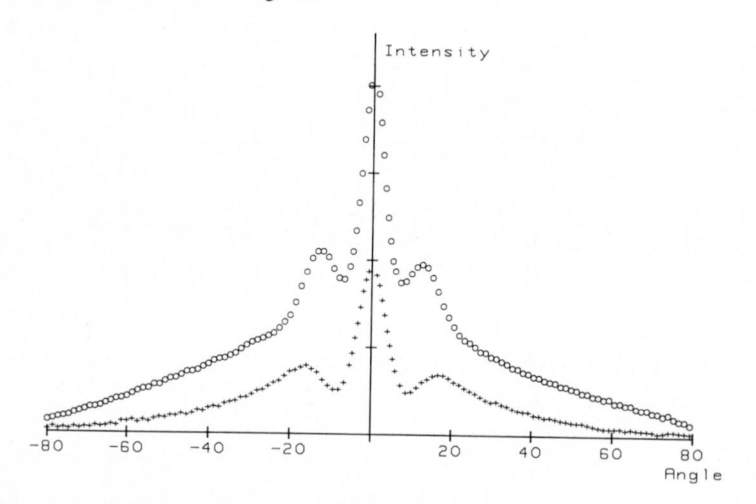

FIGURE 2. Mean scattered intensity from diffuser #83 ($a \cong 1.4$ μm and $\sigma \cong 1$ μm from exposure parameters and electron microscope estimates) as a function of detector angle. The light is normally incident, illumination wavelength 0.633 microns, S incident polarization, and the o's denote co-polarized (SS) scatter and the +'s denote cross-polarized (SP) scatter.

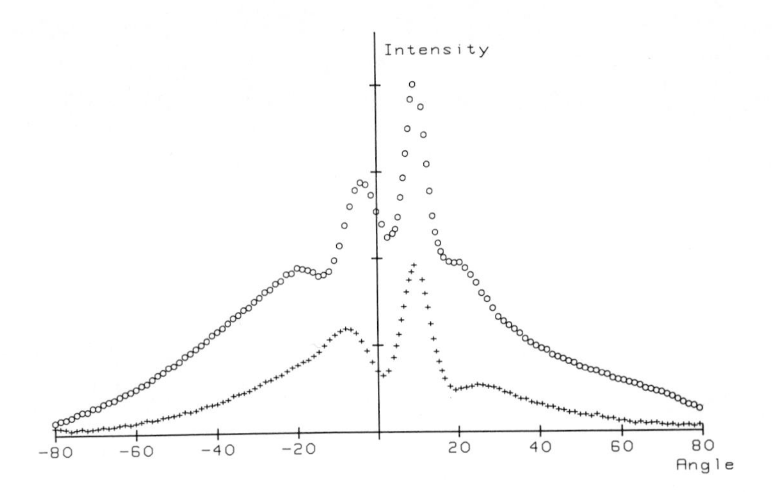

FIGURE 3. Mean scattered intensity from diffuser #83 as a function of detector angle. The angle of incidence is +10 degrees, illumination wavelength 0.633 microns, and o's denote co-polarized (*SS*) scatter and the +'s denote cross-polarized (*SP*) scatter. Enhanced backscatter is clearly seen in both scattering components as the peak at +10 degrees.

It is also possible to show that, even in normal incidence, the diffuse scatter from surface #83 is not rotationally symmetric. Figure 4 shows photographs of the two-dimensional co- and cross-polarized scattering patterns in normal incidence. These patterns very nearly have a four-fold symmetry about the optical axis, and are connected with polarization rather than any lack of statistical isotropy of the surface. It is straightforward to demonstrate this, as rotation of the incident polarization leads to similar rotation of the scattering patterns. An annular region surrounding the backscattering direction has highly polarization-dependent scatter, for if these patterns are viewed through a rotating analyzer, the four dark regions rotate about the optical axis at one-half the rate of analyzer rotation. The dark regions of the cross-polarized pattern have higher contrast than those of the co-polarized pattern, though this may not be obvious from the photographs. Figure 2 corresponds to scans of Fig. 4 along the x-axis and Fig. 12 to scans along the y-axis, and the dark spots fall at roughly the first minima of the data. Similar observations were made for all the fine-scale diffusers that were fabricated, with small variations of the backscattering strength and other details.

3. QUALITATIVE INTERPRETATION

The scattering properties of diffuser #83 (and similar surfaces) are unusual and careful interpretation of these results is important. At present, we believe that multiple scattering is the cause of the unusual backscattering effects, but these observations are not consistent in detail with any existing theoretical calculations. In the following is presented a discussion of the physical mechanisms that we believe are responsible for the observed effects; much of this discussion is based on observations of these surfaces in a high-resolution optical microscope.

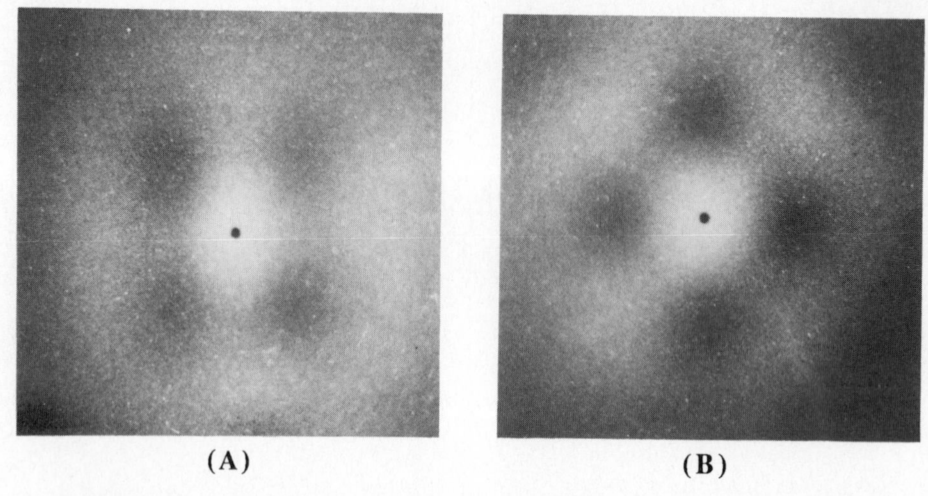

FIGURE 4. Two-dimensional scattering pattern of diffuser #83 in normal incidence and wavelength 0.633 microns; (A) is the co-polarized and (B) is the cross-polarized scattering (the field shown has a half-angle of approximately 25 degrees). The dark spot is a hole in the projection screen that lets the laser source through. For each component, it can be seen that there is a scattering maximum in the backscattering direction surrounded by four regions where depleted scattering is observed. Figures 2 and 12 correspond to scans along the x- and y-axes of these patterns.

First, it is important to consider the multiple scattering mechanism that has been found to be responsible for backscattering enhancement from other random media. In particular, it has been discussed at length that the scattered amplitudes from a multiple scattering path and its time-reversed partner in a random medium add in phase only in a small cone surrounding the backscattering direction, and these paths add essentially incoherently at significantly different angles[8,9]. To apply a similar argument to the case of a rough surface, consider light incident on a valley of a random surface as shown in Fig. 5. If the valley sides are particularly steep, a significant amount of light will follow the path 1→2 shown; moreover, light will also follow the 2→1 path. The phase difference between the 1→2 path and the 2→1 path follows from straightforward geometry as

$$\phi = (\mathbf{k_i} + \mathbf{k_s}) \cdot \Delta\mathbf{r}$$

where $\mathbf{k_i}$ and $\mathbf{k_s}$ are the incident and scattered wavevectors, and $\Delta\mathbf{r}$ is the vector between points 1 and 2. If $\mathbf{k_i}$ and $\mathbf{k_s}$ are significantly different, averaging over all such paths occurring on a random surface (that is, with stochastically varying $\Delta\mathbf{r}$) will destroy constructive interference between such pairs of paths, and they will contribute on only an intensity basis to the scattered intensity. However, near backscattering where $\mathbf{k_i} \cong - \mathbf{k_s}$, the amplitude from these pairs of paths add in phase and there should be a strong contribution to the scattered intensity.

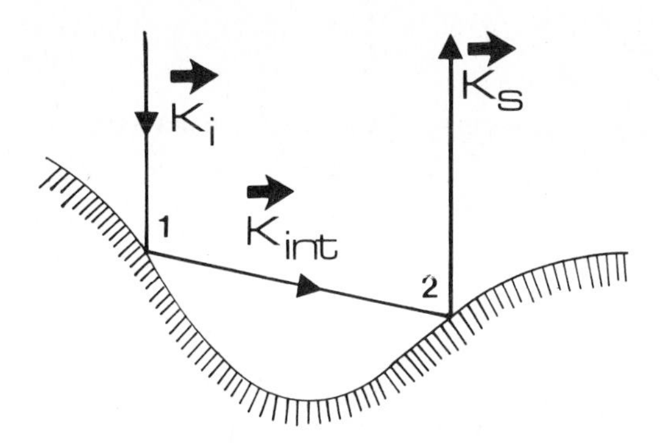

FIGURE 5. Multiple scattering in a valley of the surface.

There is direct evidence that these light paths are occurring on diffuser #83. In Fig. 6 is shown a photograph of a portion of a similar surface taken at high magnification in a reflection optical microscope. The dark shadow is a sharp edge from another plane in the optical system that is being imaged onto the surface. As this dark edge is scanned along the surface, the edge seen in reflection from the surface appears slightly jagged because of the surface roughness. When this shadow encounters what we believe to be a deep valley of the surface (Fig. 6), the direction of edge scan *inside of the valley* appears to reverse. In other words, if the edge is positioned at the center of a particular valley (as shown in the encircled area of the photograph), the left side of the valley darkens and the right half remains lit. Hence the right side of the valley is apparently being illuminated with sources that lie on the left side (and vice versa), in a manner entirely consistent with the light paths shown in Fig. 5.

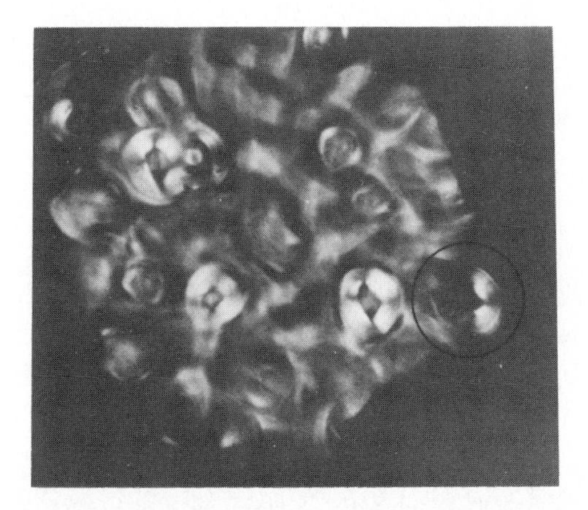

FIGURE 6. Scanning a sharp edge of intensity across the surface. The illuminated area in the photo is a field aperture in the optical system that is being imaged onto the surface. When the sharp edge of the aperture crosses deep valleys, as can be seen in the encircled area above, the direction of scan of the edge in the valley is reversed, in a manner consistent with the scattering paths of Fig. 5.

There are other unusual aspects of the light emanating from these deep valleys. An area of the surface illuminated with vertically polarized light is shown in Fig.7A and 7B photographed with co- and cross-polarized analyzers, respectively. The light originating from the deep valleys (at least deep enough to give rise to the edge reversal discussed above) has an unusual polarization structure that can be seen in the photo. If this area is viewed without an analyzer, the pattern in a given valley is nearly rotationally symmetric. In the photo with the cross-polarized analyzer, each valley (see the encircled areas) has four dark regions which are oriented at approximately 0°, 90°, 180°, and 270° in the valley, with bright areas in between. With a co-polarized analyzer, the entire pattern in a valley rotates by 45°. That is, the dark regions now appear at 45°, 135°, 225° and 315° in Fig. 7A, and there is also a reduction of contrast of this structure. These dark regions are difficult to see in Fig. 7A, but they can be quite convincingly observed.

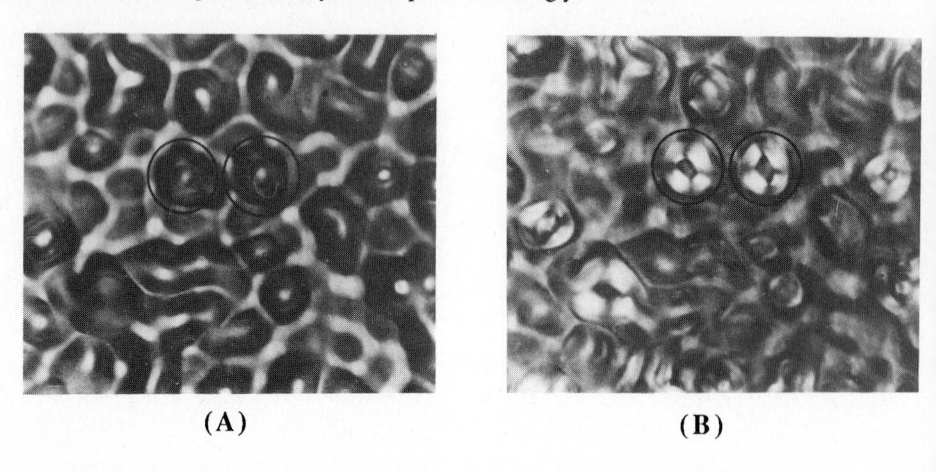

FIGURE 7. Photo of the surface taken in an optical reflection microscope (NA = .80, 50x magnification objective) with linearly polarized incident light. Photo (A) is taken with a co-polarized analyzer and (B) is taken with a cross-polarized analyzer. The encircled areas are deep valleys where the edge reversal discussed in the text can be observed, and the scattering from these valleys produces patterns in agreement with the light paths of Fig. 8.

In order to understand these observations, it is necessary to consider how the polarization is affected by multiple scattering within a valley. Consider the incidence of a vertically polarized wave onto a two-dimensional valley (shown from above in Fig. 8). If the valley is sufficiently deep, multiple scattering will occur along all possible paths from one side of the valley to the other. The interaction of a wave with such a structure would be complicated, and for simplicity we assume that the surface is locally planar. In this case the polarization of the light following any paths is determined by the (generally complex) S and P reflectivities with respect to the local plane of incidence. The light following the scattering paths of Fig. 8A and 8B will not suffer any depolarization, as the light is purely S or P polarized. For the diagonal cross-paths of Fig. 8C and 8D the light has both S and P components; the complex reflectivities will lead to elliptical polarization of the scattered return. In the case of a perfect conductor, as shown in Fig. 8, these diagonal paths cause a rotation of the polarization vector by 90° and the light is returned cross-polarized. Hence the light paths of (8A) and (8B) should be extinguished by a crossed analyzer, while (8C) and (8D) should be extinguished by a parallel analyzer; this is precisely what was found in Fig. 7. The light paths within a circular valley, at least in the case of a perfect conductor, have a four-fold symmetry about the valley center. Propagation of these patterns to the far-field also produce a pattern that has four-fold symmetry, as we apparently saw in Fig. 4. It is also worth noting that the lower contrasts of the patterns in Fig. 7A could arise because paths (8C) and (8D) are indeed not perfectly cross-polarized for a real metal with a complex reflectivity.

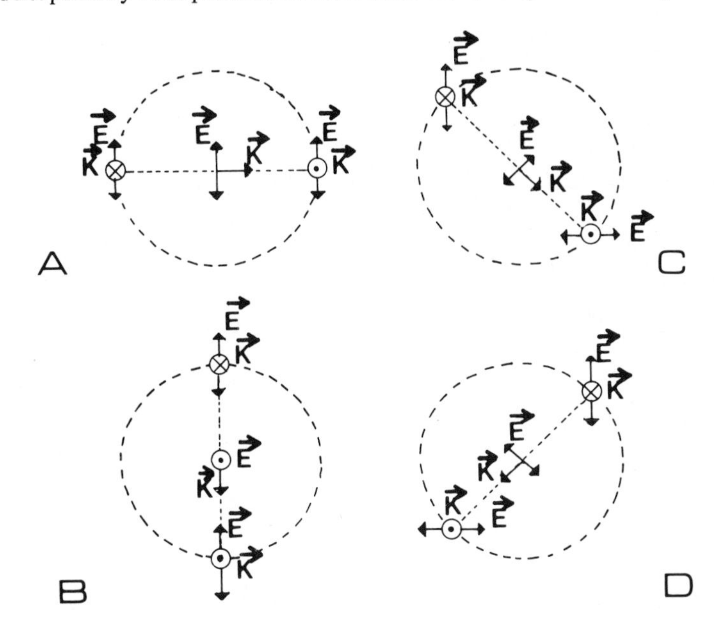

FIGURE 8. Light paths arising in a valley of the surface as viewed from above (with vertical incident polarization). For the case of a locally planar surface, multiple scattering along paths (A) and (B) involves only the S or P reflectivity and do not produce depolarization. Paths (C) and (D) produce in general elliptically polarized light, and for the case of a perfect conductor (shown) they produce cross-polarized scattering. Compare with the observations of Fig. 7.

The discussion above provides an intuitive way of understanding these observations, but a more sophisticated approach could very well lead to similar conclusions, at least in certain limits. A second comment we make here is that the valleys take on a wide variety of (non-circular) shapes on a typical microscope image; the region in the Fig. 7 shown was chosen quite carefully to have a few nearly circular regions to illustrate these points. However, the polarization structure seen in these distorted valleys usually maintains some resemblance to that of the more circular ones.

Another observation helps to substantiate these ideas. Co-polarized light is observed over most of the random surface in the microscope, but essentially all of the cross-polarization seen on the surface is associated with multiple scattering in deep valleys (the light that appears to come from other parts of the surface in Fig. 7B arises through imperfections in the polarizer). All of this cross-polarized light propagates to the far-field and forms a pattern that appears to be associated only with backscattering enhancement (see the lower curve of Fig. 2). This is highly significant in that it implies that enhanced backscattering is associated only with multiple scattering paths analogous to those of Fig. 5.

The models that we have proposed here give qualitatively correct results, but we do not claim that they are in any sense a precise way of understanding these observations. Particularly because these valleys are but a few wavelengths across, a more rigorous solution would be desirable. The main difficulty encountered in scattering problems is that the values of the field and its normal derivative along the surface are not generally known. Even for simple bowl-shaped regions such as the one shown in Fig. 5, we are not aware of any rigorous analytical solutions. On the other hand, related problems can be solved numerically; such an approach is adequate for deterministic surfaces and has recently been employed in Monte-Carlo simulations[4,5,12]. These numerical techniques can be quite rigorous in their approach and take into account effects such as shadowing and multiple scattering. Even if it is only for deterministic surfaces, exact solutions can be useful in the understanding of the physical processes involved. A specific example for a bowl-shaped surface is considered in the next section.

4. RIGOROUS NUMERICAL SOLUTION: AN EXAMPLE

Although the models discussed in the last section are in qualitative agreement with the experimental results, it is fairly obvious from the more recent literature on the subject that, at present, the mechanism responsible for the observed phenomena is considered an open question[10,11]. There are two main reasons for which there has been some reluctance for the acceptance of these models. First of all, our use of *rays*, or wavevectors, to indicate multiple scattering paths (as in Fig. 5) has been questioned by many authors; particularly in view of the length scales involved. And second, Bahar and Fitzwater[11], using the so-called full wave method in a single scattering approximation, have produced results that are indeed similar to our experimental ones.

We do not claim that our models provide a quantitative representation of the physical processes that take place on the surface, but they provide useful approximations that, at least qualitatively, account for the observed phenomena. From first principles, there is no reason to expect single scattering theories to provide adequate results for surfaces with steep slopes. We have ample evidence that multiple scattering takes place on our surfaces and believe that the fact that a single scattering theory produces similar results is fortuitous. On the other hand, the mechanism responsible for the backscattering enhancement from randomly rough surfaces is still uncertain and could well be a combination of several of the mechanisms that have been suggested in the literature. Nevertheless, we are convinced that multiple scattering is strongly connected with this effect. The purpose of this section is to support some of our naive models and comments with more rigorous arguments.

Due to the considerable difficulties for the application of exact analytical techniques, several groups have conducted numerical simulations of scattering from randomly rough two-dimensional (corrugated) surfaces[4,5,10,12]. The approach requires considerable computer power but has proven quite successful. Still, so far the simulations have given us very little explicit information about the physical processes that take place on the surface. Following this work, we have conducted some numerical experiments to investigate the boundary conditions for a relatively simple bowl-shaped surface illuminated by a Gaussian beam. We contend that the numerical solutions obtained are consistent with the mechanism we have suggested and support our belief in a multiple scattering explanation for the observations.

Since detailed accounts of the numerical techniques can be found elsewhere[4,5,12], our discussion will be rather brief and schematic. The interested reader will find more information in those references and, in particular, in the paper by Thorsos[4]. The problem we consider is that of scattering from a perfectly conducting two-dimensional surface with a Gaussian groove, when an S-polarized Gaussian beam is incident perpendicularly on it. We employ a Gaussian beam in order to avoid edge effects. Due to the two-dimensional geometry of the problem and the polarization of the incident beam, the interaction of the light with the surface will not modify its state of polarization, and the electromagnetic field can be treated as a scalar quantity[13]. We assume a surface profile given by the function

$$\zeta(x) = -\delta \, exp\{-x^2/\alpha^2\} \tag{1}$$

where, in particular, we have chosen the parameters $\delta = 1$ µm and $\alpha = 1$ µm for this numerical example. The surface profile and its first derivative are shown in Fig. 9.

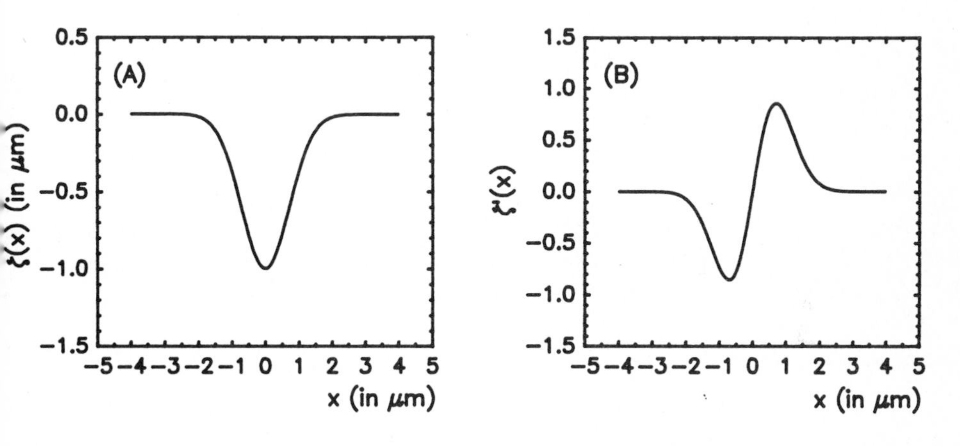

FIGURE 9. Gaussian groove with $\delta = 1$ µm and $\alpha = 1$ µm. (A) Surface profile $\zeta(x)$. (B) First derivative of the profile $\zeta'(x)$.

By the use of Green's integral theorem, it can be shown that the scattered field at a point of observation $\boldsymbol{r} = (x, z)$ above the surface S is given by [4]

$$E(\boldsymbol{r}) = E_{inc}(\boldsymbol{r}) - \frac{1}{4i}\int_S H_o^{(1)}(k|\boldsymbol{r}-\boldsymbol{r}'|)\,\frac{\partial E(\boldsymbol{r}')}{\partial n'}\,dS' \ , \qquad (2)$$

where $E_{inc}(\boldsymbol{r})$ is the incident electric field, $\partial E(\boldsymbol{r}')/\partial n'$ represents the normal derivative of the total electric field on the surface (which is still unknown at this point), $H_o^{(1)}$ is the zero-order Hankel function of the first kind and, as usual, $k = 2\pi/\lambda$; for the example we chose $\lambda = 0.5$ μm. It is convenient to define a source function $F(\boldsymbol{r}')$ as follows,

$$F(\boldsymbol{r}') = \left[\,1 + \left(\frac{\partial \zeta(x')}{\partial x}\right)^2\,\right]^{1/2} \frac{\partial E(\boldsymbol{r}')}{\partial n'} \ . \qquad (3)$$

We note that in Eq. (2) $\boldsymbol{r}' = (x', \zeta(x'))$ and thus, for a given surface, $F(\boldsymbol{r}')$ is only a function of x'. The source function can be found by solving an integral equation obtained from Eq. (2) by letting $\boldsymbol{r}$ approach a point on the surface and using the boundary condition $E(\boldsymbol{r}) = 0$. This results in the integral equation,

$$E_{inc}(\boldsymbol{r}) = \frac{1}{4i}\int_S H_o^{(1)}(k|\boldsymbol{r}-\boldsymbol{r}'|)\,F(\boldsymbol{r}')\,dx' \ , \qquad (4)$$

which uniquely determines the source function $F(\boldsymbol{r})$. To find $F(\boldsymbol{r})$, Eq. (4) was converted into a matrix equation by partitioning the region of integration into 160 equally spaced subintervals of width .05 μm. Special care needs to be taken with the diagonal elements (i.e. when $\boldsymbol{r} = \boldsymbol{r}'$), but the details are well documented in Ref. 4. The matrix equation was solved using standard numerical techniques[14].

Following Refs. 4 and 12 we employed an incident field of the form

$$E_{inc}(\boldsymbol{r}) = exp\{-ikz\,(1 + W(x)\,)\}\ \ exp\{-x^2/w^2\} \ , \qquad (5)$$

where

$$W(x) = \frac{\left(\dfrac{2x^2}{w^2}\right) - 1}{k^2\,w^2} \ ,$$

and for the example we chose $w = 2$ μm.

The result of this simulation is illustrated in Fig. 10A, where we show the modulus of the source function $F(\boldsymbol{r})$ as a function of position (x) on the surface. For comparison, the modulus of the source function obtained with the Kirchhoff approximation is shown with a dashed line in Fig. 10B. It is fairly evident that in this case the Kirchhoff approximation breaks down. We have conducted other numerical experiments that show that it fails due to multiple scattering and not to surface slope or curvature effects. For instance, the strong oscillations observed in Fig. 10A are not present for a surface consisting of only half a groove, as is the case for the surface

$$\zeta_1(x) = \zeta(x) \qquad \text{for} \quad x<0 \ ,$$

$$\zeta_1(x) = -\delta \qquad \text{for} \quad x>0 \ .$$

The period of the oscillations in the curve of Fig. 10A is approximately $\lambda/2$, suggesting the presence of standing waves in this region. This is what one would expect from our simple arguments and the light paths depicted in Fig. 5. From Figs. 9 and 10A, we see that the oscillations are more violent in the region of the surface in which the slope is close to unity, which would also be expected from geometrical optics arguments.

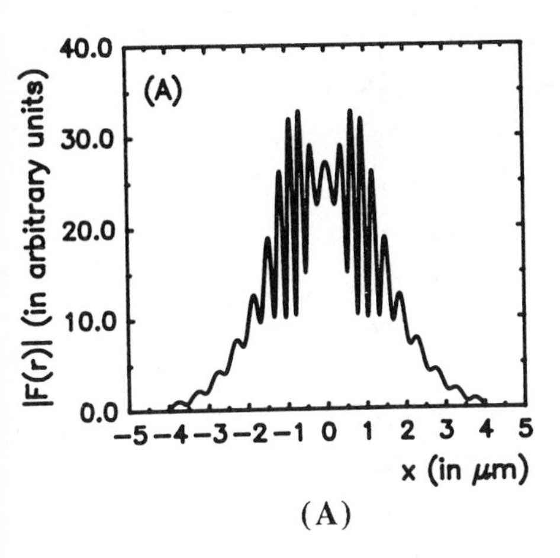

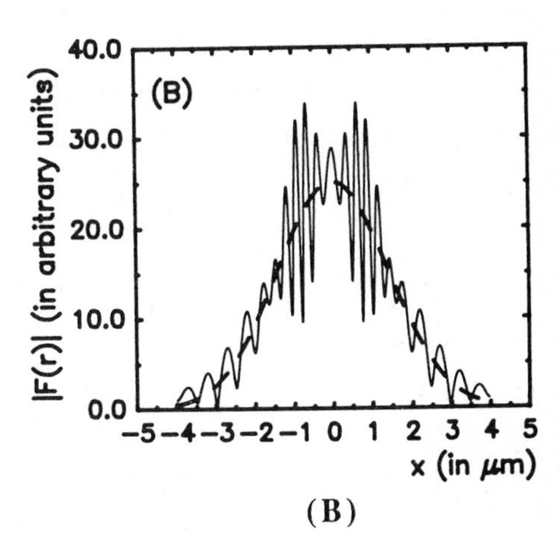

FIGURE 10. Modulus of the source function $F(\boldsymbol{r})$ as a function of position on the surface. (A) Results obtained by solving the matrix equation. (B) Results obtained with the iterative method. The dashed curve (Gaussian-like) involves the zeroth order iterate or Kirchhoff approximation, while the continuous line involves the first order iterate.

The solution obtained above is essentially exact. The only errors arise from the discretization of the integral equation. The source function $F(\boldsymbol{r})$ can also be obtained by a different technique that, although approximate, can provide useful insight into the physics of the problem. The method uses the following Fredholm integral equation of the second kind

$$\frac{\partial E(\boldsymbol{r})}{\partial n} = 2\,\frac{\partial E_{inc}(\boldsymbol{r})}{\partial n} - \frac{1}{2i}\int_S \frac{\partial}{\partial n} H_{\mathrm{o}}^{(1)}(k|\boldsymbol{r}-\boldsymbol{r}'|)\,\frac{\partial E(\boldsymbol{r}')}{\partial n'}\,dS' \ , \qquad (6)$$

which is obtained from Eq. (2). In Eq. (6), the points $\boldsymbol{r}$ and $\boldsymbol{r}'$ both lie on the surface. It appears that this expression was first derived by Meecham[15] in his two-dimensional formulation of the scattering problem. The first term on the right hand side of Eq. (6) is readily recognized as the Kirchhoff approximation. The form of the equation suggests a series of approximate solutions obtained by iteration, in which the zeroth-order iterate is the Kirchhoff approximation. The convergence and interpretation of these iterative solutions has been considered by Liszka and McCoy[16], who have shown that in the geometrical optics limit, the first-order iterate corresponds, essentially, to the case depicted in Fig. 5.

The form of the matrix elements obtained by converting Eq. (6) into a matrix equation can also be found in Ref. 4. The result for the first-order iterate in our example is shown with the continuous line in Fig. 10B. Although there are some differences, the similarity with the curve in Fig. 10A is evident. For the example considered, the curves were indistinguishable after three iterations. These results reinforce our belief in the proposed models, and demonstrate that multiple scattering takes place in these deep valleys despite their relatively small dimensions.

5. DISCUSSION

Other observations also aid in the understanding of scattering from the fine-scale surfaces. As has been discussed by Etemad, Thompson and Andrejco[17], to some degree it is possible to isolate the single scattering from the multiple scattering in the sample. This may be done with circularly polarized light. For example, if a scatterer is illuminated with right-circularly polarized (RCP) light, the single-scatter should be returned as left-circularly polarized light (LCP) and multiple scatter should be predominantly polarized as RCP. In Fig. 11 are shown scans of the scattering from surface #83 with RCP illumination. It can be seen that the RCP scattering has an enormous backscattering peak with an associated ring structure. The LCP scattering is broader and, perhaps surprisingly, has a small backscattering enhancement peak of its own. The reason for this other peak is not well-understood, but the significant disparity of the backscattered signals of the two components strongly suggests that the enhancement arises because of multiple scattering. In the weak localization experiments that have been done, it has been said that the enhancement should be exactly twice the height of the background[8,9,17], but in the case of a surface we have found that it can be significantly greater. We also note that effects which may be related to these observations have been noted in radar scattering[18].

It may also be shown that, despite the earlier discussion, the scattering in the far-field does not quite have fourfold symmetry for the case of normal incidence. This is most clearly seen in Fig. 12 which shows the scattering in normal incidence with P-polarized illumination. In comparing with Fig. 2, it can be seen that the SP and PS scattering is nearly identical. However, the PP scattering does not have secondary maxima, while the SS scattering has strong secondary maxima. For other surfaces that have been fabricated we sometimes do observe secondary maxima in the PP scattering, but they are always weaker than the corresponding maxima in the SS data. The reasons for this are not yet clear, but there is not necessarily a four-fold symmetry in the light paths of Fig. 8 if the surface is not a perfect conductor. This can be easily seen from the paths of Fig. 8A and

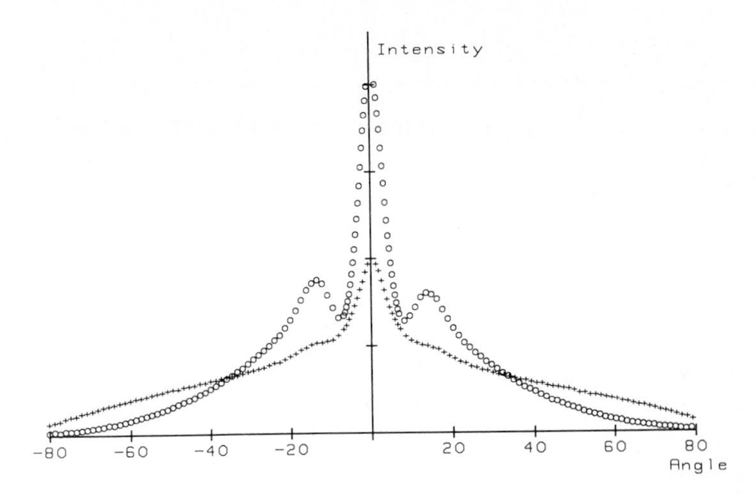

FIGURE 11. Scattering from diffuser #83 with right-circularly polarized illumination in normal incidence. The +'s denote scatter with a left-circular polarized (LCP) analyzer, and the o's are scattering with a right-circular (RCP) polarized analyzer. In this case the enhancement is considerably more than a factor of two above the background.

8B, as one involves the *S* and the other the *P* reflectivity; these are different for real metals and hence the fourfold symmetry breaks down. There is also another possibility, as it is well known that *P*-waves are more likely to excite surface waves than *S*-polarized waves do. In other words, energy is more easily coupled into the surface. The fact that the secondary maxima are less pronounced for *PP* scattering than for *SS* scattering is also consistent with the results obtained by numerical work for two-dimensional random surfaces[19].

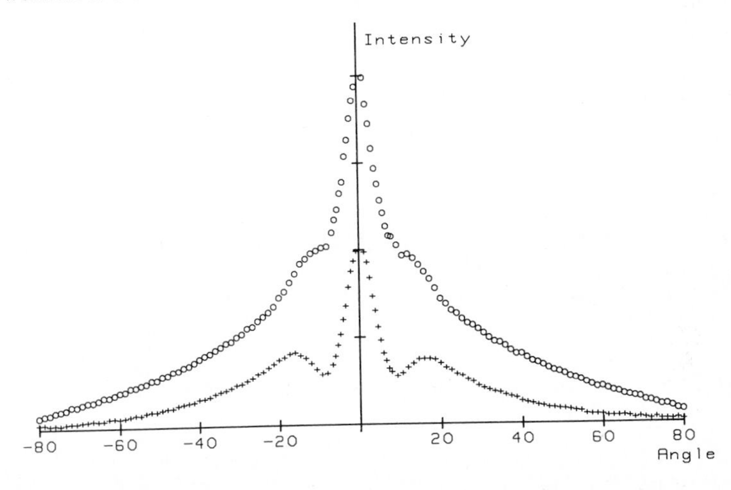

FIGURE 12. Mean scattered intensity from diffuser #83 as a function of detector angle. The light is normally incident, illumination wavelength 0.633 microns, *P* incident polarization, and the o's denote co-polarized (*PP*) scatter and the +'s denote cross-polarized (*PS*) scatter. The secondary maxima seen in the *SS* scatter of Fig. 2 are not seen in the *PP* data, while the *SP* and *PS* data are quite similar.

6. CONCLUSIONS

In this paper we have discussed some of the unusual results of our experiments with specially fabricated rough surfaces. In particular, we have found that in certain regimes these surfaces scatter in a manner that is predicted by theories based on the Kirchhoff approximation. For surfaces that have steep slopes and near wavelength-sized structure, we observe enhanced backscattering and depolarization phenomena that is not predicted in detail by any theory. In view of this situation, we have attempted to present an understanding of the physical processes occurring on a random surface so as to gain insight into the unusual observations. Specifically, we believe that these observations arise from multiple scattering; this interpretation was supported by microscopic observations of the surfaces and by some numerical calculations. At present the theoretical understanding of these phenomena is still being developed, and it is clear that considerable work remains to be done.

ACKNOWLEDGEMENTS

We are grateful to J.C. Dainty, F.C. Reavell and M.J. Kim of Imperial College, London, for discussions and help with various aspects of the experimental work. One of us (ERM) is grateful to A.A. Maradudin and T. Michel of the University of California, Irvine, for several useful discussions concerning the numerical work.

REFERENCES

1) J.A. DeSanto and G.S. Brown, Analytical techniques for multiple scattering from rough surfaces, in: Progress in Optics XXIII, ed. E. Wolf (North-Holland, Amsterdam, 1986) pp. 3-62.

2) P. Beckmann and A. Spizzichino, The Scattering of Electromagnetic Waves From Rough Surfaces (Pergamon, New York, 1963).

3) K.A. O'Donnell and E.R. Méndez, J. Opt. Soc. Am. **A4** (1987) 1194.

4) E.I. Thorsos, J. Acoust. Soc. Am. **83** (1988) 78.

5) J.M. Soto-Crespo and M. Nieto-Vesperinas, J. Opt. Soc. Am. A6 (1989) 367.

6) E.R. Méndez and K.A. O'Donnell, Opt. Commun. **61** (1987) 91.

7) P.F. Gray, Opt. Acta **25** (1978) 765.

8) M.P. Van Albada and A. Langendijk, Phys. Rev. Lett. **55** (1985) 2692.

9) P.E. Wolf and G. Maret, Phys. Rev. Lett. **55** (1985) 2696.

10) M. Nieto-Vesperinas and J.M. Soto-Crespo, Phys. Rev. B **12** (1989) 8193.

11) E. Bahar and M.A. Fitzwater, J. Opt. Soc. Am. **A6** (1989) 33.

12) A.A. Maradudin, E.R. Méndez and T. Michel, Opt. Lett. **14** (1989) 151. See also A.A. Maradudin, E.R. Méndez and T. Michel, "Backscattering effects in the elastic scattering of P-polarized light from a large amplitude random grating", in this volume.

13) M. Born and E. Wolf, Principles of Optics, 5th edition, (Pergamon, New York, 1975), p. 560.

14) See e.g. W.H. Press, B.P. Flannery, S.A. Teukolsky and W.T. Vetterling, Numerical Recipes (Cambridge University Press, New York, 1986), p. 31.

15) W.C. Meecham, J. Ration. Mech. Anal. **5** (1956) 323.

16) E.G. Liszka and J.J. McCoy, J. Acoust. Soc. Am. **71** (1982) 1093.

17) S. Etemad, R. Thompson, and M.J. Andrejco, Phys. Rev. Lett. **57** (1986) 575.

18) Similar (monostatic) observations have been made of circularly polarized radar waves scattered by the satellites of Jupiter, and a related multiple-scattering mechanism has been proposed by S.J. Ostro and G.H. Pettengill, Icarus **34** (1978) 268.

19) Unpublished results by T. Michel, A.A. Maradudin and E.R. Méndez (1989).

'cattering in Volumes and Surfaces
I. Nieto-Vesperinas and J.C. Dainty (Editors)
) Elsevier Science Publishers B.V. (North-Holland), 1990

MEASUREMENTS OF ANGULAR SCATTERING BY RANDOMLY ROUGH METAL AND DIELECTRIC SURFACES

J C DAINTY, M-J KIM and A J SANT

Blackett Laboratory, Imperial College, London SW7 2BZ, UK.

Measurements of the light scattering by three well-characterised random rough surfaces are presented. The surfaces studied include one- and two-dimensional gold-coated samples and a one-dimensional dielectric sample. Measurement wavelengths are 633 nm and 10.6μm. The purpose of these measurements is to provide a reliable database to aid theoretical understanding of multiple light scattering from rough surfaces.

1. INTRODUCTION

There is no shortage of experimental and theoretical studies of angular scattering from randomly rough surfaces. However, experiment and theory rarely have the opportunity to be compared in a critical way, for two reasons; first, theoretical analyses frequently make quite specific assumptions on the nature of the surface (e.g. gaussian probability density of height to all orders, or perhaps perfect conductivity) and on the nature of the scattering process (e.g. "shadowing" or multiple scattering ignored); second, experiments are frequently carried out on surfaces whose properties do not match the theoretical assumptions. In the optical region, there have been few experiments on controlled, well-defined randomly rough surfaces and the results of Reneau[1,2] are of particular note.

From the point of view of assisting our understanding of light scattering, there seems little point in measuring the angular scatter from low-sloped surfaces whose correlation length is much greater than the optical wavelength, since this situation is explained adequately by scalar Kirchhoff diffraction theory (except at grazing incidence). However, scattering from high-sloped surfaces is inherently more interesting, since multiple scattering occurs, and until recently no controlled surfaces were available for visible light experiments. In 1987, Mendez and O'Donnell published experimental results for the angular scatter from well characterised high-sloped metallic gaussian surfaces[3,4] which have stimulated a large number of theoretical and numerical studies[5–11]. The experiments of Mend and O'Donnell were prompted by the question: is it possible to observe, from rough surfaces, enhanced backscattering of the type observed from dense volume media and which is attributed to coherent co-operative effects resulting from multiple scattering? Enhanced backscattering was observed in the pioneering experiments of Mendez and O'Donnell (and subsequently from many other surface prepared in our laboratory) although it is still not clear whether the coherent co-operative effect that gives the effect in dense volume media is the dominant cause the enhancement for rough surfaces.

In this paper, new experimental results[12–14] are presented for two-dimension metallic rough surfaces and one-dimensional metallic and dielectric surfaces — in both cases, results at visible (633nm) and infrared (10.6μm) wavelengths are presented. The one-dimensional results (random gratings) are given to assist comparisons with analytical and numerical calculations which are frequently limited to this case; however, no detailed comparisons with theory are made here.

2. SCATTERING EQUIPMENT AND SURFACE PREPARATION

Figure 1 shows the essential features of the scatterometer used to measure the angular dependence of the scattered intensity. In the present work, wavelengths of 0.633 μm (He-Ne laser) and 10.6 μm (CO_2 laser) were used for illumination. Onl scattering in the plane of incidence is measured, the incident (θ_i) and scattered (θ) angles being both measured from the surface normal, i.e. $\theta = \theta_i$ in the specular direction. A photomultiplier was used to detect the visible wavelength and a pyroelectric detector for the infrared radiation. The measurement of intensity involved some angular averaging, approximately 0.5° for visible and 1.0° for the infrared. Since the speckle size was on the order of 5 arc seconds for the visible and 1 arc minute for the infrared, speckle noise was effectively eliminated and the results are equivalent to ensemble averages blurred by the (small) angular cone of measurement.

The measured intensity is normalised by the incident power and is thus the mean normalised differential scattering cross-section; for a perfect lambertian diffuser, the variation of intensity with angle would be $\cos(\theta)$. Measurements of a smoked Magnesium Oxide surface[13] show this cosine law is obeyed approximately

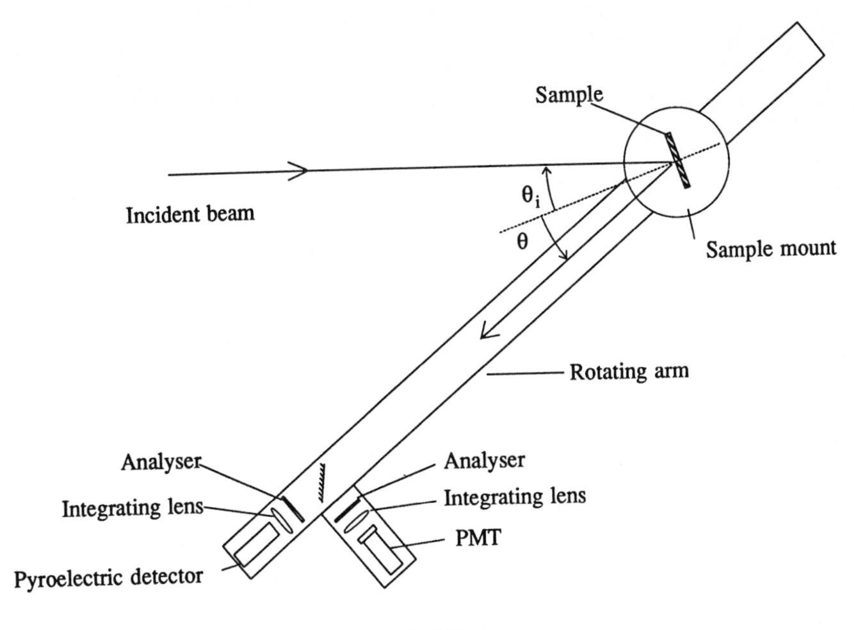

FIGURE 1

for small angles of incidence, except for the presence of the enhanced backscatter peak due to coherent effects within the scattering volume.

Random rough surfaces were prepared by exposing photoresist (Shipley S1400-37 in the most recent experiments) to laser speckle patterns, following the method of Gray[15]. Both one- and two-dimensional surfaces have been made. In the case of the "one-dimensional" surfaces, correlation lengths are typically microns in the direction of interest and millimetres in the orthogonal direction. Surfaces have been replicated[12] by first forming a copy in silicone rubber (Dow Corning Sylgard 182) and then casting in epoxy resin (Araldite MY 778 + HY 956). The resin copy is a positive replica of the original surface. Surfaces are coated with a gold layer ≈90 nm thick; the dielectric results shown in Section 4 were obtained using the silicone rubber intermediate (refractive index 1.43).

Measurements of the surface profile were made with a stylus instrument (Rank Taylor Hobson "Talystep") equipped with a special stylus. For the case of the one-dimensional surfaces, a chisel-shaped stylus was used, of approximate dimensions 0.6 by 2.0 μm. The probability density function of surface height and correlation function are both gaussian to a good approximation. Figure 2 shows an example for a one-dimensional surface #39. Surfaces are therefore characterised by their

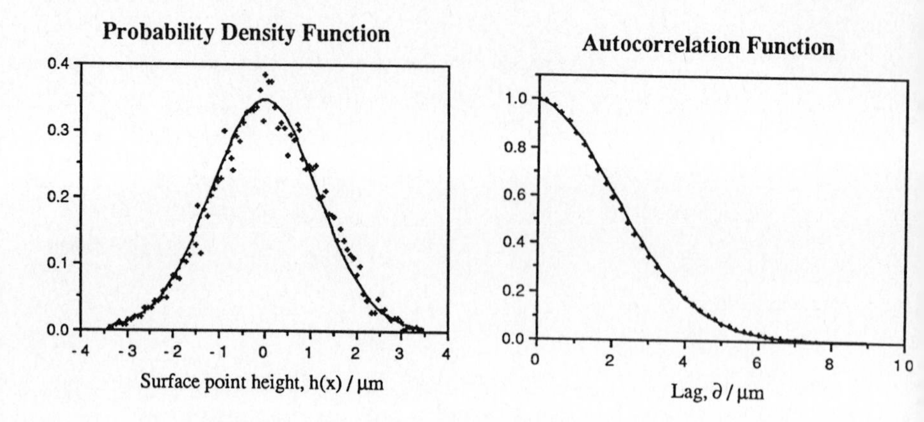

FIGURE 2. Probability density and autocorrelation function of surface height for one-dimensional surface #39.

root-mean-square (rms) surface height σ and 1/e correlation length τ. The measurements on one-dimensional surfaces are considered to be much more reliable than those made on the two-dimensional surfaces because the effect of stylus on the measurement is much smaller. In this paper, results are presented on three surfaces (note that the errors given are statistical errors):

(i) Surface #313: two-dimensional, gold-coated; $\sigma = 1.0 \pm 0.1$ μm, $\tau = 2.9 \pm 0.2$ μm

(ii) Surface #440: one-dimensional, gold-coated; $\sigma = 1.2 \pm 0.1$ μm, $\tau = 2.0 \pm 0.2$ μm

(iii) Surface #39: one-dimensional, dielectric and gold-coated, $\sigma = 1.18 \pm 0.13$ μm, $\tau = 2.97 \pm 0.05$ μm.

3. RESULTS FOR TWO-DIMENSIONAL SURFACES

Mendez and O'Donnell[3,4] have reported measurements of enhanced backscattering on gold-coated high-sloped surfaces. Apart from the backscatter peak, the scattered intensity also has an interesting polarisation structure that strongly suggests that multiple scattering plays an important rôle. To investigate this further, we measured the Stokes' parameters of the scattered intensity for a few fixed angles of scattering. For linearly polarised incident light, it was found that the scattered light had only two non-zero Stokes' parameters (to within the experimental error of 5% of the first Stokes' parameter), corresponding to components that were (a) linearly polarised in the direction of the incident light and

›) unpolarised. It should be stressed that, although a single surface was used in the xperiment, the measurement involved angular averaging over many speckles and ıus was equivalent to an ensemble average. In this context, "unpolarised" means ıat there was no preferred direction of polarisation averaged over all speckles.

As a consequence of this result, it is possible to re-plot the usual co- and cross-olarised intensity measurements as polarised, I_{pol}, and unpolarised, I_{unpol} ıtensities; for example, for s-polarised incident radiation (s-polarised means that ıe electric vector is perpendicular to the plane of incidence):

$$I_{pol} = I_{ss} - I_{sp} \quad ; \quad I_{unpol} = 2\, I_{sp} \quad . \tag{1}$$

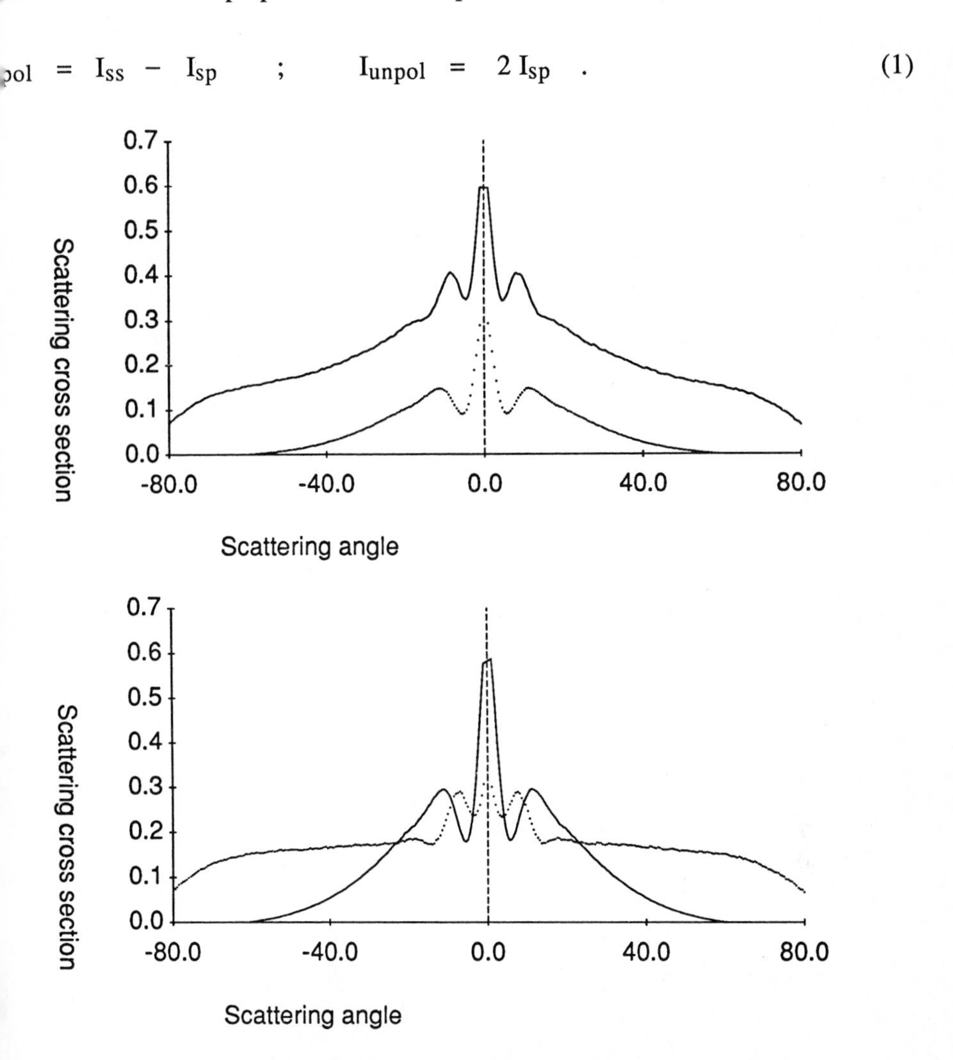

FIGURE 3. Two-dimensional surface #313, normal incidence.

The upper parts of Figures 3 and 4 show the angular scattered intensity for surfac #313 (σ = 1.0 ± 0.1 µm, τ = 2.9 ± 0.2 µm) for s-polarised incident light (λ = 633 nm) and s- and p-polarised scattered light at angles of incidence of 0° and -10° respectively (in these and subsequent Figures, the enhanced backscatter peak is situated on the right hand side and the specular peak — if any — on the left). In each case the solid line is the co-polarised return and the broken line is the cross-polarised return. These measurements clearly show the enhanced backscatter peak and side-lobe structure reported by Mendez and O'Donnell[3,4]. The lower parts of Figures 3 and 4 show these results re-plotted in terms of the polarised (broken line

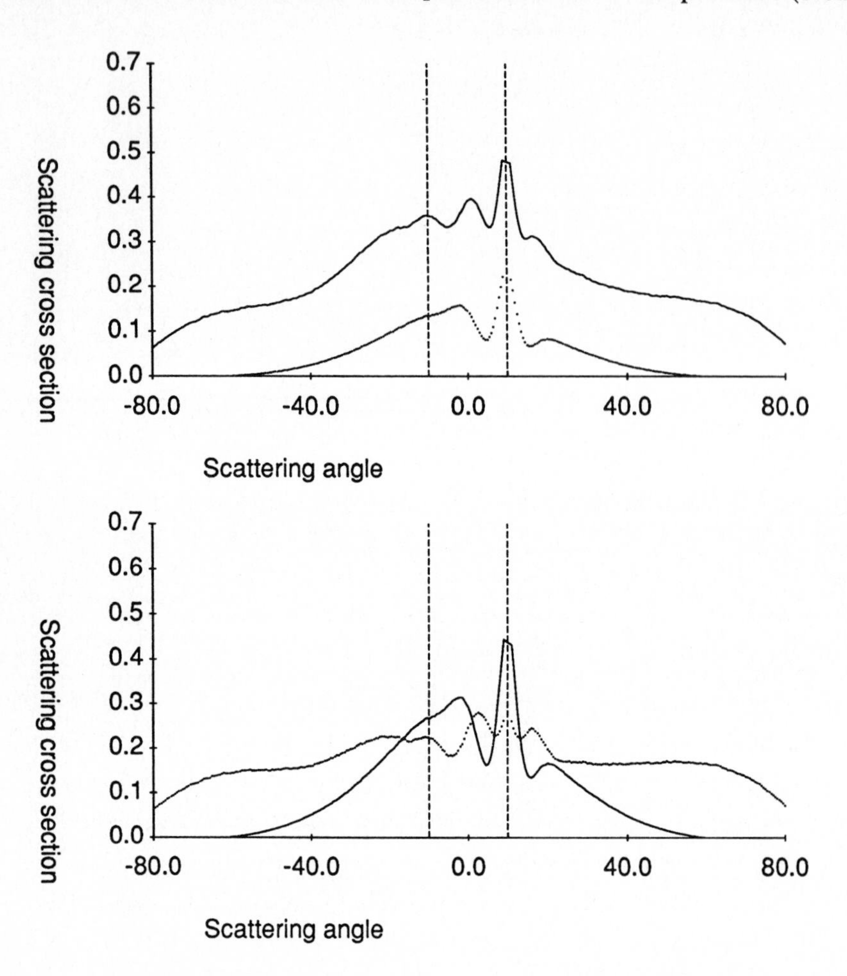

FIGURE 4. Two-dimensional surface #313, −10° incidence.

nd unpolarised (solid line) cross-sections, as defined by Eq.(1). *Note that the nhanced backscatter peak is present only in the unpolarised intensity.*

To a first approximation, the polarised component is the result of single cattering and the unpolarised component due to multiple scattering. Thus the bove result implies that it is *multiple scattering* that is the dominant cause of nhanced backscattering for these high-sloped metallic rough surfaces.

. RESULTS FOR ONE-DIMENSIONAL SURFACES

Whilst experimental measurements for two-dimensional surfaces are nteresting, particularly as regards the polarisation behaviour of the scattered light, hey are of limited value for verifying theoretical predictions since these are almost lways limited to the one-dimensional case (i.e. to random gratings). Since the rincipal aim of our work is to provide experimental data for comparison with heory[5–11], we have measured the light scattering from a number of well haracterised one-dimensional random rough surfaces.

Figures 5 – 8 show the measurements of the angular scattering for surface 440 ($\sigma = 1.2 \pm 0.1$ μm, $\tau = 2.0 \pm 0.2$ μm) for several angles of incidence. igures 5 & 6 are for visible light ($\lambda = 633$ nm) and Figs 7 & 8 for infrared adiation ($\lambda = 10.6$ μm), and Figs 5 & 7 are for s-polarised incident light and Figs 5 & 8 for p-polarised incident light. There was no measurable cross-polarised cattering. In all of the Figures, the circles represent the experimental neasurements and the solid (noisy) lines represent the results of numerical alculations (see below).

Some general comments on these experimental measurements are in order. (i) At the visible wavelength, there is a pronounced enhanced backscatter peak and idelobe structure at small angles of incidence (< 20°), similar to that exhibited by wo-dimensional surfaces. This peak is above a broad plateau of scattering as in the wo-dimensional case. (ii) The differences between the curves for incident light hat is s-polarised (Fig 5) and p-polarised (Fig 6) are small. (iii) The angular cattering curves for the infrared do not show any obvious backscattering peak; however, given that the angular width of the backscatter peak is wavelength-dependent[4,14], this is not surprising. For s-polarisation, the scattering is fairly ymmetrical around the origin with a little more than 50% of the radiation in the pecular half-plane, whereas for p-polarisation a little more than 50% goes into the ackscatter half-plane.

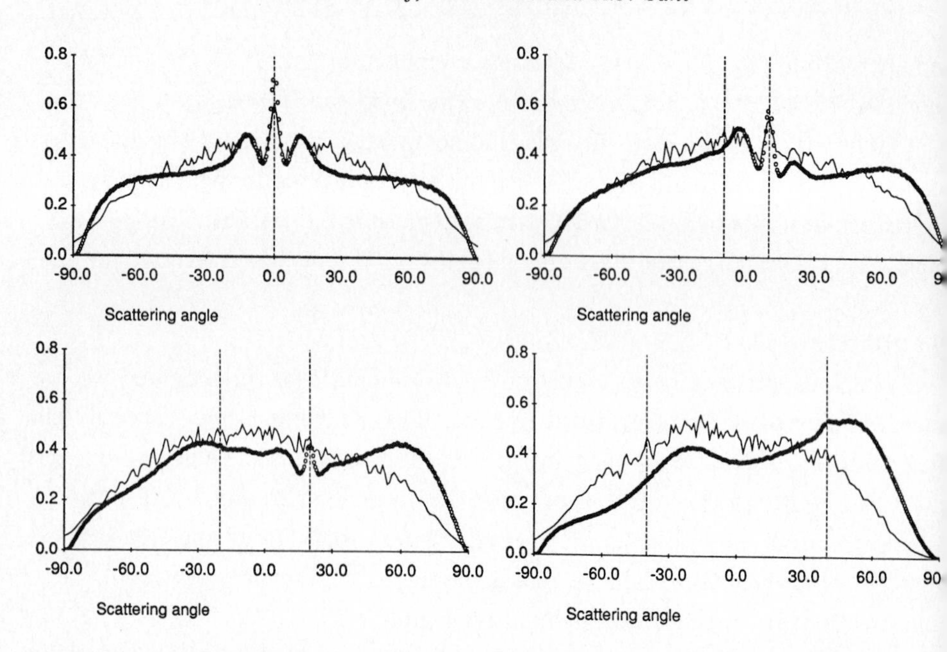

FIGURE 5. Surface #440, λ = 633 nm, gold-coated, s-polarisation, angles of incidence 0°,-10°,-20° and-40°. Circles represent measurements, solid (noisy) line is numerical calculation. Vertical axis is scattering cross-section

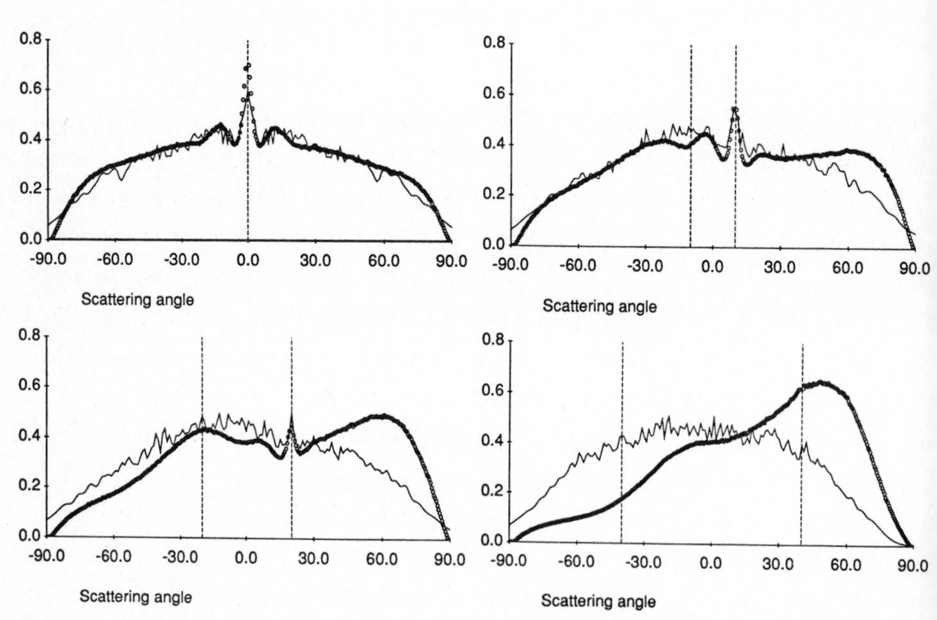

FIGURE 6. Surface #440, as Fig 5, but p-polarisation.

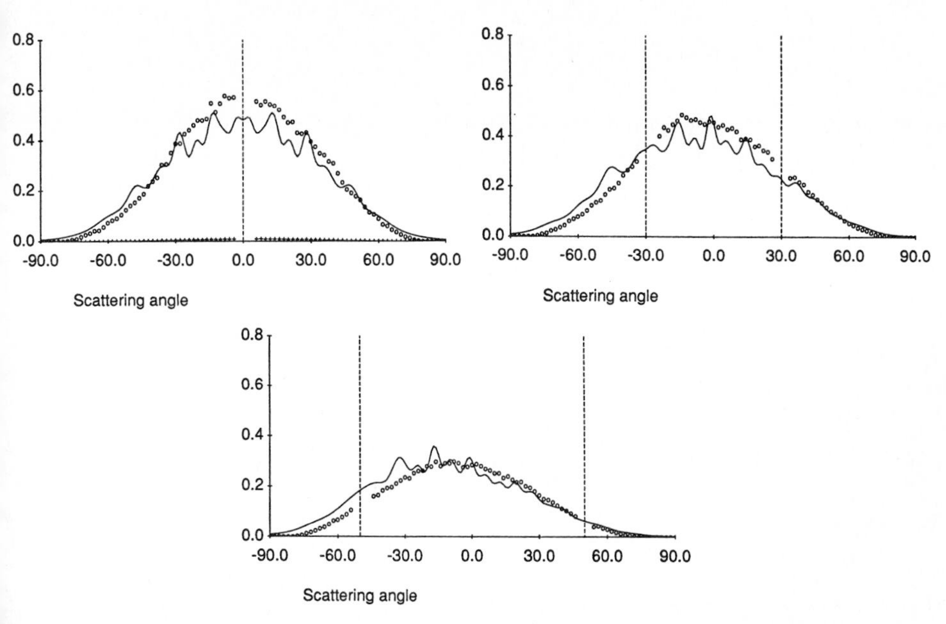

FIGURE 7. Surface #440, λ = 10.6 μm, gold-coated, s-polarisation, angles of incidence 0°,-30°, and-50°. Circles represent measurements, solid (noisy) line is numerical calculation. Vertical axis is scattering cross-section

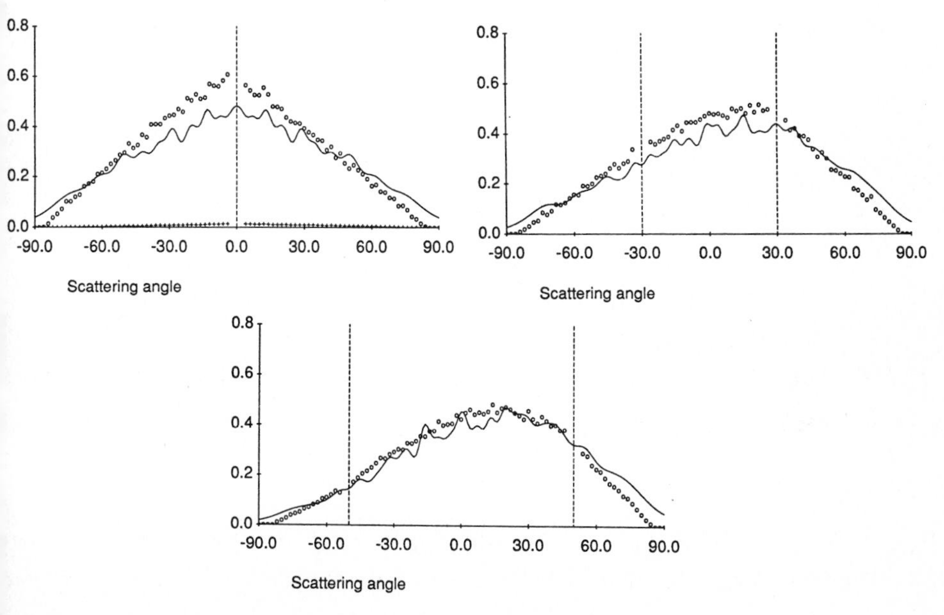

FIGURE 8. Surface #440, as Fig 7, but p-polarisation.

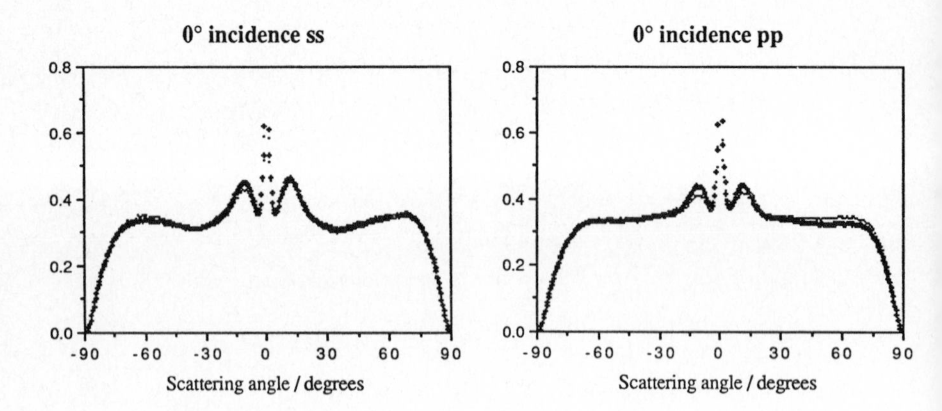

FIGURE 9. Surface #39, relative scattering cross-sections (vertical axis) for the original and epoxy replica, both gold-coated, for normal incidence (λ = 633 nm).

Also plotted in Figs 5 – 8 are the results of numerical calculations *for a perfe conductor* based on the method of Nieto-Vesperinas[6] (solid noisy lines). Good agreement is obtained for the infrared wavelength and for small angles of incidenc at the visible wavelength. It is not known at the present time whether the discrepancies observed in the other cases are due to deficiencies in the calculation or to the assumption of a perfect conductor.

In order to compare the scattering by metallic and dielectric surfaces, replicas of surface #39 (σ = 1.18 ± 0.13 μm, τ = 2.97 ± 0.05 μm) were made. Figure 9 compares the scattered intensity for the original and epoxy replica, for s-polarised and p-polarised normally incident light of wavelength λ = 633 nm; in both cases the surfaces were gold-coated. The agreement between the light scattering curves for the original and replica is excellent, indicating that the replicating process faithfully reproduces the surface structure of the original. Figure 10 shows the angular scattering of the original surface for angles of incidence of-20° and-40° for both incident polarisations. Both Figs 9 & 10 show features (i) and (ii) of Figs 5 & 6 for surface #440 described above — in particular, note that the light scattering is not strongly dependent on the incident polarisation.

Figure 11 shows the angular scattering by the dielectric intermediate used to cast the epoxy replicas; since the surface has a gaussian probability density functio of height, the statistics of this intermediate are identical to those of the original anc

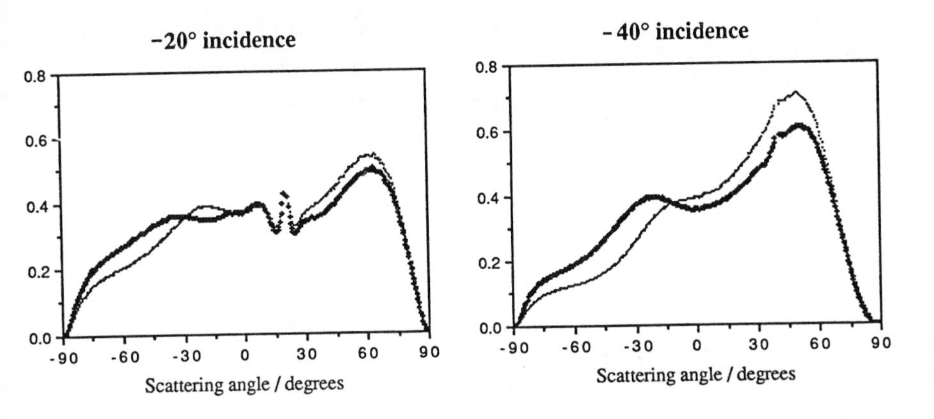

FIGURE 10. Surface #39, relative scattering cross-sections (vertical axis) for the gold-coated original for angles of incidence of -20° and -40° for s-polarisation (bold) and p-polarisation (light). λ = 633 nm.

epoxy replica. The most important feature of Fig 11 is the large difference between the angular scattering for s- and p-polarisations of the incident light. As might be expected from the Fresnel reflection coefficients for a flat surface, the scattered light is always greater for s-polarised incident light. For s-polarised incident light, most of the scattered energy lies in the specular half-plane, whereas it lies in the backscatter half-plane for the p-polarised case.

5. CONCLUSIONS

Experimental measurements of the light scattering by three well-characterised high-sloped random rough surfaces at two wavelengths have been presented. For the gold-coated two-dimensional surface, measurement of the Stokes' parameters has shown that for linearly polarised incident radiation, the scattered light has two components, linearly polarised and unpolarised. The enhanced backscatter is exhibited only by the unpolarised component, thus supporting the hypothesis that the enhancement is due to multiple scattering. One-dimensional gold-coated surfaces also exhibit enhanced backscattering. For the gold-coated surfaces, the scattering is similar for both s- and p-polarisations of the incident light, whereas quite different behaviours are observed for dielectric surfaces. The measurements presented for the one-dimensional surfaces are of particular usefulness for comparisons with theory, since the theory is usually only possible for the one-dimensional case.

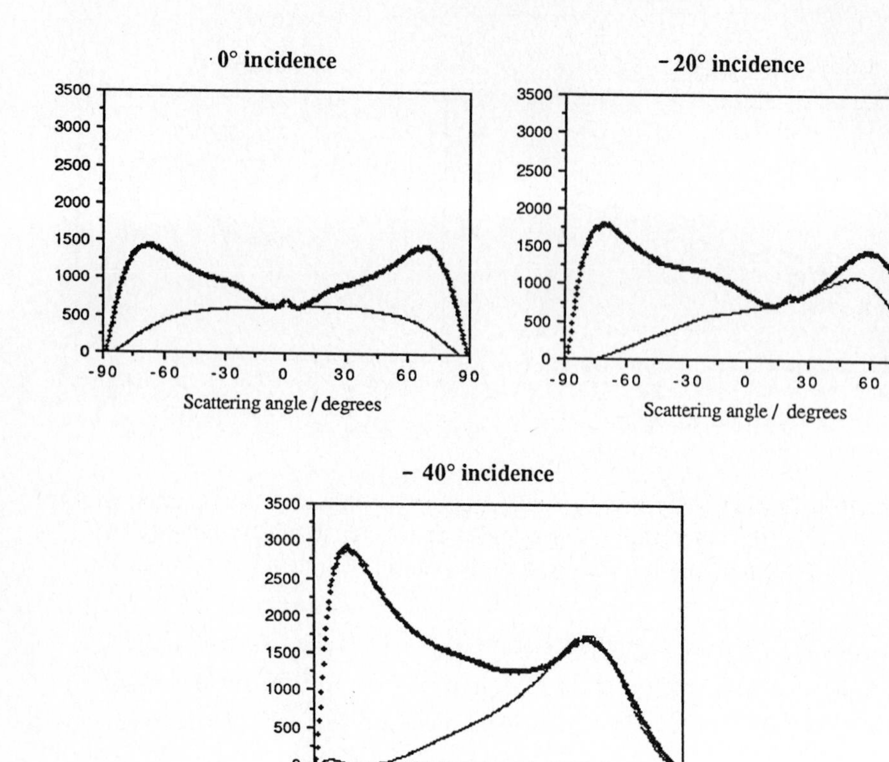

FIGURE 11 Surface #39, relative scattering cross-sections (vertical axis) for a *dielectric* replica (refractive index ≈ 1.43) for angles of incidence of 0°,-20° and -40° for s-polarisation (bold) and p-polarisation (light). $\lambda = 633$ nm.

ACKNOWLEDGEMENT

We are grateful to Dr A T Friberg for his assistance in the interpretation of the experimental data. This work was supported by the US Army (DAJA45-87-C-0039) and UK Science and Engineering Research Council (GR/E 40910).

REFERENCES

1 J Reneau and J A Collinson, "Measurements of electromagnetic backscattering from known rough surfaces", Bell Syst Tech J, 44, 2203 – 2226 (1965)

2 J Reneau, P K Cheo and H G Cooper, "Depolarisation of linearly polarised electromagnetic waves backscattered from rough metals and inhomogeneous dielectrics", J Opt Soc Am, 57, 459 – 467 (1967)

E R Mendez and K A O'Donnell, "Observation of depolarisation and backscattering enhancement in light scattering from gaussian rough surfaces", Opt Commun, 61, 91-95 (1987)

K A O'Donnell and E R Mendez, "Experimental study of scattering from characterised random surfaces", J Opt Soc Am A, 4, 1194-1205 (1987)

E Bahar and M A Fitzwater, "Depolarisation and backscatter enhancement in light scattering from random rough surfaces: comparison of full wave theory with experiment", J Opt Soc Am A, 6, 33-43 (1989)

6 M Nieto-Vesperinas and J M Soto-Crespo, "Monte Carlo simulations for scattering of electromagnetic waves from perfectly conducting randomly rough surfaces", Opt Lett, 12, 979-981 (1987)

7 J M Soto-Crespo and M Nieto-Vesperinas, "Electromagnetic scattering from very rough random surfaces and deep reflection gratings", J Opt Soc Am A, 6, 367–384 (1989)

8 A A Maradudin, E R Mendez and T Michel, "Backscattering effects in the elastic scattering of p-polarised light from a large-amplitude random metallic grating", Opt Lett, 14, 151-153 (1989)

9 E Jakeman, "Enhanced backscattering through a deep random phase screen", J Opt Soc Am A, 5, 1638–1648 (1988)

10 M Saillard and D Maystre, "Scattering from metallic and dielectric rough surfaces", submitted to J Opt Soc Am A.

11 C Makaskill, "Geometric optics and enhanced backscatter from very rough surfaces", submitted to J Opt Soc Am A.

12 A J Sant, J C Dainty and M J Kim, "Comparison of surface scattering between identical, randomly rough metal and dielectric diffusers", Opt Lett, (in press)

13 M J Kim, J C Dainty, A T Friberg and A J Sant, "Experimental study of enhanced backscattering from one and two dimensional random rough surfaces", J Opt Soc Am A, (submitted)

14 M J Kim, "Light Scattering from Characterised Random Rough Surfaces", Thesis, University of London, 1989

15 P F Gray, " A method of forming optical diffusers of simple known statistical properties", Optica Acta, 25, 765-775 (1978)

Scattering in Volumes and Surfaces
M. Nieto-Vesperinas and J.C. Dainty (Editors)

BACKSCATTERING EFFECTS IN THE ELASTIC SCATTERING OF P-POLARIZED LIGHT FROM A LARGE AMPLITUDE RANDOM GRATING

A. A. Maradudin, E. R. Méndez*, and T. Michel

Department of Physics and the Institute for Surface and Interface Science, University of California, Irvine, CA 92717, USA

By the use of Green's second integral theorem we have written an exact expression for the scattered electromagnetic field produced by a p-polarized beam of finite width incident from the vacuum side onto a random grating whose grooves are perpendicular to the plane of incidence. The scattered field is expressed in terms of the values of the total magnetic field and its normal derivative on the surface of the grating. The coupled pair of integral equations satisfied by these functions is solved numerically for each of several hundred (~ 1000) realizations of the surface profile, which are generated numerically and possess a Gaussian spectrum. The diffuse component of the differential reflection coefficient averaged over these realizations of the surface profile displays a well-defined peak in the retroreflection direction in both the small roughness and strong roughness limits for a metallic random grating and for a large amplitude, perfectly conducting, random grating, but not for a large amplitude random grating on a dielectric surface.

1. INTRODUCTION

Since the prediction of a narrow peak in the retroreflection direction in the angular distribution of the intensity of the diffuse component of the light scattered from a randomly rough metal surface[1-3], this effect has been studied both experimentally[4-7] and theoretically[8,9] by several authors, who have confirmed its existence.

However, all of the theoretical studies of this effect to date have contained approximations of one sort or another. Thus, in the perturbation-theoretic approach of Refs. 1-3 only a selected subset of diagrams that gives the dominant contribution to the backscattering peak was summed. In the work of Bahar and Fitzwater[8] only the single-scattering approximation to their

---

* Permanent Address: CICESE, División de Fisica Aplicada, Apdo. Postal 2732, Ensenada, Baja California, MEXICO

full-wave solution was used, and averaging over the slopes of the surface profile was carried out independently of the averaging over its heights. Finally, the Kirchhoff approximation, even the double-scattering version used by Jin and Lax[(9)], is not always valid.

In a recent letter[10] we presented the results of a numerical simulation of the scattering of light from a randomly rough dielectric surface that, in principle, is free from the kinds of approximations made in the more analytical treatments cited above. The present paper is an expansion of Ref. 10. In it we present more of the theoretical analysis underlying the calculations whose results were reported there, and present several new results.

The kind of numerical approach to the scattering of light or of a scalar plane wave from a randomly rough surface described in Ref. 10 and in the present paper has been carried out by several authors in the past, but mostly for perfect conductors or hard walls[11-15]. Comparatively few such calculations have been carried out for finitely conducting surfaces.

The work that ours most closely resembles, in that it too is devoted to penetrable media, is that of García and Stoll[16,17] and Tran and Celli[18]. In both cases a random grating of length L was generated numerically on the surface of a metal of finite conductivity, and was then illuminated from the vacuum side by a plane wave. Because of the infinite extent of the plane wave along the grating surface the random grating of length L was replicated periodically, and grating theory was used in calculating the scattering from the resulting structure. Garcia and Stoll[17] did not find a peak in the retroreflection direction in the angular dependence of the intensity of the diffuse component of the scattered light because the angular resolution of their calculation was insufficiently small. Tran and Celli[18] obtained this peak, but the scattering theory they used is applicable only to surface that are weakly rough.

The calculations reported here differ from those of Refs. 16-18 in our assumption that the incident electromagnetic field is a beam of finite width rather than a plane wave[19]. By choosing the width of this beam suitably, we make its amplitude at the ends of the segment of the dielectric surface that is rough sensibly zero. This has the consequence that it is not necessary to replicate the rough surface periodically. A second difference between our calculation and those of Refs. 16-18 is that we formulate the scattering problem exactly through the use of Green's theorem[20] and invoke neither the extended boundary condition[16,17] nor the Rayleigh hypothesis[18] in obtaining the scattered field. As a result, we are able, at least in principle, to calculate the scattering from a surface of arbitrary roughness, not just from weakly rough surfaces.

2. THE DIFFERENTIAL REFLECTION COEFFICIENT

The physical system we consider consists of vacuum in the region $x_3 > \zeta(x_1)$ and a dielectric medium, characterized by an isotropic, frequency-dependent, complex dielectric constant $\epsilon(\omega) = \epsilon_1(\omega)+i\epsilon_2(\omega)$ $(\epsilon_2(\omega) > 0)$, in the region $x_3 < \zeta(x_1)$. The surface profile function $\zeta(x_1)$ is assumed to be a stationary, Gaussian, stochastic process defined by the properties $\langle\zeta(x_1)\rangle = 0$, $\langle\zeta(x_1)\zeta(x_1')\rangle = \delta^2\exp(-(x_1-x_1')^2/a^2)$, where the angular brackets denote an average over the ensemble of realizations of the surface profile, while $\delta^2 = \langle\zeta^2(x_1)\rangle$ is the mean-square departure of the surface from flatness.

The surface $x_3 = \zeta(x_1)$ is illuminated from the vacuum side by a p-polarized beam of light whose plane of incidence is the x_1x_3-plane[21]. The amplitude of the single, nonzero component of the magnetic vector of this beam is written initially in the form[19]

$$H_2^>(x_1,x_3|\omega)_{inc} = \frac{w\omega}{2\sqrt{\pi}c}\int_{-\frac{\pi}{2}}^{\frac{\pi}{2}} d\theta\, e^{-\frac{w^2\omega^2}{4c^2}(\theta-\theta_o)^2}\, e^{i\frac{\omega}{c}(x_1\sin\theta-x_3\cos\theta)}, \qquad (2.1a)$$

which is an exact solution of Maxwell's equations in vacuum, where θ_o is the mean angle of incidence measured from the normal to the mean surface $x_3 = 0$, and w is the half-width of the beam in the plane perpendicular to the direction of incidence that passes through the point $x_1 = x_3 = 0$. We will also use the relation $w = g\cos\theta_o$, where g is the half-width of the intercept of the beam with the plane $x_3 = 0$. The width of the beam w must satisfy the inequality $w >> 2c/\omega$ in order that Eq. (2.1a) represent a well-defined beam. Rather than work with the integral expression (2.1a), we have chosen to work with the analytic expression derived from it by an approximate integration valid in the limit $w >> 2c/\omega$, viz.

$$H_2^> (x_1,x_3|\omega)_{inc} = \exp\left\{i\frac{\omega}{c}(x_1\sin\theta_o-x_3\cos\theta_o)[1+w(x_1,x_3)]\right\} \times$$
$$\times \exp\left[-(x_1\cos\theta_o+x_3\sin\theta_o)^2/w^2\right], \qquad (2.1b)$$

where $w(x_1,x_3) = (c^2/(w^2\omega^2))\left\{\left[2(x_1\cos\theta_o+x_3\sin\theta_o)^2/w^2\right]-1\right\}$.

The differential reflection coefficient (drc) $\partial R/\partial\theta_s$ can be written in the form

$$\frac{\partial R}{\partial \theta_s} = \frac{1}{2(2\pi)^{3/2}} \frac{c}{\omega w} \frac{|r(\theta_s)|^2}{\left[1-c^2(1+2\tan^2\theta_o)/(2\omega^2 w^2)\right]} \qquad |\theta_s| < \frac{\pi}{2}, \tag{2.2}$$

where θ_s is the scattering angle, measured from the normal to the plane $x_3 = 0$, and the scattering amplitude $r(\theta_s)$, obtained by the use of Green's second integral theorem[20], is given by

$$r(\theta_s) = \int_{-\infty}^{\infty} dx_1 e^{-i\frac{\omega}{c}(x_1 \sin\theta_s + \zeta(x_1)\cos\theta_s)} \times$$
$$\times \left\{ i\frac{\omega}{c}\left[\zeta'(x_1)\sin\theta_s - \cos\theta_s\right] H(x_1|\omega) - L(x_1|\omega) \right\}. \tag{2.3}$$

The source functions $H(x_1|\omega)$ and $L(x_1|\omega)$ are $H_2^>(x_1 x_3|\omega)\big|_{x_3=\zeta(x_1)}$ and $[1+(\zeta'(x_1))^2]^{1/2}\, \partial H_2^>(x_1 x_3|\omega)/\partial n\big|_{x_3=\zeta(x_1)}$, respectively, where $H_2^>(x_1 x_3|\omega)$ is the amplitude of the total magnetic field in the vacuum region, and $\partial/\partial n$ denotes the derivative along the normal to the surface $x_3 = \zeta(x_1)$ at each point, directed from the dielectric into the vacuum. They satisfy the pair of coupled integral equations

$$H(x_1|\omega) = H(x_1|\omega)_{inc} + \int_{-\infty}^{\infty} dx_1' \left[H_o(x_1|x_1')H(x_1'|\omega) - L_o(x_1|x_1')L(x_1'|\omega)\right] \tag{2.4a}$$

$$0 = \int_{-\infty}^{\infty} dx_1' \left[H_\epsilon(x_1|x_1')H_1(x_1'|\omega) - \epsilon(\omega)L_\epsilon(x_1|x_1')L(x_1'|\omega)\right], \tag{2.4b}$$

where $H(x_1|\omega)_{inc} = H_2^>(x_1,\zeta(x_1)|\omega)_{inc}$, and the kernels are given by

$$H_\epsilon(x_1|x_1') = \lim_{\epsilon\to 0+} -\frac{i}{4} n_c^2 \frac{\omega^2}{c^2} \frac{H_1^{(1)}\left(n_c \frac{\omega}{c}\left[(x_1-x_1')^2+(\zeta(x_1)-\zeta(x_1')+\epsilon)^2\right]^{1/2}\right)}{n_c \frac{\omega}{c}\left[(x_1-x_1')^2+(\zeta(x_1)-\zeta(x_1')+\epsilon)^2\right]^{1/2}} \times$$
$$\times \left[(x_1-x_1')\zeta'(x_1') - (\zeta(x_1)-\zeta(x_1')+\epsilon)\right] \tag{2.5a}$$

$$L_\epsilon(x_1|x_1') = \lim_{\epsilon\to 0+} \frac{i}{4} H_o^{(1)}\left(n_c \frac{\omega}{c}\left[(x_1-x_1')^2+(\zeta(x_1)-\zeta(x_1')+\epsilon)^2\right]^{1/2}\right). \tag{2.5b}$$

In Eqs. (2.5) $H_{0,1}^{(1)}(z)$ are Hankel functions of the first kind, while $n_c(\omega) = (\epsilon_1(\omega)+i\epsilon_2(\omega))^{1/2}$ is the complex refractive index of the dielectric medium. We require that Re $n_c(\omega) > 0$, Im $n_c(\omega) > 0$. The kernels $H_o(x_1|x_1')$ and $L_o(x_1|x_1')$ are obtained from Eqs. (2.5) by replacing n_c by unity.

Equations (2.4) were converted to matrix equations by replacing the nfinite range of integration by the finite range (-L/2, L/2) and replacing ntegration by summation. In the latter step care has to be taken in passing o the limit as $\epsilon \to 0+$ in Eqs. (2.5) because of the singularities of the lankel functions at vanishing arguments. The resulting equations can be ritten as

$$H(x_m|\omega) = 2H(x_m|\omega)_{inc} + \sum_{n=1}^{N} \left[\mathcal{H}_{mn}^{(o)} H(x_n|\omega) - \mathcal{L}_{mn}^{(o)} L(x_n|\omega)\right] \quad (2.6a)$$

$$H(x_m|\omega) + \sum_{n=1}^{N} \left[\mathcal{H}_{mn}^{(\epsilon)} H(x_n|\omega) - \epsilon(\omega) \mathcal{L}_{mn}^{(\epsilon)} L(x_n|\omega)\right] = 0, \quad (2.6b)$$

where $m = 1,2,\ldots,N$, $x_n = -L/2+(n-\frac{1}{2})\Delta x$, $\Delta x = L/N$, and

$$\mathcal{H}_{mn}^{(\epsilon)} = \Delta x\left(-\frac{i}{2}\right) n_c^2 \frac{\omega^2}{c^2} \frac{H_1^{(1)}\left(n_c \frac{\omega}{c}\left[(x_m-x_n)^2+(\zeta(x_m)-\zeta(x_n))^2\right]^{1/2}\right)}{n_c \frac{\omega}{c}\left[(x_m-x_n)^2+(\zeta(x_m)-\zeta(x_n))^2\right]^{1/2}} \times$$

$$\times \left[(x_m-x_n)\zeta'(x_n) - (\zeta(x_m)-\zeta(x_n))\right] \quad m \neq n \quad (2.7a)$$

$$= \Delta x \frac{\zeta''(x_m)}{2\pi\gamma_m^2} \quad m = n \quad (2.7b)$$

$$\mathcal{L}_{mn}^{(\epsilon)} = \Delta x\left(\frac{i}{2}\right) H_o^{(1)}\left(n_c \frac{\omega}{c}\left[(x_m-x_n)^2+(\zeta(x_m)-\zeta(x_n))^2\right]^{1/2}\right) \quad m \neq n \quad (2.8a)$$

$$= \Delta x\left(\frac{i}{2}\right) H_o^{(1)}\left(n_c \frac{\omega}{c} \frac{\gamma_m \Delta x}{2e}\right) \quad m = n, \quad (2.8b)$$

with $\gamma_m = \left[1 + (\zeta'(x_m))^2\right]^{1/2}$. The expressions for $\mathcal{H}_{mn}^{(o)}$ and $\mathcal{L}_{mn}^{(o)}$ are obtained from Eqs. (2.7) and (2.8) by replacing n_c by unity.

The values of the surface profile function $\zeta(x_1)$ and of its derivatives at the points $\{x_n\}$ were calculated by the method described by Thorsos[19]. The Hankel functions of complex argument appearing in Eqs. (2.7) and (2.8) were calculated by means of an algorithm due to Amos[22].

When $H(x_n|\omega)$ and $L(x_n|\omega)$ have been obtained by solving Eqs. (6), the amplitude $r(\theta_s)$ that appears in the expression (2.2) for the differential reflection coefficient is calculated from Eq. (2.3) according to

$$r(\theta_s) = \Delta x \sum_{n=1}^{N} e^{-i\frac{\omega}{c}\left(x_n \sin\theta_s + \zeta(x_n)\cos\theta_s\right)} \times$$

$$\times \left\{ i\frac{\omega}{c}\left[\zeta'(x_n)\sin\theta_s - \cos\theta_s\right]H(x_n|\omega) - L(x_n|\omega)\right\}. \tag{2.9}$$

We now note that scattering from a surface defined by the profile function $\zeta(-x_1)$ for an angle of incidence θ_o yields a scattering amplitude $r(\theta_s)$ that is the mirror image of the scattering amplitude obtained in scattering from a surface defined by the profile function $\zeta(x_1)$ for an angle of incidence $-\theta_o$. The surface profile function $\zeta(-x_1)$ is a Gaussianly distributed random variable with the same statistical properties as $\zeta(x_1)$. The calculation of the scattering amplitude for any two angles of incidence, in particular for θ_o and $-\theta_o$, can be carried out virtually simultaneously. This is because the rate determining step in the calculation is the calculation of the coefficient matrices $\overleftrightarrow{\mathcal{H}}^{(0,\epsilon)}$ and $\overleftrightarrow{\mathcal{L}}^{(0,\epsilon)}$ in the equations (2.6) for the source functions, and these are independent of the angle of incidence. The latter enters the calculation only through the inhomogeneous term $2H(x_m|\omega)_{inc}$. Thus, we calculated an averaged differential reflection coefficient in the following way. We constructed N_p different surface profiles $\zeta(x_1)$, calculated the scattering amplitude $r(\theta_s)$ for each profile for angles of incidence θ_o and $-\theta_o$, added the squared modulus of the mirror image of the latter to the squared modulus of the former, summed the result over the N_p realizations of $\zeta(x_1)$, and divided the sum by $2N_p$. The function of θ_s obtained in this way was denoted by $\langle|r(\theta_s)|^2\rangle$. The mean differential reflection coefficient was then obtained from

$$\left\langle\left(\frac{\partial R}{\partial\theta_s}\right)\right\rangle = \frac{1}{2(2\pi)^{3/2}}\frac{c}{\omega w}\frac{\langle|r(\theta_s)|^2\rangle}{[1-c^2(1+2\tan^2\theta_o)/(2\omega^2 w^2)]} \qquad |\theta_s| \le \frac{\pi}{2}. \tag{2.10}$$

The computer time required to calculate $\langle(\partial R/\partial\theta_s)\rangle$ in this way was only about half of what would be required for a calculation based on $2N_p$ realizations of $\zeta(x_1)$.

The mean differential reflection coefficient given by Eq. (2.10) contains the contribution from specular reflection as well as the contribution from diffuse scattering. In a study of enhanced backscattering it is the contribution from the diffuse scattering that is of interest. In all of the results that are displayed below we have therefore subtracted off the contribution to the mean differential reflection coefficient from the specular reflection. This is the differential reflection coefficient calculated by the use of the mean scattered field, and is given by

$$\left\langle\left(\frac{\partial R}{\partial \theta_s}\right)\right\rangle_{spec} = \frac{1}{2(2\pi)^{3/2}} \frac{c}{\omega w} \frac{|\langle r(\theta_s)\rangle|^2}{\left[1-c^2(1+2\tan^2\theta_o)/(2\omega^2 w^2)\right]} \qquad |\theta_s| \le \frac{\pi}{2} . \quad (2.11)$$

he function $\langle r(\theta_s)\rangle$ was calculated from the same N_p surface profiles used in alculating $\langle|r(\theta_s)|^2\rangle$ by calculating $r(\theta_s)$ for each profile for angles of ncidence θ_o and $-\theta_o$, adding the mirror image of the latter to the former, umming the result over the N_p realizations of $\zeta(x_1)$, and dividing the sum by N_p. It should be noted, however, that in all the cases except the one epicted in Fig. 1, the surface from which the scattering occurred was ufficiently rough that the specular contribution to the mean differential eflection coefficient was small enough that its retention or elimination nfluenced the latter's form negligibly.

The corresponding results for scattering from a perfect conductor are btained by setting $L(x_n|\omega) \equiv 0$ in Eqs. (2.6a) and (2.9), and omitting Eq. (2.6b).

3. RESULTS

Several results of our calculations are presented in Figs. 1-5. For comparison with the perturbation-theoretic results of Refs. 1-3 we present in Fig. 1 the differential reflection coefficient as a function of the scattering angle for the scattering of p-polarized light incident on a weakly rough random grating ruled on the surface of silver. The wavelength of the incident light and the parameters characterizing the surface roughness have the same values as were used in plotting Fig. 2 of Ref. 1. In obtaining these figures we used N_p = 2590 different surface profiles. This large number of profiles was used because the contribution to the differential reflection coefficient from the diffuse component of the scattered light in this case is the small difference between two large and nearly equal quantities given by Eqs. (2.10) and (2.11). A well-defined peak in the retroreflection direction is present in this reflection coefficient for each angle of incidence, and the amplitude of the peak decreases with increasing angle of incidence. A decrease in the amplitude of the peak in the backscattering direction with an increase in the angle of incidence is observed in the results of the perturbation-theoretic calculations in Refs. 1-3. The height of the peak above the background at its position observed in Fig. 1 is consistent with the results of Ref. 1, as is the magnitude of the differential reflection coefficient (if one corrects the results presented in Fig. 2 of Ref. 1 by dividing them by a factor of 2π which, regrettably, was omitted in the latter calculations). The widths of the peaks in Fig. 1 are broader than

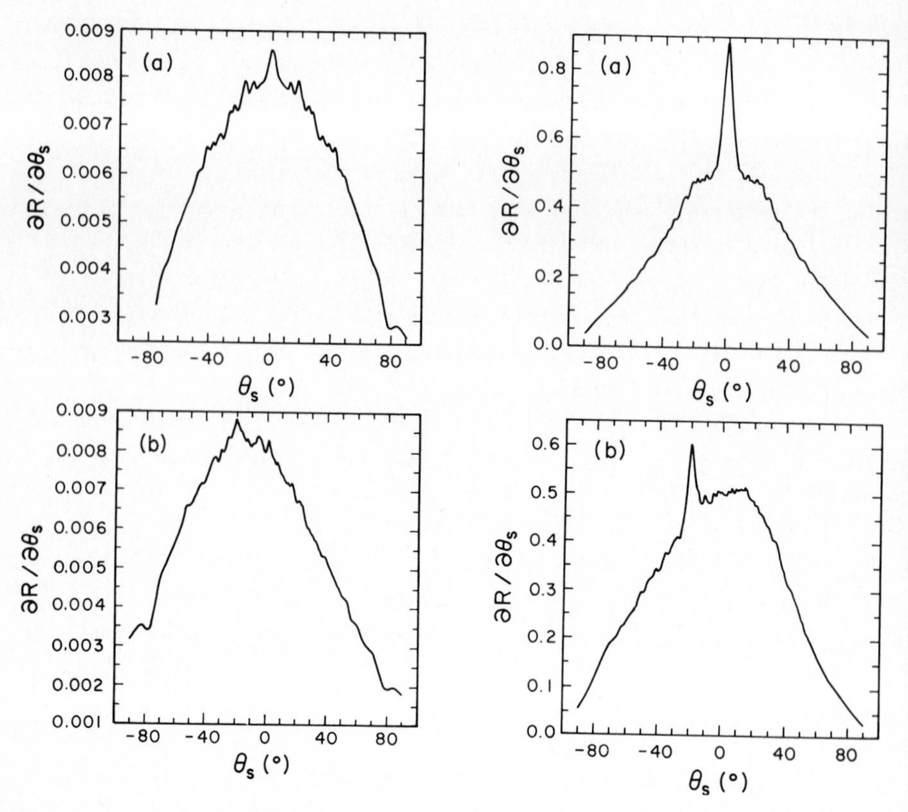

Fig. 1. The drc for a silver surface characterized by δ = 50Å and a = 1000Å. λ = 4579Å, $\epsilon(\omega)$ = -7.5 + i0.24, while g = 2.75 μm, L = 13.7 μm, and N = 300. (a) $\theta_o = 0^o$; (b) $\theta_o = 20^o$.

Fig. 2. The drc for a silver surface characterized by δ = 1.4142μm and a = 2 μm. λ = 6127Å, $\epsilon(\omega)$ = -17.2+i0.498, while g = 6.4 μm, L = 25.6 μm, and N = 300. (a) $\theta_o = 0^o$; (b) $\theta_o = 20^o$.

the corresponding widths in Fig. 2 of Ref. 1, however. This may be due to our use of a beam of finite width as the incident field in the present work rather than the plane wave used in Ref. 1.

We now turn to results obtained for large amplitude gratings. In Fig. 2a we present the differential reflection coefficient as a function of the scattering angle for the scattering of p-polarized light incident normally on a random grating ruled on the surface of silver. In obtaining this figure we have used N_p = 2000 different surface profiles. A well-defined peak in the retroreflection direction ($\theta_s=0^o$) is present in this reflection coefficient. Its form, even to the presence of the subsidiary maxima on both sides of the backscattering peak, closely resembles that of the experimental reflection coefficient depicted in Fig. 9 of Ref. 5. A detailed comparison between our theoretical curve and the result of Ref. 5, as well as of Ref. 6, is not possible, because the latter were obtained in the scattering of light from a randomly rough surface characterized by a surface profile function $\zeta(x_1,x_2)$ that was a function of both coordinates in the plane of the mean surface. A quantitative comparison between our theoretical result and the experimental results of Ref. 7 for one-dimensional rough surfaces is also not possible at this time, because although they indicate an enhanced backscattering, data points close enough to the backscattering direction for such a comparison to be made are not available, and because in the experiments gold-coated surfaces were employed. In Fig. 2b we present the differential reflection coefficient coefficient for the scattering of light from a random grating ruled on a silver surface when the angle of incidence is 20°. A peak in the backscattering direction ($\theta_s = -20^o$) whose amplitude is smaller than in the case of normal incidence, is visible in this reflection coefficient. A peak in the backscattering direction is also observed in the experimental results presented in Fig. 10 and 19 of Ref. 5, as well as in the results of Ref. 7.

The method described here can also be applied to the scattering of light from random gratings ruled on the surface of nearly transparent dielectric media, i.e. dielectric media characterized by a complex refractive index whose real part is positive and whose imaginary part is small. In Fig. 3 we present the differential reflection coefficient for the scattering of p-polarized light from a random grating ruled on the surface of $BaSO_4$, for two angles of incidence. In obtaining these results $N_p = 2000$ realizations of the surface profile were used. No peak in the retroreflection direction is observed for either angle of incidence, although the scattering is predominantly in the backward direction in the case of an angle of incidence of 20°.

In Fig. 4 we display the differential reflection coefficient for the scattering of p-polarized light from a random grating ruled on the surface of a perfect conductor for light incident normally on the grating and in the case that the angle of incidence is 20° ($N_p = 1000$ realizations of the surface were used in both of these cases). A well-defined peak in the backscattering direction is observed for each angle of incidence. Again, the height of this peak relative to the background at its position decreases with increasing angle of incidence. The results presented in this figure were tested for energy conservation, which was found to be satisfied to within about 1%.

The results displayed in Fig. 4 are also of interest in a rather different context from the one to which this paper is devoted. They give the differential reflection coefficient for the scattering of a finite beam of bulk acoustic waves of shear horizontal polarization, incident from inside a semi-infinite, isotropic elastic medium onto the stress-free random grating surface bounding it. It is only necessary to replace the speed of light c by the speed of bulk transverse waves c_t to go from the electromagnetic to the acoustic case. Thus, enhanced backscattering should also be observed in the scattering of acoustic waves from randomly rough surfaces. The corresponding effect for acoustic waves of sagittal polarization had been predicted earlier

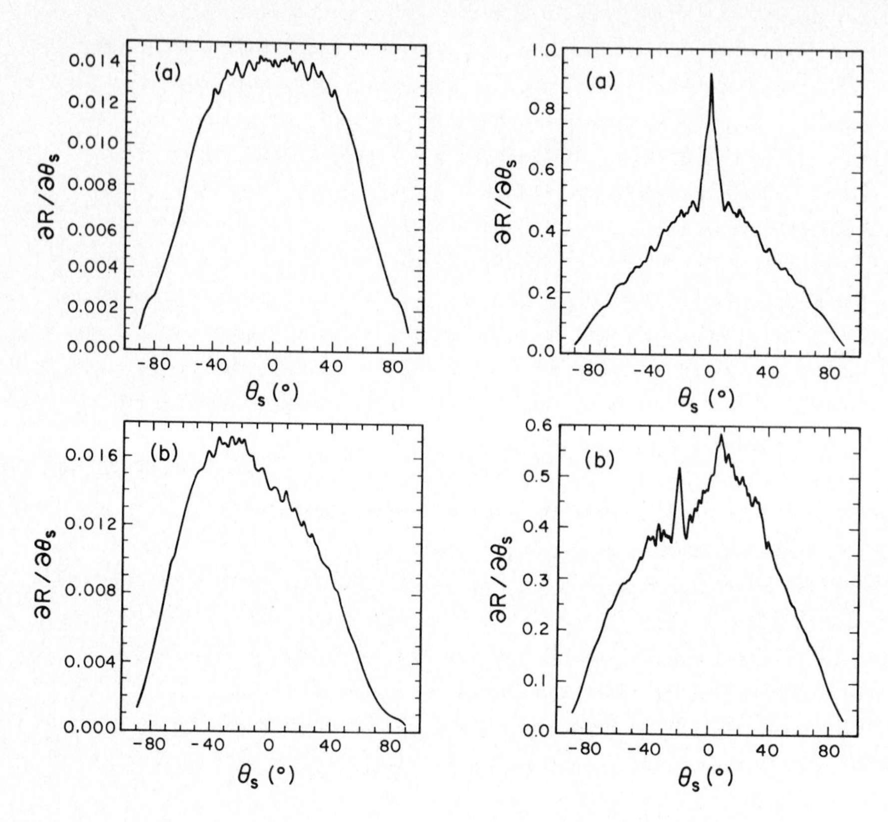

Fig 3. The drc for a $BaSO_4$ surface characterized by $\delta = 1.2\ \mu m$ and $a = 2\ \mu m$. $\lambda = 6328Å$, $n_c(\omega) = 1.628 + i0.0003$, while $g = 6.4\ \mu m$, $L = 25.6\mu m$, and $N = 300$. (a) $\theta_o = 0^o$; (b) $\theta_o = 20^o$.

Fig. 4. The drc for a perfect conductor characterized by $\delta/a = 0.6$. $k_o a = (\omega/c)\ a = 20$, $g = L/4.6875$, and $N = 300$. (a) $\theta_o = 0^o$; (b) $\theta_o = 20^o$.

on the basis of a perturbation-theoretic calculation of such scattering[23].

The analysis presented in Refs. 1-3 and 9 indicates that enhanced back-scattering is a multiple scattering effect, and is already contained in a ouble-scattering approximation. To examine whether the results presented in his paper are consistent with this picture, we have carried out the following alculation. Starting from the version of Eq. (2.6a) that applies to cattering from a perfect conductor,

$$H(x_m|\omega) = 2H(x_m|\omega)_{inc} + \sum_{n=1}^{N} \mathcal{H}_{mn}^{(o)} H(x_n|\omega), \tag{3.1}$$

e construct a sequence of iterates $H^{(1)}(x_m|\omega)$, $H^{(2)}(x_m|\omega), \ldots$ ccording to

$$H^{(s)}(x_m|\omega) = \sum_{n=1}^{N} \mathcal{H}_{mn}^{(o)} H^{(s-1)}(x_n|\omega) \qquad s \geq 2 \tag{3.2a}$$

$$H^{(1)}(x_m|\omega) = 2H(x_m|\omega)_{inc} . \tag{3.2b}$$

t is clear that

$$H(x_m|\omega) = \sum_{s=1}^{\infty} H^{(s)}(x_m|\omega) . \tag{3.3}$$

he physical significance of $H^{(s)}(x_m|\omega)$ is that when it is substituted into q. (2.9), (with $L(x_m|\omega) \equiv 0$), and the result used in Eqs. (2.10) and (2.11), e obtain the contribution to the diffuse component of the differential eflection coefficient from s-fold scattering processes[24].

In Fig. 5a we present the diffuse contribution to the differential reflec- ion coefficient from single-scattering processes ($H(x_m|\omega) = H^{(1)}(x_m|\omega)$) for n angle of incidence of 20°. (This approximation is exact when $\zeta(x_1) \equiv 0$.) his result is to be compared with that given in Fig. 4b for the "exact solu- ion." No evidence of an enhanced backscattering peak is observed. In con- rast, in Fig. 5b we present the diffuse contribution to the differential eflection coefficient in the double-scattering approximation ($H(x_m|\omega) = H^{(1)}(x_m|\omega) + H^{(2)}(x_m|\omega)$). In this figure a well-defined enhanced backscat- ering peak is observed.

4. DISCUSSION AND CONCLUSIONS

In this paper we have presented the results of formally exact calculations f the angular distribution of the intensity of the diffuse component of p- olarized light scattered from a random grating ruled on the surfaces of both enetrable and impenetrable dielectric media. We describe these calculations s formally exact because their accuracy is limited both by the numerical uadrature method employed in the solution of the integral equations and in he evaluation of $r(\theta_s)$, and by the finite number of surface profile functions used in obtaining the averaged differential reflection coefficients. The accuracy of the results can be improved, however, by the expenditure of more computer time. In contrast with the perturbation-theoretic calculations of Refs. 1-3, which are restricted to weakly corrugated random surfaces, the calculations presented here are not so restricted, and indeed have been carried out for large-amplitude random gratings as well as for weakly rough random gratings. The results obtained display a peak in the backscattering direction in the cases of a metal surface and a perfectly conducting surface, but not in the case of a nearly transparent dielectric surface. Thus, the results presented here support the results obtained in Refs. 1-3 in predicting backscattering enhancement in the scattering of p-polarized light from a weakly rough random metallic grating.

The question that has now to be addressed is, what is the origin of the enhanced backscattering that is observed in the results presented here? On the basis of these results we can offer the following partial answers to this question.

From the results presented in Figs. 4b and 5 we conclude that the enhanced backscattering is a multiple-scattering phenomenon, which occurs already in the double-scattering approximation. This result is in agreement with the perturbation theoretic results of Refs. 1-3, and with the second-order Kirchhoff approximation result of Jin and Lax[9]. It is in disagreement with the results of Bahar and Fitzwater[8], who claim to obtain enhanced backscattering already in the single-scattering approximation to their full-wave theory of the scattering of light from a randomly rough metal surface. The difference between these two sets of results is not understood at the present time. We note, however, that since shadowing is taken into account in their treatment, multiple-scattering effects are built into their theory.

On the basis of the conclusion that the enhanced backscattering is due to multiple scattering, it can be understood, at least qualitatively, as a coherent interference effect in a manner that is analogous to the manner in which enhanced backscattering in volume scattering is explained (see, e.g. Refs. 25, 26). Let us consider just one of the plane wave components of the incident field (Eq. 2.1a) and represent it by its wave vector $\vec{k}_i$. For the parameters that we employ the angular width of the incident field is small and, for the sake of argument, the plane wave with wave vector $\vec{k}_i$ can be considered a good approximation to the incident field. When the incident wave hits the surface element ds, a scattered field is produced, which can also be written as a superposition of plane waves. We can restrict our attention to one of these components and label it $\vec{k}_s$. We consider a sequence of such cattering events in which the incident light strikes the surface s-1 times efore being scattered into the vacuum away from the surface at the sth cattering event. This picture is formally correct but it represents only art of the total scattered field. How important, or representative, one of hese multiple scattering paths is, depends on the local slope and curvature f the surface, as compared to the wavelength of the radiation. We assume hat all such s-order sequences are uncorrelated due to the random nature of he surface profile. However, any given sequence and its time-reversed artner, where the light is scattered from the same points on the surface but n the reverse order, interfere constructively if the wave vectors of the ncident and final waves are oppositely directed. Such a pair of scattering equences is illustrated in the double-scattering case in Fig. 6. These two

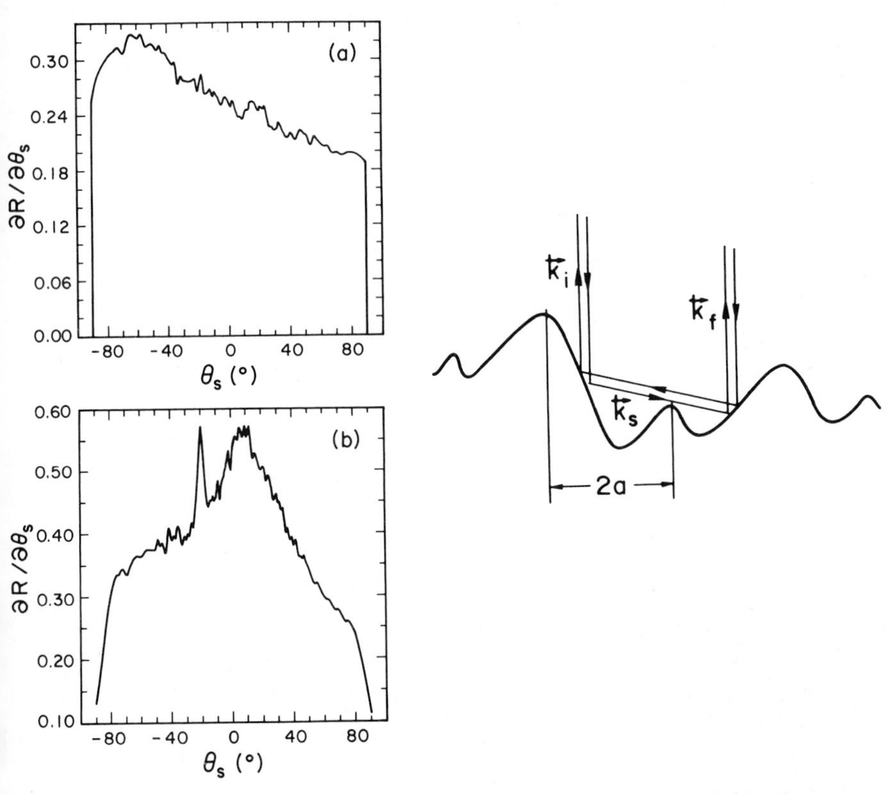

Fig. 5. The drc for a perfect conductor characterized by $\delta/a = 0.6$. $k_o a = (\omega/c)a = 20$, $g = L/4.6875$, and $N = 300$. $\theta_o = 20^o$. (a) The single-scattering approximation; (b) the double-scattering approximation.

Fig. 6. A schematic depiction of a light path and its coherent, time-reversed partner in a typical double-scattering event that contributes to enhanced backscattering.

partial waves have the same amplitude and phase and add coherently in forming the intensity of the scattered light. For scattering into directions other than the retroreflection direction the different partial waves have a non-zero phase difference, and very rapidly become incoherent, so that only their intensities add. Thus, the intensity of scattering into the retroreflection direction is a factor of two larger than the intensity of scattering into other directions. The contribution of the single-scattering processes has to be subtracted off in obtaining this factor of 2 enhancement because it is not subject to coherent backscattering.

For normal incidence, the phase difference ϕ between a given light path and its time-reversed partner is proportional to $(2\pi/\lambda)\theta_s d$, where θ_s is the scattering angle, d is the distance between the first and last scattering points on the surface, and λ is the wavelength of the light. With these arguments, one would expect subsidiary maxima whenever the average phase shift $\langle\phi\rangle$ is a mulitple of 2π, that is, at angles of observation given by

$$\theta_s \simeq n\lambda/\langle d\rangle, \tag{4.1}$$

where n is the order of interference. However, in the results that we present here, as well as in the experimental ones[4,5,7] only one pair of subsidiary maxima is observed, corresponding to $n = \pm 1$. Moreover, in the experiments reported in Ref. 25 no subsidiary maxima are observed. The explanation for this lies in the statistics of the random quantity d. A straightforward calculation shows that the standard deviation σ_ϕ of the phase difference ϕ at an angle of observation given by expression (4.1) is proportional to $n\sigma_d/\langle d\rangle$, where σ_d is the standard deviation of d. Thus, these results suggest that for $n = 2$, σ_ϕ is sufficiently large to destroy all interference effects in surface scattering, while these effects are already washed out for $n = 1$ in volume scattering.

The coherence of the light path of Fig. 6, and its time-reversed partner, is lost for angles of observation greater than $\lambda/\langle d\rangle$. Since the average value of d for the shortest sequence ($s = 2$) is the mean distance between two scattering events, i.e. the elastic mean free path of the light interacting with the surface ℓ, one expects the reflected intensity to increase by up to the above-mentioned factor of 2 inside a region of angular width of order λ/ℓ. The mean distance between consecutive peaks on our random surface is of the order of 2a[27]. If we then take a as a measure of ℓ, we expect the angular width of each backscattering peak to be of the order of λ/a [5]. This estimate is reasonably well satisfied by our results.

We also suggest that the decrease of the height of the backscattering peak relative to the background at its position with increasing angle of incidence is due to the effects of shadowing on both the incident and scattered beams. This refers to the fact that a smaller fraction of the surface is illuminated as the angle of incidence is increased due to the lengthening shadows cast by the ridges on the surface; at the same time, an illuminated portion of the surface can scatter the light into an increasingly restricted range of angles for the same reason. The latter effect is already seen away from the retroreflection direction in the vanishing of the differential reflection coefficient as the scattering angle approaches $\pm 90^{\circ}$, for a fixed angle of incidence.

As it stands, the picture does not explain the absence of enhanced backscattering that we observe in the case of a nearly transparent dielectric. The same coherence of a given light path and its time-reversed partner exists here as in the case of a metallic surface or a perfectly conducting surface, so that enhanced backscattering would be expected in this case as well. It may be that, since the dielectric we have studied is so highly transparent, so little light is thrown back into the vacuum in each scattering event that even the double-scattering contribution to the differential reflection coefficient is overwhelmed by the single-scattering contribution, and the peak it contributes cannot be observed with the resolution of our calculations. This suggestion is supported by the results presented in Fig. 7. In this figure we

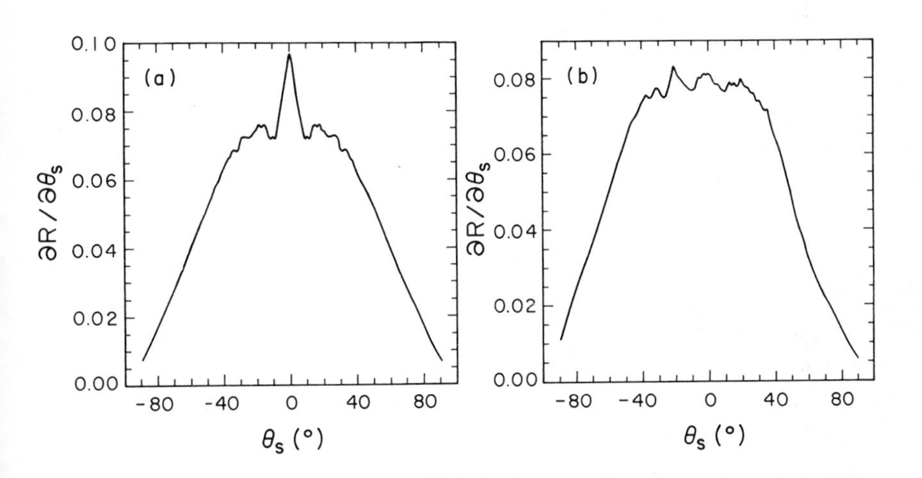

Fig. 7. The drc for an artificial dielectric surface characterized by $\delta = 1.2\ \mu m$ and $a = 2\ \mu m$. $\lambda = 6137$ Å, $n_c(\omega) = 3.256+i0.0006$, while $g = 6.4\ \mu m$, $L = 25.6\ \mu m$, and $N = 300$. (a) $\theta_0 = 0^{\circ}$; (b) $\theta_0 = 20^{\circ}$.

show the differential reflection coefficient as a function of the scattering angle for the scattering of p-polarized light from a random grating ruled on the surface of a dielectric medium whose complex index of refraction has been increased artificially by a factor of 2 from the value for $BaSO_4$ used in obtaining the results depicted in Fig. 3. A total of $N_p = 4000$ realizations of the surface profile was used in obtaining these results. In contrast with the results displayed in Fig. 3, the results shown in Fig. 7 for normal incidence (Fig. 7a) and for an angle of incidence of 20° (Fig. 7b), display a well-defined peak in the retroreflection direction. This is accompanied by an overall increase in the scattered intensity relative to that in Fig. 3, due to the higher reflectivity of the dielectric medium. These results suggest that enhanced backscattering occurs from the randomly rough surface of even a transparent medium provided that its surface is sufficiently reflective. The mechanism responsible for it is then the coherent addition of the scattering amplitudes for a light path and its time-reversed partner, as in the scattering of light from metallic and perfectly conducting surfaces.

The enhanced backscattering observed in the scattering of light from a nearly mirror-like rough metal surface of the kind giving rise to the results depicted in Fig. 1 is unlikely to arise from the multiple scattering of the incident light from the surface. This is because the surface is so weakly rough that the probability of multiple scattering from it is very low. Instead, it is more likely due to the multiple scattering of the surface electromagnetic waves excited by the incident beam, through the surface roughness, from the ridges and valleys on the surface. This is the mechanism underlying the perturbation theoretic predictions of enhanced backscattering in reflection presented in Refs. 1-3. It may also play a role in the enhanced backscattering from strongly corrugated surfaces, but it appears to become the dominant mechanism for weakly corrugated surfaces.

ACKNOWLEDGMENTS

We are grateful to Dr. D. E. Amos of the Sandia National Laboratories for furnishing us the computer code for calculating Hankel functions of complex arguments. We also acknowledge the award of time on the CRAY XMP-48 at the Ballistic Research Laboratory of the Aberdeen Proving Ground, Maryland. This work was supported in part by Army Research Office Grant No. DAAL-88-K-0067.

REFERENCES

1) A. R. McGurn, A. A. Maradudin, and V. Celli, Phys. Rev. B31, (1985) 4866.

2) V. Celli, A. A. Maradudin, A. M. Marvin, and A. R. McGurn, J. Opt. Soc. Am. A2, (1985) 2225.

3) A. R. McGurn and A. A. Maradudin, J. Opt. Soc. Am. B4, (1987) 910.

4) E. R. Méndez and K. A. O'Donnell, Optics Commun. 61, (1987) 91.

5) K. A. O'Donnell and E. R. Méndez, J. Opt. Soc. Am. A4, (1987) 1194.

6) Zu-Han Gu, R. S. Dummer, A. A. Maradudin, and A. R. McGurn, Appl. Optics (in print).

7) J. C. Dainty, M.-J. Kim, and A. J. Sant, "Measurements of Enhanced Back-scattering of Light From One and Two Dimensional Random Rough Surfaces," Notes for the Workshop "Recent Progress in Surface and Volume Scattering," Madrid, September 14-16, 1988, and for the International Working Group Meeting on "Wave Propagation in Random Media," Tallinn, September 19-23, 1988.

8) E. Bahar and M. A. Fitzwater, Optics Commun. 63 (1987) 355.

9) Ya-Qiu Jin and M. Lax, Phys. Rev. B (in print).

10) A. A. Maradudin, E. R. Méndez, and T. Michel, Optics Lett. (in print).

11) R. R. Lentz, Radio. Sci. 9, (1974) 1139.

12) R. M. Axline and A. K. Fung, IEEE Trans. Antennas Propag. AP-26, (1978) 482; AP-28, (1980) 949.

13) H. L. Chan and A. K. Fung, Radio Sci. 13, (1978) 811.

14) A. K. Fung and M. F. Chen, J. Opt. Soc. Am. A2, (1985) 2274.

15) M. Nieto-Vesperinas and J. M. Soto-Crespo, Optics Lett. 12, (1987) 979.

16) N. García and E. Stoll, Phys. Rev. Lett. 52, (1984) 1798.

17) N. García and E. Stoll, J. Opt. Soc. Am. A2, (1985) 2240.

18) P. Tran and V. Celli, J. Opt. Soc. Am. A5 (1988) 1635.

19) This approach has been used effectively in several recent papers. See, for example, E. I. Thorsos, J. Acoust. Soc. Am. 83, (1988) 78; S. L. Broschat, E. I. Thorsos, and A. Ishimaru, J. Electromagnetic Waves and Applications (in print).

20) See, for example, J. D. Jackson, Classical Electrodynamics (John Wiley and Sons, Inc., New York, 1962) pp. 14-15.

21) Our results for an incident beam of s-polarization will be presented elsewhere.

22) D. E. Amos, ACM Trans. on Math. Software 12, (1986) 265.

23) A. R. McGurn and A. A. Maradudin, Localization Effects in the Scattering of Acoustic Waves from a Random Stress-free Grating Surface, in Recent Developments in Surface Acoustic Waves, D. F. Parker and G. A. Maugin, Eds. (Springer-Verlag, Berlin, 1988) p. 135.

24) This interpretation has been justified in the high frequency limit, and for the case of the Dirichlet boundary condition, by E. G. Liszka and J. J. McCoy, J. Acoust. Soc. Am. 71 (1982) 1093.

25) E. Akkermans, P. E. Wolf, and R. Maynard, Phys. Rev. Lett. 56 (1986) 1471.

26) E. Akkermans, P. E. Wolf, R. Maynard, and G. Maret, J. Phys. France 49 (1988) 77.

27) A. A. Maradudin and T. Michel (unpublished work).

Scattering in Volumes and Surfaces
M. Nieto-Vesperinas and J.C. Dainty (Editors)

ELECTROMAGNETIC SCATTERING FROM VERY ROUGH RANDOM SURFACES AND ITS CONNECTION WITH BLAZES FROM REFLECTION GRATINGS

M.Nieto-Vesperinas and J.M.Soto-Crespo

Instituto de Optica C.S.I.C., Serrano 121, 28006 Madrid Spain

We present theoretical results on the scattering of electromagnetic waves from perfectly conductive very rough random rough one-dimensional surfaces. The dependence of the mean scattered intensities on the state of polarization and parameters of the surface are discussed. The phenomenon of enhanced backscattering is obtained and interpreted in connection with the blaze effect observed in reflection gratings by constructive interference in the antispecular direction. The enhancement in the specular direction for symmetric profiles is also discussed.

1. INTRODUCTION

The phenomenon of enhanced backscattering from deep random rough surfaces has been experimentally observed[1]. This effect appears to be related to that of weak localization of photons in random media[2,3] and has also been predicted in shallow surfaces with finite conductivity under p-polarization[4].

Thus far, this phenomenon has not been adequately studied in deep random surfaces, apart from the case of perfect conductors[5,6].

In the study that follows we shall discuss the effect of enhanced backscattering for perfectly conductive very rough surfaces. This will be done by solving numerically the scattering equations which are derived from the extinction theorem[7,9]. The surfaces will be generated by means of a Monte Carlo procedure, and they will be one-dimensional, thus, no depolarization effects will be obtained. The reason of considering one-dimensional surfaces is based on the limited number of points available to sample the data and at which the source samples are to be determined.

We shall show that for correlation length, T, of the surface heights smaller than the wavelength λ, as the root mean square σ of the surface increases, the surface behaves differently; first, for small values of σ/λ, the scattering of S and P waves is quite different. Then as σ is larger, the surface acquires a near Lambertian regime within moderate scattering angles. Further, as σ is still larger, the phenomenon of enhanced backscattering appears, even when T is slightly larger than λ.

Another important aspect that will be emphasized in this study is the intimate connection between the enhanced backscattering from perfectly conductive random rough surfaces and the blaze effect from reflection gratings[10,11]. Both effects being due to a constructive interference on reflection[12,13]. In fact, deep gratings have a marked tendency to reflect in the antispecular direction under Littrow mount. Also, when the grating profile is symmetric, either the specular or antispecular orders tend to concentrate the larger amount of reflected energy. It is shown, in addition, that in connection with this behaviour, one-dimensional random rough surfaces which present a center of symmetry produce simultaneously enhanced reflected intensity both in the backscattering and in the specular directions.

2. SCATTERING EQUATIONS FOR RANDOM ROUGH SURFACES

Let us consider a linearly polarized electromagnetic wave incident under an angle θ_o upon a surface $z=D(x)$ separating vacuum from a perfect conductor (See Fig.1). Since the variation takes place only in the x-coordinate, one has to consider only the plane of incidence and there is no depolarization.

In the case of s-polarization the incident electric vector is:

$$\mathbf{E}^{(i)}(\mathbf{r})=\mathbf{j}E^{(i)}\exp[i(K_o x-q_o z)], \tag{1}$$

where $\mathbf{r}=(x,z)$, $\mathbf{j}$ is the unit vector along OY, $E^{(i)}$ is a complex constant amplitude and:

$$K_o=k_o\sin\theta_o, \tag{2a}$$

$$q_o=k_o\cos\theta_o, \tag{2b}$$

are the components of the incident wave vector, $k_o=2\pi/\lambda$. The waves are assumed monochromatic with a time dependent factor exp(-iwt).

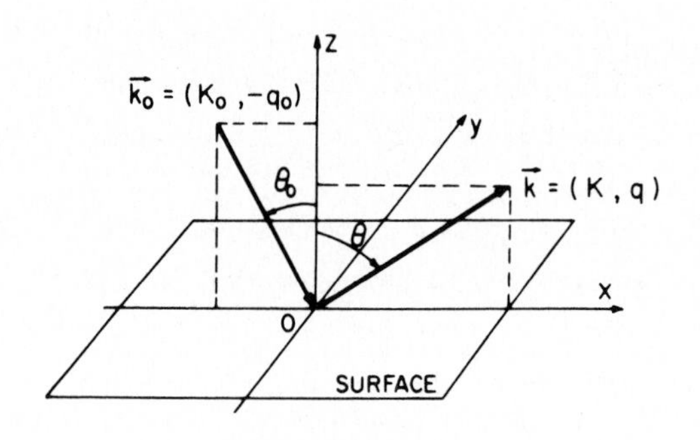

FIGURE 1
Scattering Geometry

The normalized mean scattered intensity in the far zone at angle of observation θ is:

$$\langle I^{(s)}(\theta)\rangle = \frac{2\pi k}{c^2}\frac{1}{|E^{(i)}|^2 L\cos\theta_0}\langle\left|\int_{-\infty}^{\infty}dx' J_y(x',D(x'))\right.$$
$$\left.\times\exp[-ik(x'\sin\theta+D(x')\cos\theta)]\right|^2\rangle. \quad (3)$$

In Eq.(3) J_y is the y-component of the induced electric current density. L is the length of the illuminated surface. Therefore in calculations, the x-integrals are restricted to this interval.

For p-polarization the incident wave is represented by the magnetic vector:

$$\mathbf{H}^{(i)}(\mathbf{r}) = \mathbf{j}\ H^{(i)}\exp[i(K_0x-q_0z)], \quad (4)$$

$H^{(i)}$ being a complex constant amplitude.

And the normalized mean intensity for p-waves in the far zone reads:

$$\langle I^{(p)}(\theta)\rangle = \frac{2\pi k}{c^2}\frac{1}{|H^{(i)}|^2 L\cos\theta_0}$$
$$\times\langle\left|\int_{-\infty}^{\infty} J(x',D(x'))\exp\{-ik[x'\mathrm{sen}\theta+D(x')\cos\theta]\}\cdot\right.$$
$$\left.\times(\cos\theta-\frac{dD}{dx'}\mathrm{sen}\theta)\sqrt{1+(dD/dx')^2}\,dx'\right|^2\rangle. \quad (5)$$

In Eq.(5) J_x is the x-component of the electric current density.

Both expressions(4) and (5) must satisfy the unitarity condition:

$$\frac{1}{I_0}\int_{-\pi/2}^{\pi/2}\langle I^{s}_{p}(\theta)\rangle d\theta = 1, \quad (6)$$

where I_0 represents the intensity of the incident field.

The sources that enter in Eqs.(4) and (5) have to be found from appropriate boundary conditions. Here we use the extinction theorem[7] that for perfect conductors reads for s-waves[6]:

$$E^{(i)}\exp\{i(K_0x-q_0z)\} = \frac{\pi k}{c}\int_{-\infty}^{\infty} J_y(\mathbf{r}')H_0^{(1)}(k|\mathbf{r}-\mathbf{r}'|)\sqrt{1+(dD/dx')^2}\,dx', \quad (7)$$

where $\mathbf{r}'=(x',D(x'))$ and $H_0^{(1)}$ is the Hankel function of the first kind and zero order.

Analogously, for p-waves the extinction theorem is expressed as:

$$H^{(i)}\exp\{i(K_0x-q_0z) =$$
$$\frac{-\pi i k_0}{c}\left[\int_{-\infty}^{\infty}dx' J_x(\mathbf{r}')H_1^{(1)}(k_0|\mathbf{r}-\mathbf{r}'|)\frac{(z-z')-\{dD/dx'\}(x-x')}{\sqrt{(x-x')^2+(z-z')^2}}\sqrt{1+\{dD/dx'\}^2}\right] \quad (8)$$

In Eq.(8) $H_1^{(1)}$ is the Hankel function of the first kind and order one. The integration must be done by isolating the singularity at $\mathbf{r}=\mathbf{r}'$ according to the method in Ref.14.

3. SURFACE GENERATION AND NUMERICAL CALCULATIONS

The surface profiles are generated by using a Monte Carlo procedure. A sequence of random numbers (around 100,000) with Normal statistics, zero mean, and variance equal to unity is constructed from another series of random numbers directly generated by the computer. Then the former sequence is scaled in order to obtain a desired variance σ^2 and, further, correlation with a Gaussian $(2/\sqrt{\pi}T\exp[-2(k/T)^2])$ is performed in order to obtain a profile with a Gaussian correlation function. In this way we generate a statistically homogeneous random process with zero mean and normal statistics. The covariance function $C(\tau)$ being also Gaussian and given, as usual, by:

$$C(\tau)=\langle D)(x)D(x+\tau)\rangle=\sigma^2\exp(-\tau^2/T^2). \tag{9}$$

T being the correlation length and σ^2 being the mean square deviation of the profile from the line $z=0$.

The mean scattered intensities are obtained by averaging far-field intensities over 200 samples of length L. Each sample has typically 221 sampling points. By sampling both the field and the surface, Eqs.(7) and (8) become a linear system of equations from which the samples of the sources are found. For angle of incidence $\theta_o=0$, this procedure yields a slightly asymmetric distribution of mean scattered intensity, which is symmetrized by averaging the values at θ and $-\theta$. Further smoothing can be done by averaging every three sampling points of these mean intensities. In addition, a Gaussian profile can be introduced in the incident wave, (thus being an incident Gaussian beam).

An idea of the sampling interval can be obtained from the following figures. For $T\approx\lambda$ the sample lenght L is $L\approx22\lambda$, although when $\sigma/T\ll1$ L can be made $L\approx66\lambda$. For $T=0.2\lambda$, $L=12\lambda$. For $T>3\lambda$, $L=60\lambda$. Therefore the sampling interval is about 0.1λ when $T\approx\lambda$, and near 0.05λ and 0.25λ when $T<\lambda$ and $T>\lambda$, respectively.

4. DEPENDENCE OF THE MEAN SCATTERED INTENSITY ON THE POLARIZATION

There is a regime of parameters σ and T of the surface in which there is an interesting markedly difference between the scattered intensities under s or p-polarization. This happens when σ is very small; then a response of the surface takes place depending on whether we deal with s-waves or with p-waves, but at the same time T is also very small so that the Kirchhoff approximation (KA) is far from being valid. Figs. 2(a) and 2(b) show two of such examples. Here we see the mean scattered intensity versus the observation angle θ at three different angles of incidence: $\theta_o=0^\circ$, 30° and 60°, for $\sigma=0.1\lambda$ and $T=0.5\lambda$ (Fig.2(a)) and $T=0.25\lambda$ (Fig.2(b)); the specular peak has been removed.

(The tip marks on the top show the specular direction to the right of $\theta_o=0°$ and the antispecular (backscattering) direction to the left of $\theta_o=0°$). As seen, s-waves (continuous line) display a smaller diffuse halo than p-waves (dotted line). Also at oblique incidences, contrary to s-waves, p-waves are skewed towards the backscattering direction and produce a bigger diffuse halo than the s-waves. This scattering halo is both for s and for p-waves broader as T decreases.

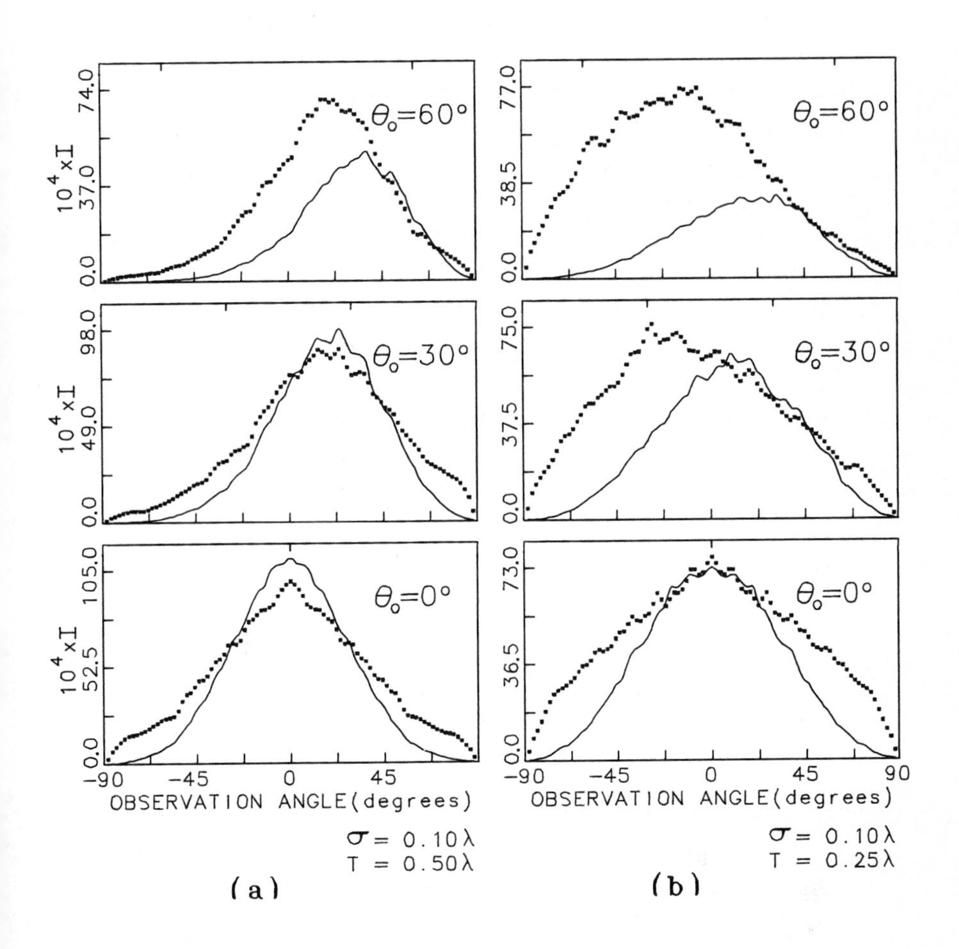

FIGURE 2
Mean scattered intensities for s (solid lines) and p polarization (dotted lines) for a surface with $\sigma=0.1\lambda$. (a) $T=0.5\lambda$. (b) $T=0.25\lambda$.

5. BEHAVIOUR FOR LARGE VALUES OF σ

For values of σ larger than, or equal to, about the wavelength λ, the difference between s-polarization and p-polarization is negligible.

As a general role, we can say that at given T, even comparable to λ, when σ is very low the KA is valid, and thus no difference exists for the results between s and p-polarization (fig.3a). As σ increases at the same value of T, one enters the domain in which the KA is no longer valid, the halo broadens and then appear some differences between s and p-waves. For σ large enough, (see Fig.3b), there appears a regime in which the surface behaves approximately as a Lambertian difusser, (at least for angles of observation θ not too large). This range of angles θ at which the Lambertian regime takes place decreases with the angle of incidence θ_o. As σ increases further (fig.3c) a peak appears in the backscattering direction of the mean scattered intensity. This peak is higher the larger σ is (Fig.3d). As θ_o increases, the highness of this peak decreases. This is the enhanced backscattering phenomenon observed in experiments on rough surfaces[1]. It is important to

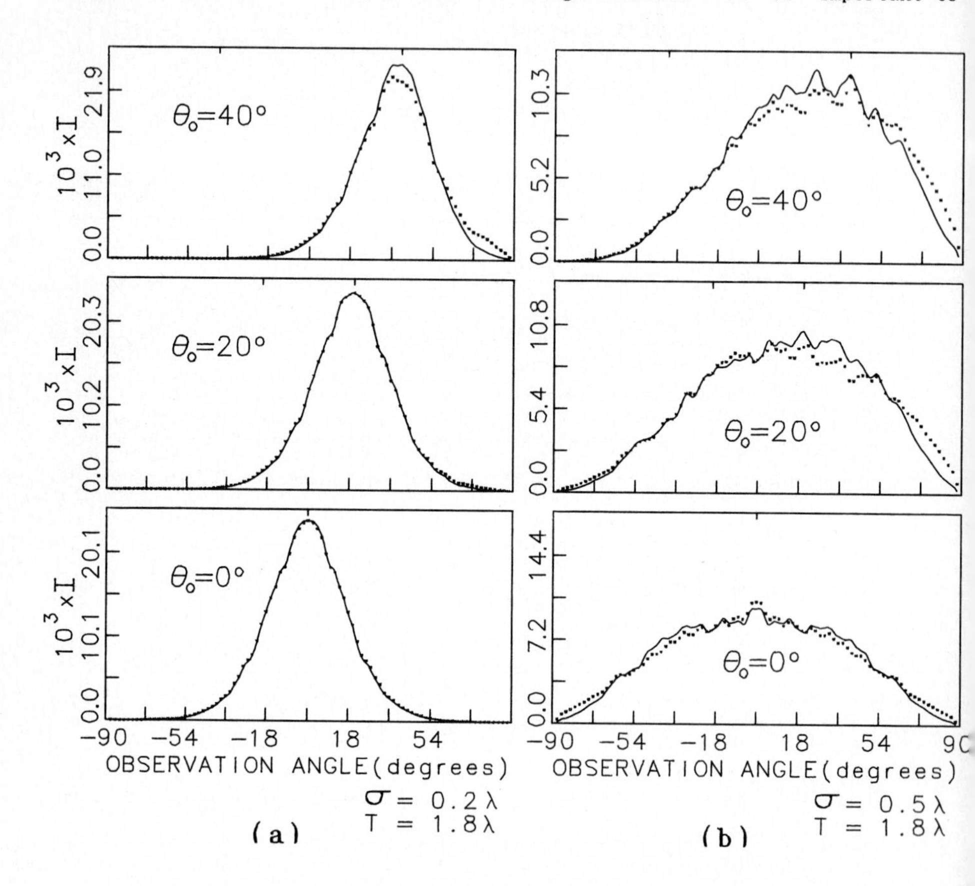

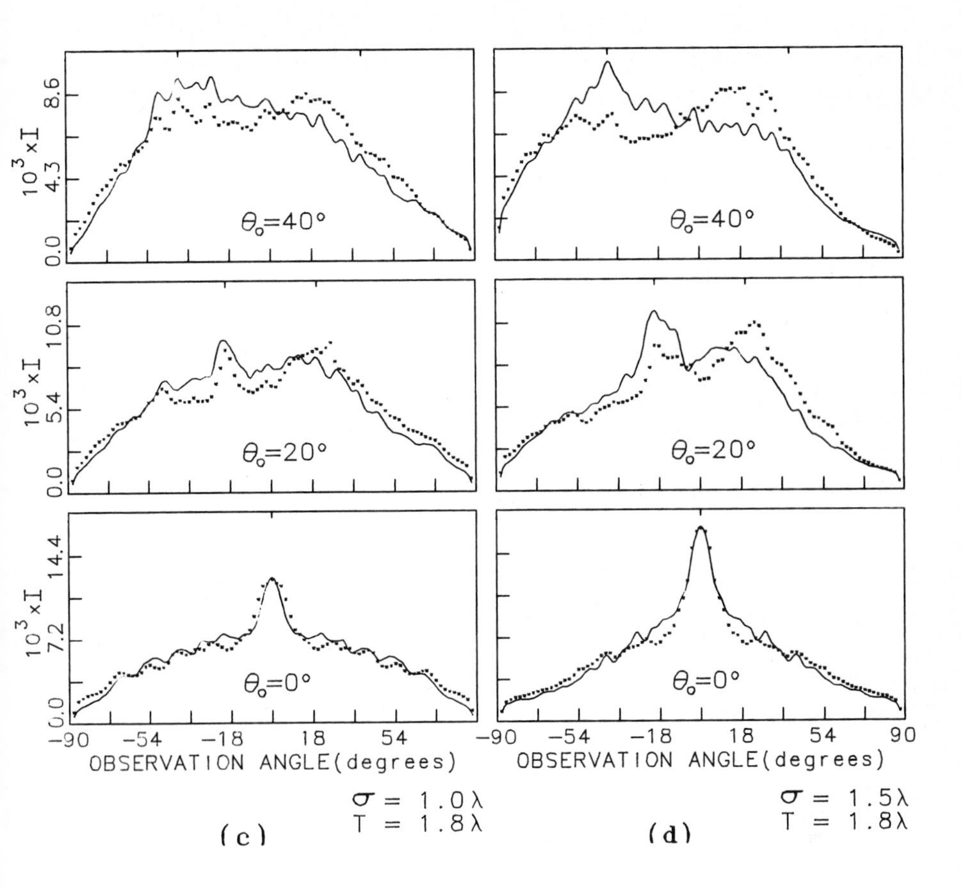

FIGURE 3
Same as Fig.2 for $T=1.8\lambda$ and $\theta_o=0°$, 20° and 40°. (a) $\sigma=0.2\lambda$. (b) $\sigma=0.5\lambda$. (c) $\sigma=\lambda$. (d) $\sigma=1.5\lambda$.

stress that this peak can be observed only when averages are made in the scattered intensities. Otherwise, if will appear swamped by the speckle fluctuations[5,15]. To illustrate this point, Fig. 4(a) and 4(b) show the scattered intensities for s and p-waves respectively from just one sample (Fig.4(a)) and averaging over 10 samples (Fig.4(b)) at $\theta_o=40°$. The average over 200 samples is shown in Fig.5.

With the same ratio σ/T, as λ decreases the enhanced backscattering peak increases. This is illustrated in Figs. 6(a)-6(c), where the mean scattered intensity is shown for s-waves at two angles of incidence, $\theta_o=0°$ and 10°, and for three different wavelengths.

We believe that the enhanced backscattering effect observed in perfectly conductive random rough surfaces is due to a pure interference effect in the reflected field. In fact, as it is seen in the next section, the blaze effect observed in reflection gratings is at the root of the enhanced backscattering in random perfectly conducting structures.

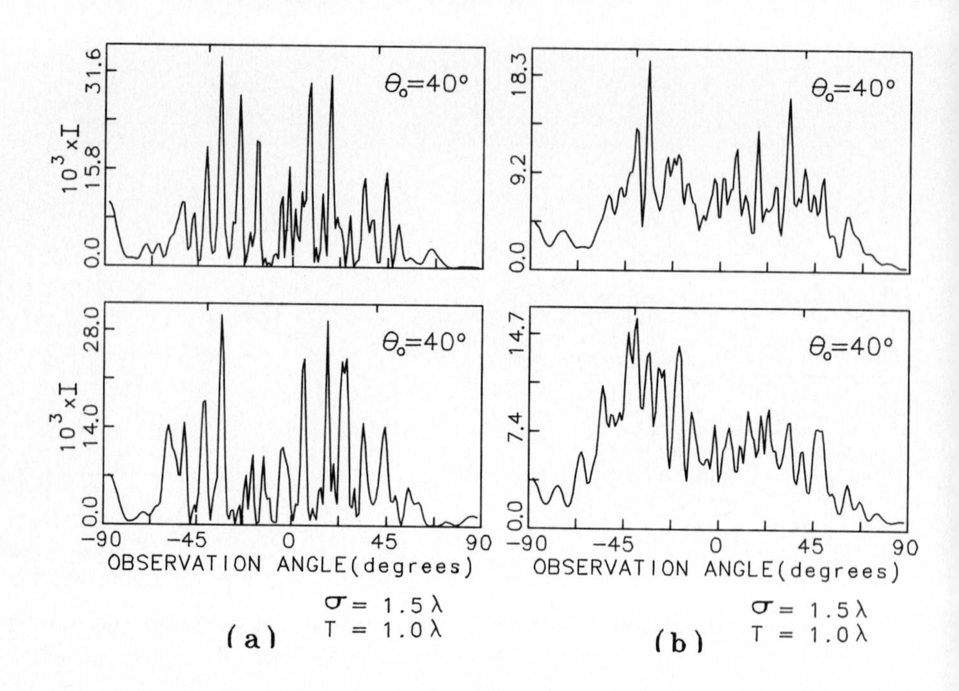

FIGURE 4
(a) Scattered intensity from one sample with T=λ, σ=1.5λ for s waves (lower curve) and p waves (upper curve). (b) Mean scattered intensity from ten samples with T=λ, σ=1.5λ for s waves (lower curve) and p waves (upper curve).

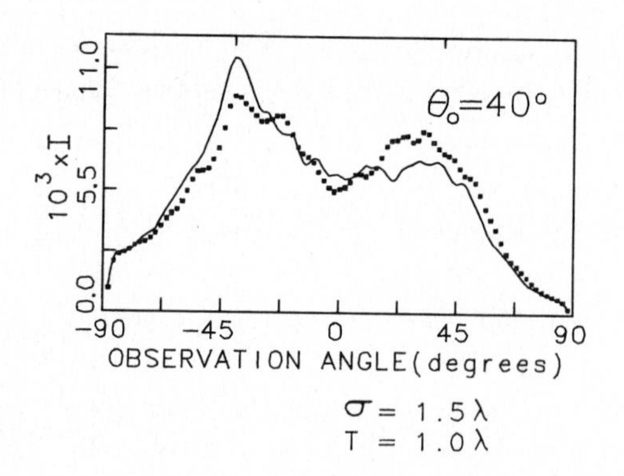

FIGURE 5
Same as Fig.2 for T=λ, θ_o=40° and σ=1.5λ.

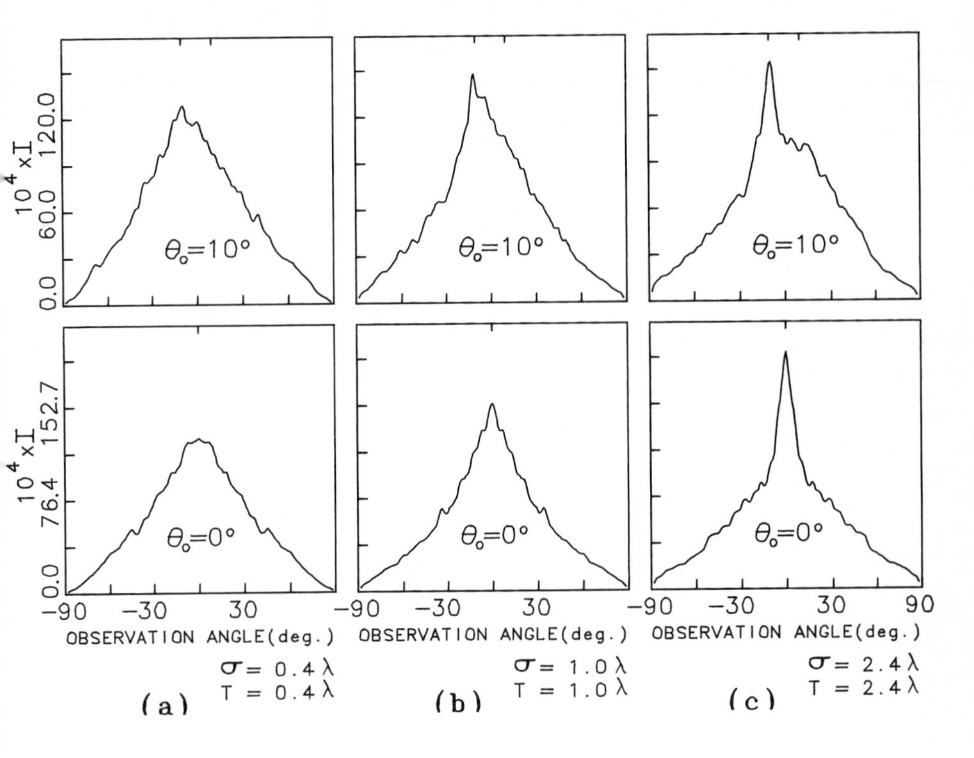

FIGURE 6
Same as Fig.2 for a surface with σ=T illuminated with three different wavelengths. (a) $\lambda=2.5\sigma$. (b) $\lambda=\sigma$. (c) $\lambda=0.42\sigma$.

6. DEEP REFLECTION GRATINGS

By using the pseudoperiodicity of the electric current density:

$$J(x+a)=J(x)\exp\{iK_oa\}; \tag{10}$$

where a is the grating period, the extinction theorem for s-waves adopts the form:

$$E^{(i)}(\mathbf{r}) = \frac{\pi k_o}{c}\int_0^a dx' J_y(x')G^s(k_o|\mathbf{r}-\mathbf{r}'|)\sqrt{1+(dD/dx')^2} \tag{11}$$

$$\mathbf{r}=(x,z)$$

Where the Green function for s-waves is given by the following expansion:

$$G^s(k_o|\mathbf{r}-\mathbf{r}'|) = \sum_l \left[H_o^{(1)}(k_o\sqrt{[x-(x'-la)]^2+[z-D(x')]^2})\exp(iK_ola) \right] \tag{12}$$

And the diffracted amplitude at the nth-order is given by:

$$A_n=\frac{2\pi k_o}{cak_{nz}}\int_0^a J_y(x')\exp\{-i[K_nx'+k_{nz}D(x')]\}\sqrt{1+(dD/dx')^2}\,dx' , \tag{13}$$

where

$$K_n^2+k_{nz}^2 = k^2 , \tag{14}$$

and

$$K_n=K_o+2\pi n/a . \tag{15}$$

Analogously, for p-waves, the extinction theorem reads:

$$H^{(i)}(r)=\frac{-\pi i k_o}{c}\int_0^a J_x(x')G^p(k_o|r-r'|)\sqrt{1+(dD/dx')^2}dx' , \tag{16}$$

where the Green function for p-waves reads:

$$G^p(k_o|r-r'|)=\sum_l\left[H_1^{(1)}(k_o|\sqrt{[x-(x'-la)]^2+(z-D(x'))^2}|)\exp\{iK_ola\}\cdot\frac{(z-D(x'))-dD/dx'[x-(x'+la)]}{\sqrt{[x-(x'+la)]^2+[z-D(x')]^2}}\right] . \tag{17}$$

And the corresponding diffracted amplitudes are:

$$B_n=\frac{2\pi}{cak_{nz}}\int_0^a dx'J_x(x')\exp\{-i[K_nx'+k_{nz}D(x')]\}\left[\frac{dD}{dx'}K_n-k_{nz}\right]\sqrt{1+(dD/dx')^2}. \tag{18}$$

The diffracted intensities (efficiencies) are given by:

$$I_n^w=\frac{k_{nz}}{k_{oz}}\begin{cases}|A_n|^2 & \text{for s-waves,} \quad (19a)\\ |B_n|^2 & \text{for p-waves.} \quad (19b)\end{cases}$$

(w=s or p)

And satisfy the unitarity condition:

$$\sum_n I_n^w=1 , \tag{20}$$

where the sum is extended over the propagating orders.

For a given profile D(x) which is deep enough, there is a marked tendency to obtain enhancement (blaze) of the antispecular order when the direction of incidence of the incoming field is such that this order exists. In addition, if the profile is symmetric, then the specular order has also high probability of being the enhanced one.

Fig.7 shows the diffracted efficiencies versus the angle of incidence for the profile $z=D(x)=h\cos(2\pi x/a)$, showing enhancement of all orders in the directions at which they become antispecular. (For deep gratings the result is independent on polarization). On the other hand, Fig.8 shows the behavior of the diffracted intensities versus h/λ for the same sinusoidal profile. The character "B" indicates the antispecular order. As seen, there is a typical "rainbow" oscillatory behaviour[16,17]. And the specular and antispecular orders are those that exhibit the largest amplitude.

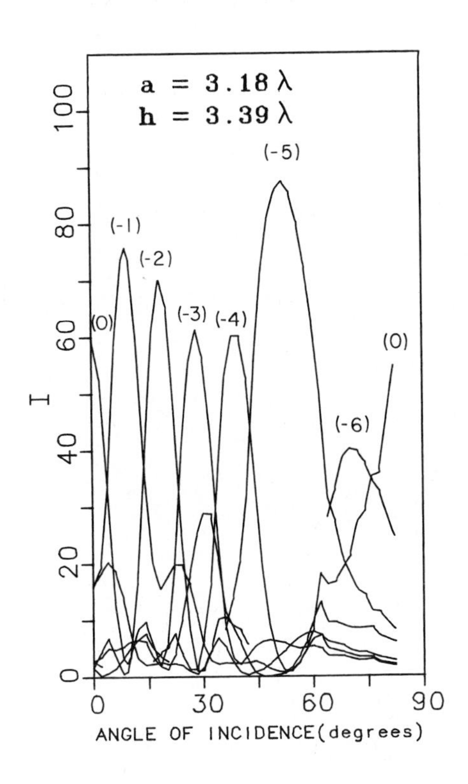

FIGURE 7
Diffracted intensities for s-polarization versus angle of incidence for a sinusoidal grating with a=3.18λ and h=3.39λ.

The tendency of the antispecular order to become enhanced (and also of the specular if the profile is symmetric) is responsible for the enhancement of the mean scattered intensity in the backscattering direction from random rough surfaces. For a particular profile it can happen that neither of those two orders is the enhanced one; like in a rough surface, for a particular sample, the speckle effect can be such that the maximum of the scattered intensity is not in the backscattering direction. But given the higher probability of the antispecular order to be enhanced, this order is the one that dominates when one averages. To illustrate this remark, Fig.9 shows the diffracted orders for three different angles of incidence obtained by averaging over 200 gratings of period a=8λ. Each of these gratings has been obtained by taking a portion of length "a" from a random profile with T=0.7λ and σ=1.2λ (left bars) and σ=1.3λ (right bars) and repeating periodically this portion. As seen, the antispecular order is the one that appears enhanced, and the enhancement is larger (right bars) for the deeper gratings.

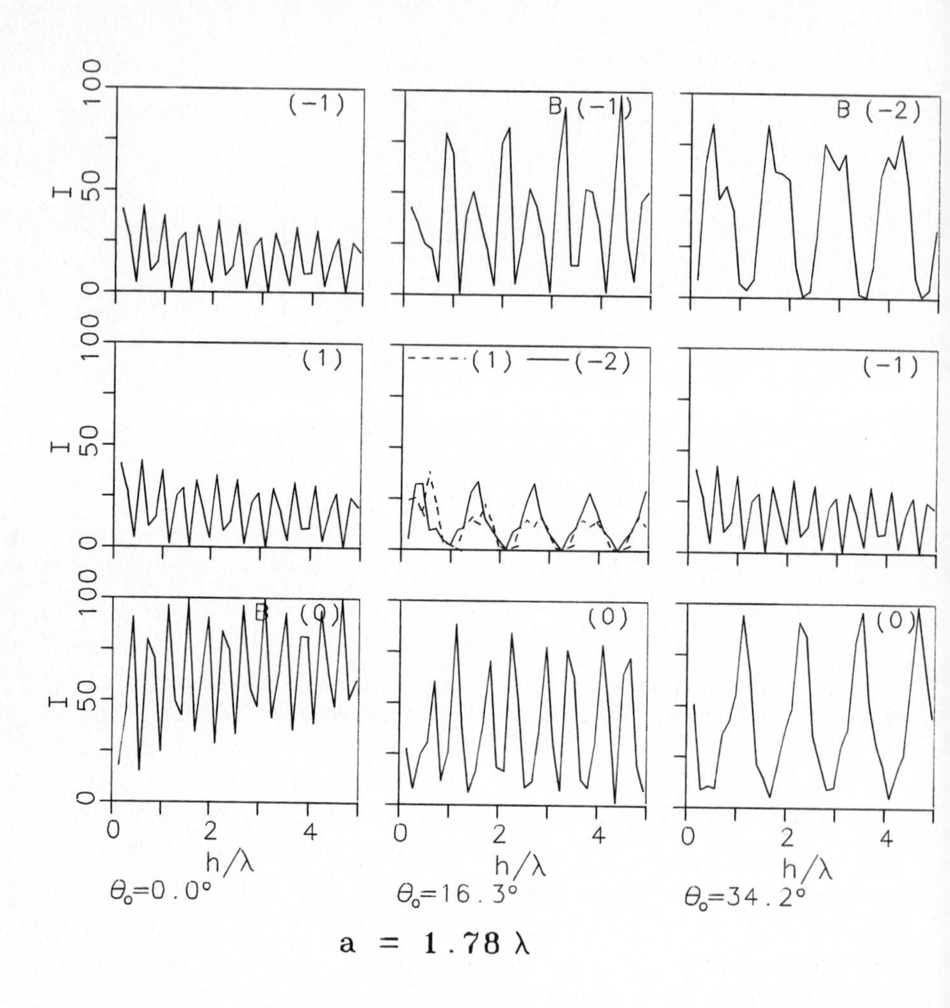

FIGURE 8
Diffracted intensities versus h/λ for a sinusoidal grating with a=1.78λ and three angles of incidence. The character "B" indicates the antispecular order.

The mean diffracted efficiencies versus angle of incidence for the gratings corresponding to $\sigma=1.2\lambda$ are shown in Fig.10. As seen the maxima of the averaged orders occur exactly at those angles at which they become antispecular.

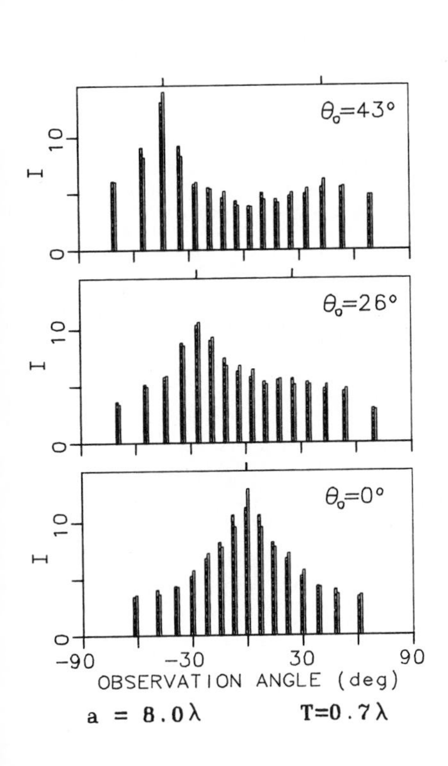

FIGURE 9
Mean diffracted intensities from 200 gratings with random profile. Left bars:$\sigma=1.2\lambda$, right bars:$\sigma=1.3\lambda$.

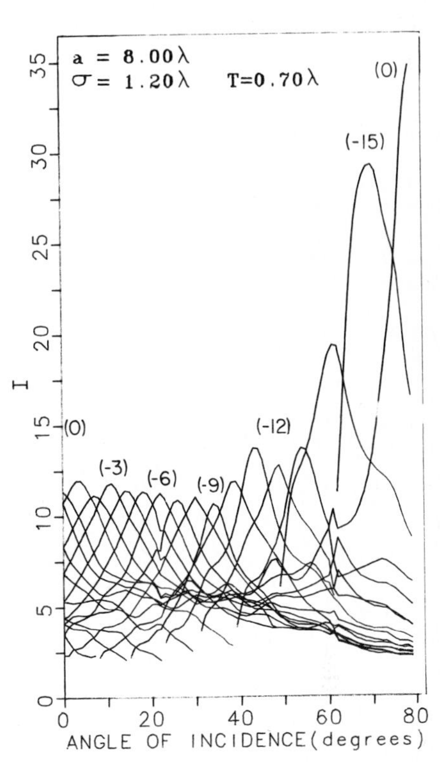

FIGURE 10
Mean diffracted efficiencies versus θ_o averaging over the same gratings as those of Fig.9 with $\sigma=1.2\lambda$.

When the profile is symmetric, the specular order appears also enhanced on performing the average over many gratings of the same period[13]. By extension of this argument, one would expect that for random rough surfaces which were artificially constructed so that their illuminated portion were symmetric with respect to a center, the mean scattered intensity would appear enhanced not only in the backscattering direction, but also in the specular direction. Fig.11 confirms this remark[13], it shows the mean scattered intensity from 200 samples of a rough random surface with $T=\lambda$ and $\sigma=\lambda$ at angles of incidence

$\theta_o=0°$, 20° and 40°. The dotted line corresponds to random samples without symmetry showing the usual backscattering enhancement. On the other hand, the continuous line represents the result for random samples in which a center of symmetry has been introduced, showing also enhancement in the specular direction. The explanation is similar to the one used for interpreting the enhanced backscattering phenomenon[18]: The coherent addition of amplitude from a ray trajectory and its reversed taking into account the symmetry properties of the scattering structure[13].

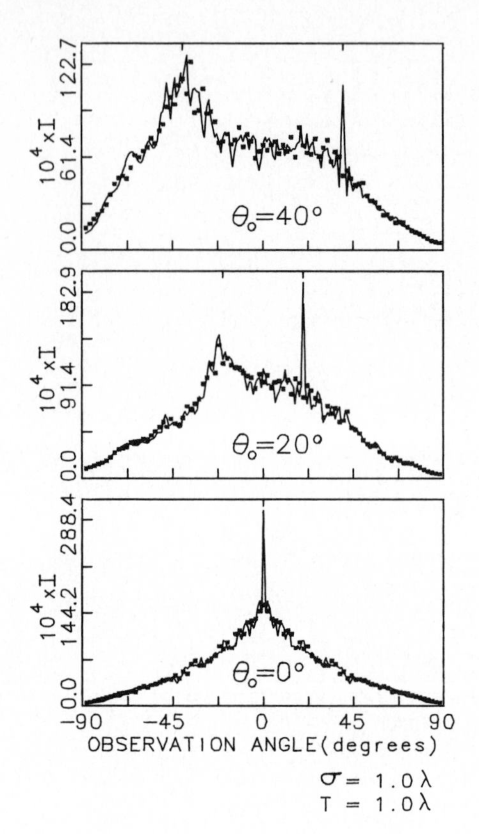

FIGURE 11
Mean scattered intensities from random surfaces without symmetry (dotted line), and with a center of symmetry included in the illuminated portion (continuous line)

. CONCLUSIONS

Blaze effects[19] and their associated Bragg anomalies[20,21] have been known for sometime in reflection gratings when they operate under Littrow mount, (i.e. when the direction of a negative order coincides with the direction of incidence of the illuminating wave). What we have shown here is that those effects are at the root of the phenomenon of enhanced backscattering from perfectly conductive random surfaces. Even in the cases with symmetry, we have pointed out the connection between the enhancement of the specular order in gratings and the strong specular peak from very rough random surfaces.

ACKNOWLEDGMENTS

Work supported by the CICYT under grant PB0278. J.M.S-C. acknowledges a grant from Ministerio de Educacion y Ciencia.

REFERENCES

1) K.A. O'Donnell and E.R. Méndez, J. Opt. Soc. Am. A4, 1194-1205 (1987).

2) H.P.Van Albada and A.Lagendijk, Phys.Rev.lett.**55**, 2692 (1983). See also P.E.Wolf and G.Maret, Phys.Rev.Lett.**55**, 2696 (1983). Y. Kuga and A. Ishimaru, J. Opt. Soc. Am. A1, 831-835 (1984).

3) M. Kaveh, M. Rosenbluth, M. Edrei and I. Freund, Phys. Rev. Lett. **57**, 2049-2052 (1986).

4) V.Celli, A.A.Maradudin, A.M.Marvin and A.R.McGurn, J.Opt.Soc.Am. A **2**, 2225-2239 (1985).

5) M.Nieto-Vesperinas and J.M.Soto-Crespo, Opt.Lett. **12**, 979-981 (1987).

6) J.M.Soto-Crespo and M.Nieto-Vesperinas, J.Opt.Soc.Am.A **6**, 367-384 (1989).

7) D.N. Pattanayak and E. Wolf, Opt. Comm. **6**, 217-220 (1972).

8) F. Toigo, A. Marvin, V. Celli and N.R. Hill, Phys. Rev.B **15**, 5618-5626 (1977).

9) M. Nieto-Vesperinas, J. Opt. Soc. Am. **72**, 539-547 (1982).

10)R. Petit, ed. **Electromagnetic Theory of Gratings** (Topics in Current Physics 22, Springer Verlag, Berlín 1980).

11)D.Y.Tseng, A.Hessel and A.A.Oliner, U.R.S.I. Symposium on Electromagnetic waves. Alta Freq. **38** special issue, 82-88 (1969).

12) M.Nieto-Vesperinas and J.M.Soto-Crespo, Phys.Rev.B **38**, 7250 (1988).

13) M.Nieto-Vesperinas and J.M.Soto-Crespo, Phys.Rev.B **39**, 8193 (1989).

14)M.G.Andreasen, IEEE Trans.Antennas Propag. **AP-12**,746 (1964).

15) S.Etemad, R,Thompson and M.J.Andrejco, Phys.Rev.Lett. **57**, 575-578 (1986).

16) V.Garibaldi, A.C.Levi, R.Spadacini and G.E. Tommei, Surf.Sci **48**,649 (1975).

17) J.M.Soto-Crespo and M.Nieto-Vesperinas, Opt.Comm. **69**,185-188 (1989).

18) E.Akkermans, P.E.Wolf and R.Maynard, Phys.Rev.Lett. **56**, 1471-1474 (1986)

19) D.Maystre, M.Neviere and R.Petit, Chapter 6 of Ref.10.

20) N.Garcia and A.A.Maradudin, Opt.Comm. **45**,301(1983).

21) A.Hessel and A.A.Oliner, Opt.Comm. **59**,327 (1986).

Scattering in Volumes and Surfaces
M. Nieto-Vesperinas and J.C. Dainty (Editors)

RIGOROUS SOLUTION OF PROBLEMS OF SCATTERING BY LARGE SIZE OBJECTS

Daniel MAYSTRE and Marc SAILLARD

Laboratoire d'Optique Electromagnétique, U.R.A. au CNRS n° 843, Faculté des Sciences et Techniques, Centre de Saint-Jérôme, 13397 Marseille Cedex 13, France

It is generally considered that the use of rigorous theories of scattering in Electromagnetism, Optics or Acoustics is restricted to scattering objects having dimensions of the order of the wavelength of the incident waves, or less. It is shown that this opinion is too pessimistic and we propose two methods able to solve rigorously the problem of scattering by very large objects. They are implemented to the calculations of radar cross-sections, scattering by echelle gratings and scattering by random rough surfaces.

1. INTRODUCTION

Solving accurately the problems of scattering by large size objects has a vital importance in Optics, Acoustics and Electromagnetism. For example, in Optics, this problem is encountered in the study of scattering by random rough surfaces, or in the calculation of the efficiency of echelle gratings, viz. gratings having a large pitch. In Electromagnetism, the same problem occurs in the calculation of Radar cross-sections of large objects.

Up to now, the methods used to solve these problems are generally asymptotic methods, for instance Physical Optics or Geometrical Theory of Diffraction. These methods are valuable but in many cases, their domains of validity are not well known and the precision of the numerical calculations is difficult to predict.

On the other hand, the rigorous methods based on the fundamental laws of Electromagnetism lead generally to the inversion of complex matrices the size of which increases linearly with the dimension of the object. Thus, they must be abandoned when the dimension/wavelength ratio exceeds a certain value (typically of the order of 20 for two dimensions problems).

Even though the progress in computer speed and memory storage have allowed one to store and invert very large complex matrices, phenomena of round-off errors, in practice, limit the size to about 500, at least if the matrix is arbitrary and if a classical method is used.

This is the reason why one consider generally that rigorous methods are unable to solve high frequency problems of diffraction. It is shown in this paper that this opinion is probably too much pessimistic. Two methods are presented in order to overcome this limitation. The first one is implemented to the calculation of Radar cross-sections and to the problem of scattering by echelle gratings. The second one is used to deal with large random rough surfaces. In that case, we give numerical results about phenomena of absorption by metallic random rough surfaces. Speckle diagrams show enhanced backscattering effects for dielectric surfaces.

2. DETERMINATION OF RADAR CROSS-SECTIONS

2.1. Rigorous integral theory

Here, we consider a perfectly conducting two-dimensional cylindrical object of surface $\mathcal{S}$ (see figure 1) described in polar coordinates by the equation

$$\rho = f(\theta) \tag{1}$$

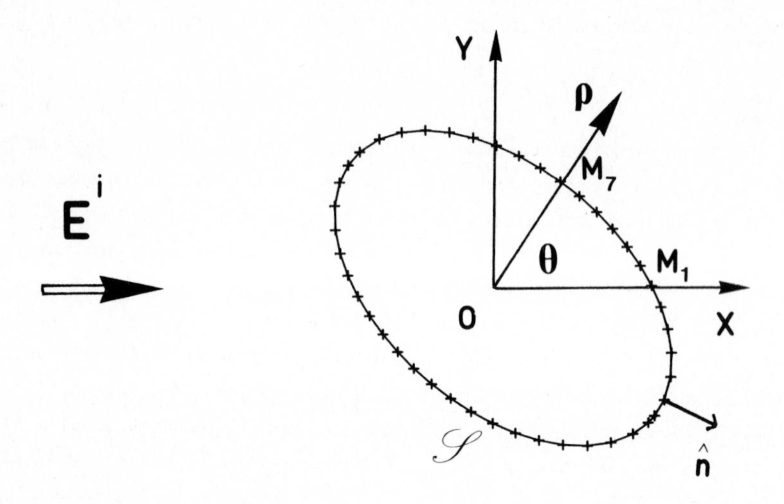

Figure 1 : Presentation of the problem of calculation of a differential cross section.

It is illuminated by a monochromatic homogeneous plane wave $E^i\hat{z}$ of wavelength $\lambda = 2\pi/k$, the time dependance being in $\exp(-i\omega t)$.

The problem is to find the scattered field E^d, difference between the total and incident fields E and E^i, and we are mainly interested in the differential bistatic cross section $D(\theta)$ which gives the scattered intensity at infinity in the direction θ. It is well known that the calculation of $D(\theta)$ reduces to that of the surface current density, which is proportional to the normal derivative

$$\psi(\theta) = dE/dn , \tag{2}$$

$\hat{n}$ being the unit normal to S, oriented towards the exterior of the object.

The function ψ satisfies a rigorous integral equation[1]

$$\int_0^{2\pi} N(\theta,\theta')\ \psi(\theta')\ d\theta' = \Gamma(\theta)\ , \tag{3}$$

with

$$N(\theta,\theta') = \frac{i}{4}\, H_0^{(1)} \left\{k_0\ [f^2(\theta) + f^2(\theta') - 2f(\theta)f(\theta')\cos(\theta-\theta')]^{1/2}\right\}\ , \tag{4}$$

$$\Gamma(\theta) = \exp[ik_0\ \cos(\theta)\ f(\theta)]\ . \tag{5}$$

In this integral equation, the kernel N, the right hand member Γ and the unknown ψ have a period 2π. The kernel $N(\theta,\theta')$ is a Hankel function of the first kind and zeroth order, the term between brackets in (4) is the distance between two points M and M′ of $\mathcal{S}$ with polar angles θ and θ'. The right hand side $\Gamma(\theta)$ is the value of the incident field $E^i(M)$ at a point M of $\mathcal{S}$ of polar angle θ.

2.2. Numerical implementation

One of the methods currently used to solve equation (3) is the discretisation. The integral equation is written at a certain number P of points M_i of $\mathcal{S}$, characterized by their polar angles θ_i

$$\forall i \in [1,P], \qquad \int_0^{2\pi} N(\theta_i,\theta')\ \psi(\theta')\ d\theta' = \Gamma(\theta_i)\ . \tag{6}$$

After making use of an integration method in order to transform the integral in a finite sum, a linear system of P equations with P unknowns is obtained

$$\forall i \in [1,P], \qquad \sum_{j=1}^{P} N_{i,j}\ \psi(\theta_j) = \Gamma(\theta_i)\ . \tag{7}$$

This linear system can be inverted on a computer, using for example the Gauss-Jordan algorithm. The main difficulty of the numerical implementation is the precise computation of the coefficients $N_{i,j}$, since the Hankel function has a logarithmic singularity. Thus the integration requires sophisticated methods to extract the singularity[2]. The singularity being removed, the integration may be performed using the trapezoidal rule, which gives a very good precision since the integrand is periodic. The numerical experiment shows that the discretisation must be made in such a way that the distance M_iM_{i+1} between two consecutive points be less than about $\lambda/10$. Under these conditions, the energy balance criterion is satisfied to within 10^{-6} and the relative precision on the scattering cross section is of the order of 10^{-3}.

2.3. Numerical limitations

As mentioned before, the inversion of arbitrary complex matrices having a size larger than about 500 using classical methods may lead to significant round-off errors. Since one must use about 10 points per wavelength on the profile, the length of this profile cannot exceed about 50 wavelengths. For a circular profile, this limitation corresponds to a radius R of about 8 wavelengths.

In order to understand the origine of the limitation, it is necessary to explain why we must use at least 10 points per wavelength in the discretisation on the profile. To this end, we have investigated the behaviour of the unknwon $\psi(\theta)$ of the integral equation where the size of the scattering object is increased.

The top of figure 2 shows the modulus M (left hand side) and argument A (right hand side) of $\psi(\theta)$ for $R = 5$, the wavelength λ being equal to 6. Just below, the same functions are drawn for $R = 10$. Obviously, the modulus $M(\theta)$ is not modified in a significant manner when the radius is multiplied by 2. The maximum value is close to 2 (the value predicted by the physical optics approximation) and the only change which can be noticed is the decrease of ψ in the shadow region ($\theta \notin [\pi/2, 3\pi/2]$) when R increases, a very intuitive fact. The same conclusion does not hold for the argument A. If the shape of the curve is conserved when R increases, the variation of $A(\theta)$ between the center of the shadow region ($\theta = 0$) and the center of the illuminated region ($\theta = \pi$) goes from 16 to 32 when R is multiplied by two.

This provides us a valuable information. Obviously, it is not necessary to increase the number P of discretisation points in order to represent accurately $M(\theta)$ when R increases. On the other hand, a good representation of the phase needs the number P to be increased with R. Indeed, the contribution of the phase $A(\theta)$ to ψ is equal to $\exp(iA(\theta))$. An accurate representation of this exponential requires a variation of $A(\theta)$ between two consecutive points less than about $\pi/6$. For $R = 5$, the total variation of $A(\theta)$ between $\theta = 0$ and $\theta = \pi$ is about 16 rd, the same between $\theta = \pi$ and $\theta = 2\pi$. Thus, a number P of the order of $2 \times 16 \times 6/\pi \simeq 60$ is needed. For $R = 10$, the total variation of phase has been multiplied by a factor 2 and thus about 120 points are necessary. These numbers P are exactly those used in our calculations and the conclusion is that, at least for a precise representation of the argument, the number P of points must be increased proportionnally to the variation ΔA of argument of ψ between $\theta = 0$ and $\theta = \pi$, and other computations have shown that ΔA is proportionnal to the size of the object (here the radius).

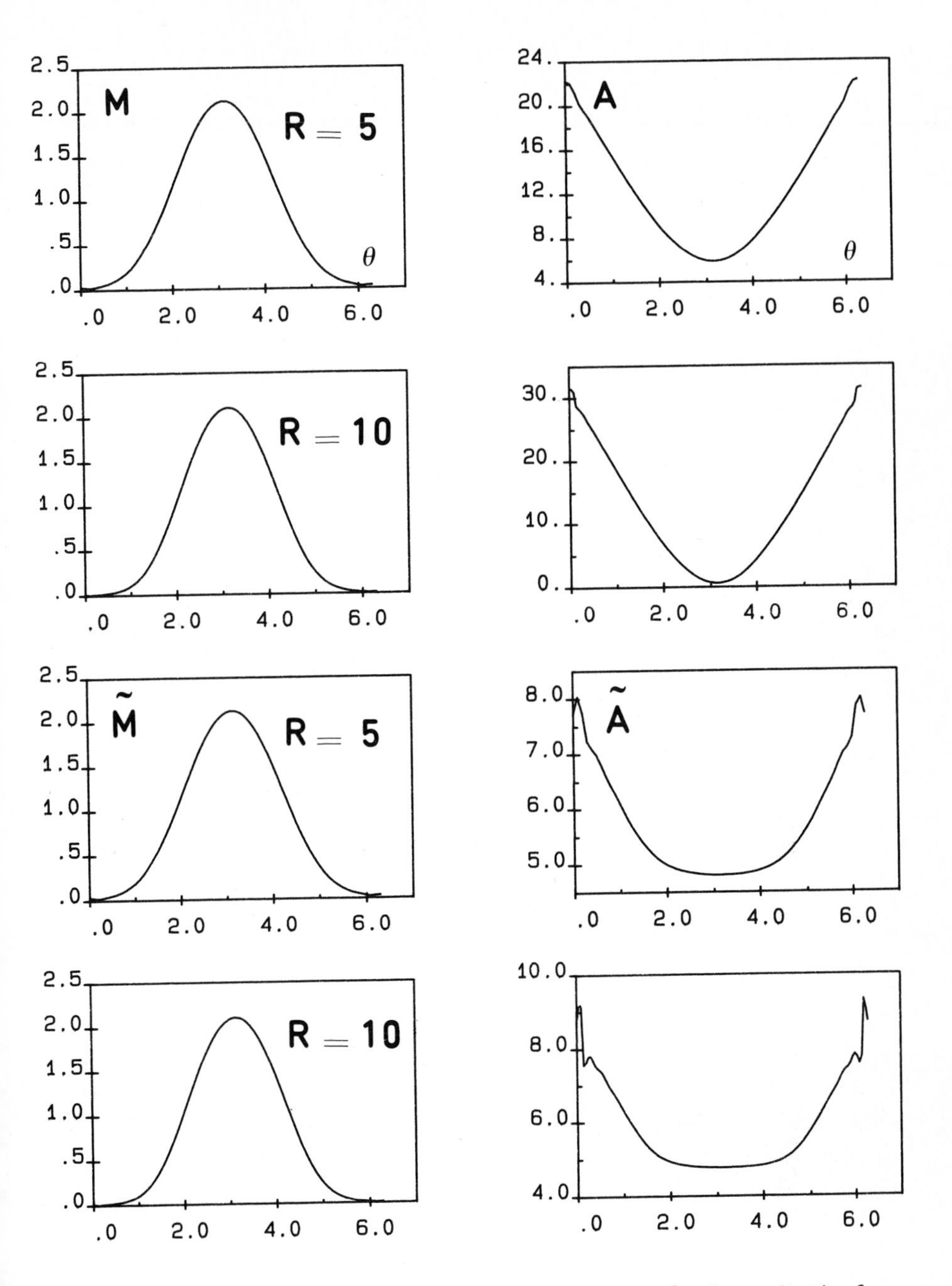

Figure 2 : At the top, modulus $M(\theta)$ and argument $A(\theta)$ (in radius) of the unknown function $\psi(\theta)$ for circular cylinder of ratio $R = 5$ and $R = 10$ (with $\lambda = 6$) and, at the bottom, modulus $\tilde{M}(\theta)$ and argument $\tilde{A}(\theta)$ of the regularized unknown function $\tilde{\psi}(\theta)$ for the same radii.

The explanation of this variation of argument along the profile is very simple. Let us consider the simple case where the object is replaced by a perfectly plane mirror (figure 3). In that case, the incident field of unit amplitude,

$$E^i = \exp\,[ik(x \sin\varphi - y \cos\varphi)] \tag{8}$$

generates a specularly reflected diffracted field E^d

$$E^d = -\exp\,[ik(x \sin\varphi + y \cos\varphi)]\ , \tag{9}$$

in such a way that the total field vanishes on the x-axis. Thus, the normal derivative $\psi(x)$ of the total field is given by

$$\psi(x) = -\,2ik \cos(\varphi)\, \exp(ikx \sin\varphi)\ . \tag{10}$$

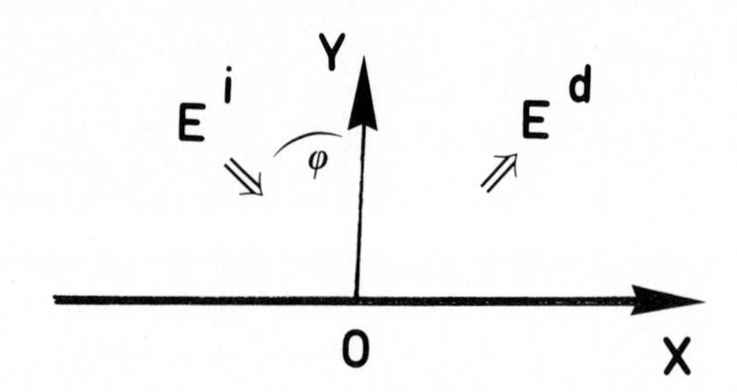

Figure 3 : Elementary explanation of the behaviour of the phase of ψ

If, for example, $\varphi = \pi/4$, one can see that a variation of x equal to 1 wavelength will result in a variation of the argument of $\psi(x)$ equal to $2\pi \sin(\varphi) = \pi\sqrt{2}$. As a consequence, an accurate representation of the phase will need about $\pi\sqrt{2}/(\pi/6) \simeq 8.5$ points per wavelength on the mirror. On the other hand, it is very important to notice that the ratio

$$\tilde{\psi}(x) = \psi(x)/E^i(x,0) = -\,2ik \cos\varphi \tag{11}$$

has a constant argument on the x axis. This is the basic idea of our "phase regularization method".

2.4. The phase regularization method

First, let us show that the conclusion drawn for a perfectly conducting plane holds for a scattering object. For such an object (fig.4), we define a "phase regularization function" $r(\theta)$ defined at a point M of plane angle θ by

$$r(\theta) = E^i \quad \text{in the illuminated region,} \tag{12}$$

$$r(\theta) = E^i(Q)\exp(ik\,\widehat{QM}) \quad \text{in the shadow region,} \tag{13}$$

where Q is the closest point of $\mathcal{S}$ in the illuminated region and $\widehat{QM}$ the distance between M and Q on the profile $\mathcal{S}$. The "phase regularized function" $\tilde{\psi}(\theta)$ is defined by

$$\tilde{\psi}(\theta) = \psi(\theta)/r(\theta) \; . \tag{14}$$

The choice of $r(\theta)$ in the shadow region is justified both by numerical experience and by intuitive considerations.

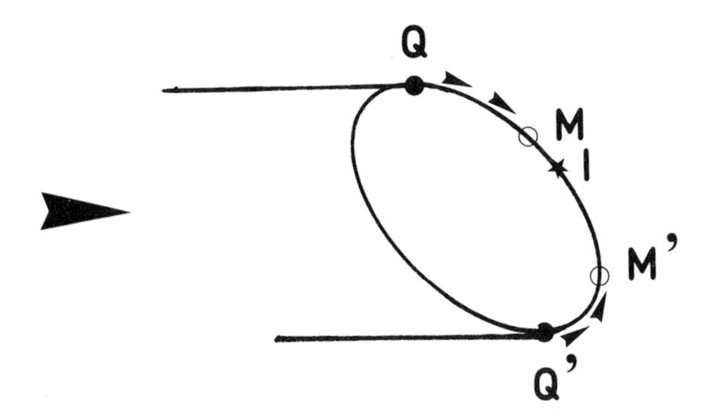

Figure 4 : Definition of the phase regularization function

Indeed, in the shadow region, the point M cannot receive directly the incident wave and so, the actual incident field comes from Q to M thanks to a creeping wave propagating along the profile. Assuming this creeping wave having a propagation constant equal to k leads to eq(13). It is worth noticing that considerations based on the Geometrical Theory of Diffraction could lead to more sophisticated phase regularization functions in the future.

It must be noticed that the definition of $r(\theta)$ leads to a discontinuity of this function. Indeed, if Q is the closest point of the illuminated region for the point M, Q' is the closest point for M' and, as M' and M tend to the point I of the shadow region equidistant to Q and Q', a discontinuity occurs. In fact, this is not an actual problem since the "phase regularization method" has an interest when the size of the object is large compared to the wavelength. Under these conditions, the function ψ at the point I is negligible and has no influence on the scattered field. Thus, the discontinuity of r does not play any role.

Using the phase regularization function and the phase regularized function, the integral equation (3) can be written in the form

$$\int_0^{2\pi} \tilde{N}(\theta,\theta') \, \tilde{\psi}(\theta') = \tilde{\Gamma}(\theta) \ , \tag{15}$$

where

$$\tilde{N}(\theta,\theta') = N(\theta,\theta') \times \frac{r(\theta)}{\Gamma(\theta)} \ , \tag{16}$$

$$\tilde{\Gamma}(\theta) = 1 \ . \tag{17}$$

In this new integral equation, not only the unknown function ψ but also the right hand member Γ have been regularized. Thus, it can be expected that $\tilde{\psi}$, as well as the two members of (15) can be projected on the Fourier basis with a small number of terms. By setting

$$\tilde{\psi}(\theta) = \sum_{n=-L}^{+L} \psi_n \exp(in\theta) \ , \tag{18}$$

and by projecting the two sides of (15) on the Fourier basis yields

$$\forall n \in [-L, +L], \qquad \sum_{m=-L}^{+L} T_{n,m} \, \psi_m = \delta_{n,0} \tag{19}$$

with

$$T_{n,m} = \frac{1}{2\pi} \int_0^{2\pi} \int^{2\pi} \tilde{N}(\theta,\theta') \exp(im\theta' - in\theta) \, d\theta' \, d\theta \ . \tag{20}$$

Finally, the size of the system is equal to 2L + 1. If L can be chosen very small, the size of the matrix to invert will be very small too. Furthermore, the calculation of the T matrix does not need the storage of the N matrix. On the other hand, the calculation of the double integrals for coefficients $T_{n,m}$ needs a discretization with a number of points which increases with the size of the object. In conclusion, it can be expected that the phase regularization method does not need the storage or the inversion of large size matrices, but the computation time will increase with the size of the object.

In order to check numerically the validity of the new method, we represent in figure 5, the modulus and the phase of the unknown function for a radius of 50, the wavelength remaining equal to 6. At the top, M and A have been obtained from the classical integral equation. At the middle, M' and A' have been calculated by solving the new integral equation using L = 5. At the bottom, we have chosen L = 2. As predicted, the modulus $\tilde{M}$ remains very close to M and the variation of the regularized arguments $\tilde{A}$ (about 10 rd) is much smaller than that of A (about 110 rd).

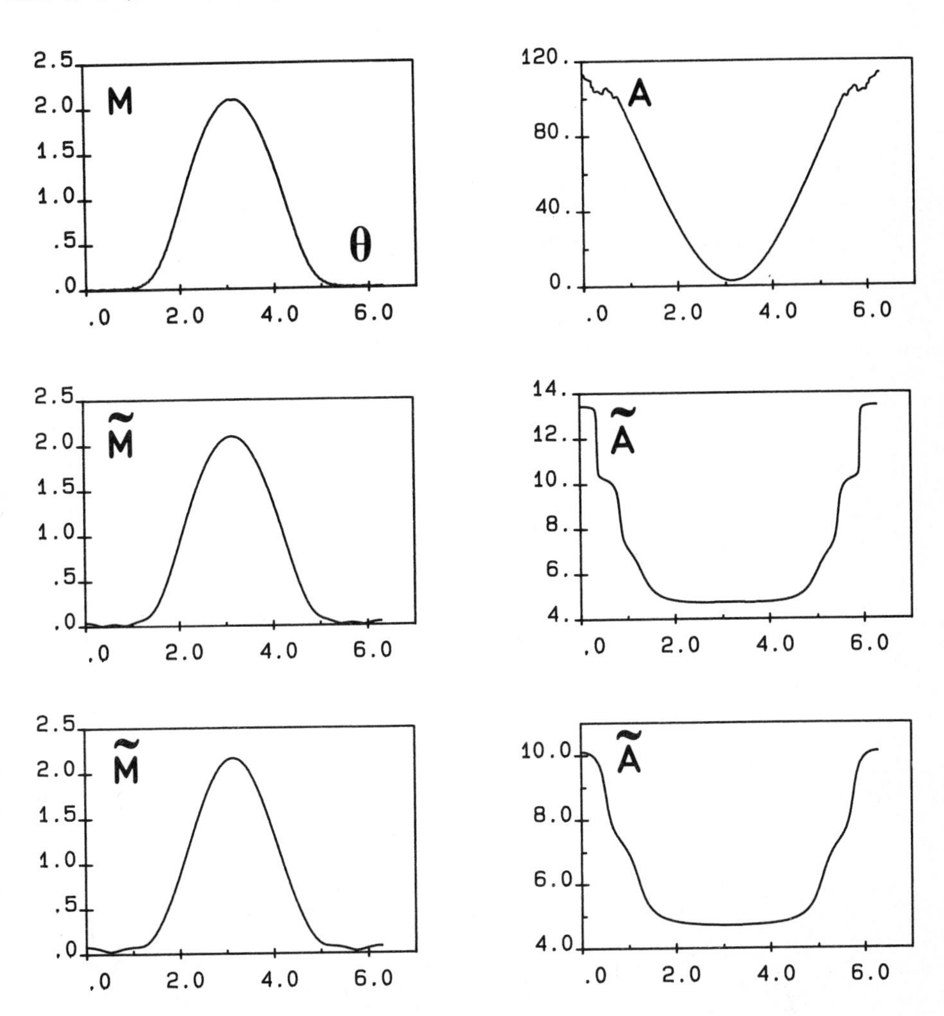

Figure 5 : Modulus and argument (in radians) of the unknown functions of the integral equations versus the angle of diffraction (in radians) for a circular cylinder of radius 50 (with λ = 6)

Figure 6 gives, from the top to the bottom, the differential cross sections corresponding to the three calculations of Figure 5. Obviously, the results are very close to each other, even for L = 2. More quantitatively, the difference between the classical method and L = 5 does not exceeds 1 % while the difference between the classical method and L = 2 never exceeds 5 %. This could be surprising if we compare the values of $\tilde{A}$ for L = 2 and L = 5. In fact, it must be noticed that the values of $\tilde{A}$ are significantly different only in the region where the modulus $\tilde{M}$ is very small, in such a way that this fact has not drastic consequences on the scattered field at infinity.

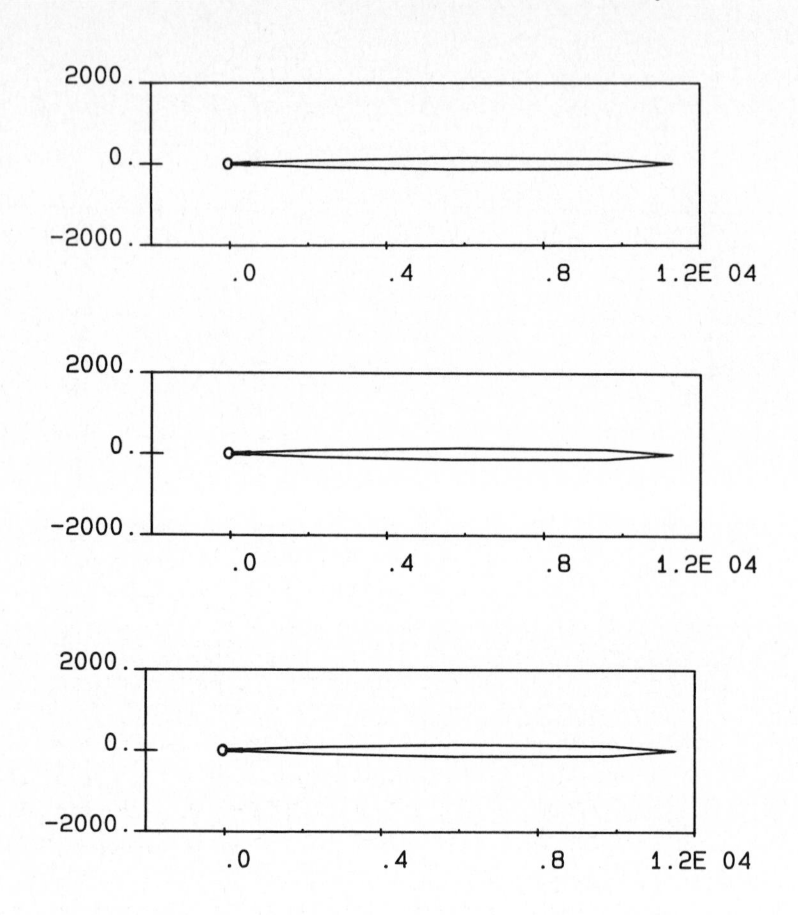

Figure 6 : Differential cross sections corresponding to the curves of figure 5.

It is quite astounding to remark that the computation by such a large object was performed with a relative precision better than 5 % by inverting a matrix of size 5! Let us notice also that the scattered intensity has a very thin peak in the direction of the incident wave.

2.5. Limits of the phase regularization method with projection on the Fourier basis.

For a cylindrical object like that represented in figure 4, it can be expected that the phase regularization method, as presented in the last section, has practically no limit, at least if we accept large computation times. On the other hand, it can be published, without any computation, that this method have no interest as soon as multiscattering phenomena occur (Figure 7). Indeed, in that case, obviously the phase of $\psi(\theta)$ in the multiscattering region (heavy line) will not follow the argument of the incident field but will be influenced by all the secondary waves reflected by the object.

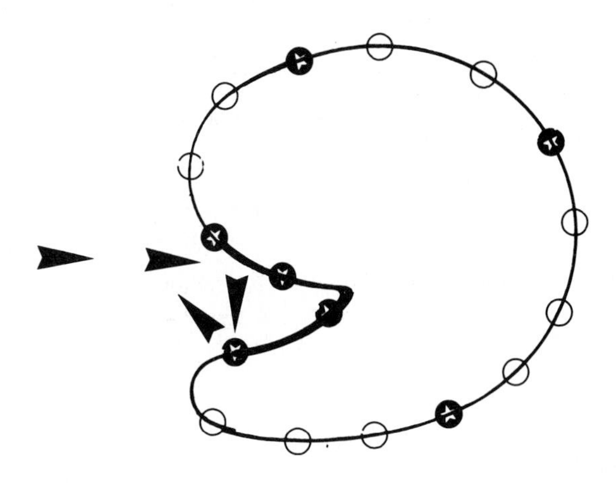

Figure 7 : Phenomenon of multiscattering

Another shortcoming of the projection on the Fourier basis is that the generalization to 3D scattering problems is not obvious.

2.6. The phase regularization method with double discretization

This method has not been numerically implemented yet and thus, we shall not describe it in detail. It uses the same regularized integral equation (20). The integral will be achieved using a classical discretization with a first set $\mathcal{P}$ of P points (circles) using the "rule of $\lambda/10$" but the unknown function $\tilde{\psi}$ will be described by its value in a subset $\tilde{\mathcal{P}}$ of $\mathcal{P}$ containing $\tilde{P}$ points. The points of $\tilde{\mathcal{P}}$ (stars) identify with those of $\mathcal{P}$ in the region when the phase regularization cannot be achieved (heavy line in figure 7) while in the region where multiscattering phenomena do not occur, these points can be placed at very large distances from each other since $\tilde{\psi}$ does not vary rapidly. However, the integration needs the values of ψ at any point of $\mathcal{P}$, a fact which apparently makes problem in the region where $\mathcal{P}$ and $\tilde{\mathcal{P}}$ differ. In fact, since $\tilde{\psi}$ is slowly varying in this region, its value at any point of $\mathcal{P}$ can be easily deduced from its values at the points of $\tilde{\mathcal{P}}$ placed on each side from an interpollation (the linear interpollation will use the two points of $\tilde{\mathcal{P}}$ on each side only). Finally, the left hand side of (15) can be calculated numerically as a linear function of the values of $\tilde{\psi}$ on $\tilde{\mathcal{P}}$. Writing this equation at the points of $\tilde{\mathcal{P}}$ will lead to a linear system of $\tilde{P}$ equations with $\tilde{P}$ unknown.

This method seems to be better than the projection on the Fourier basis since it is more general and also it can be generalized to 3D problems.

3. SCATTERING BY ECHELETTE GRATINGS

This kind of grating, very important in particular for astronomical observations, has a groove spacing very large compared to the wavelength. Typically, an echelle grating used in the visible or near infrared region has 30 or 50 grooves/mm, in such a way that more than 100 orders are diffracted. The rigorous calculation of the efficiency of such gratings can lead to the inversion of matrices having size greater than 1000, in such a way that, in general, it is impossible to achieve the precise computation.

The phase regularization method has been applied to this problem. It is not the purpose of this paper to give the first results of this study. They will be described in a next paper, and compared with those obtained from the physical optics approximation. Nevertheless, we can indicate that, using the Fourier projection method, we have been able to solve precisely this problem, for instance with 30 grooves/mm gratings used in the visible region. Unfortunately, the Fourier projection method leads to high computation times and the double discretization will be implemented in the next months.

4. SCATTERING BY RANDOM ROUGH SURFACES

4.1. Origine of the difficulty

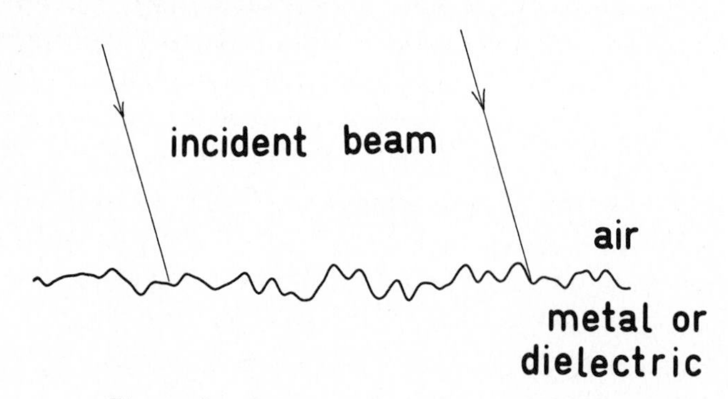

Figure 8 : Presentation of the problem

Figure 8 shows a random rough surface illuminated by a light beam. Under these conditions, the surface generates a speckle. The direct problem is to find the scattered field when the shape of the surface is given. On the other hand, the inverse problem consists in computing the deterministic or statistical features of the random rough surface from the speckle. Due to the tremendious difficulty of these problems, it is almost impossible at the present time to solve the inverse statistical problem, even in the simplest case of the two-dimensions problem.

In the direct problem, one of the main difficulties is the following : the incident beam must illuminate a large number of asperities, at least if a good investigation of the statistical properties of the speckle is needed. In our laboratory, this problem has been solved using a discretization method for solving our integral equation [3,4]. As in the problem of the calculation of the differential cross-section described in section 1, an accurate calculation of the speckle requires a discretization with about 10 points/wavelength on the surface profile. Assuming that the mean width of the asperities (correlation length) has the same order of magnitude as the wavelength, we are led to the conclusion that about 10 A points or more are necessary (thus a linear system of size greater than 10 A must be inverted), A denoting the number of illuminated asperities. In fact, this conclusion is too optimistic since the integral equation must be solved in an interval larger than the width of the illuminated region. Indeed, the unknown of the integral equation (for instance the surface current density when the material is considered as a perfectly conducting one), does not vanish immediately outside the illuminated region. So, obviously we are led to the inversion of matrices of large size and, when the width of the beam is increased, it becomes impossible to solve the problem. Unfortunately, it seems that conversely to what happens for the calculation of the cross section achieved in section 1, it does not exist a phase regularization function for the unknown of the integral equation, due to the rapid variation of the slope of the surface on a wavelength. This shows that this problem is more complicated that those we were dealing with in the first two sections.

4.2. The "Beam Simulation Method"

In order to overcome this difficulty, we have used a beam simulation method. This method is described in reference 5. Briefly, the basic idea is that a large beam can be rigorously represented as the sum of narrow gaussian beams. Since each of these gaussian beams illuminates a very small number of asperities, the elementary problem of scattering of each narrow beam can be solved precisely. It suffices to add the contribution of all the narrow beams to obtain the field scattered by the wide beam.

We must emphasize the fact that in our calculations, we actually deal with a random rough surface. Another approach, developped for instance by Nieto-Vesperinas et al [6,7], consists in modelizing a random rough surface by a grating having a groove spacing containing a large number of randomly computed asperities.

Even though Nieto-Vesperinas was able to obtain valuable results from this representation, it seems better to use our representation for a good description of the speckle since we are able to compute a continuous representation of the scattered field at infinity.

We are able to deal with perfectly conducting, metallic or dielectric materials, even though considerable numerical difficulties arise for highly conducting metallic surfaces (Aluminium or Silver surfaces for example) in the visible region and for TM incident polarization, due to the propagation of plasmon surface waves at very long distances from the illuminated region. The integral theory we have used is a generalization of the integral theory of gratings developped in our Laboratory [8]. It leads to a single integral equation, even for metallic or dielectric materials. The generalization of the theory to random rough surfaces is described in [9]. It is to be noticed that all our computations deal with gaussian correlation functions.

4.3. Inadequacy of the perfectly conducting model for metals in the visible domain.

Figure 9 shows the profile of a random rough surface (top) illuminated by a narrow beam of wavelength $\lambda = 0.4$ μm in normal incidence. The middle and the bottom of the figure show the intensity I of the speckle obtained for TE (middle) and TM (bottom) polarization, versus the diffraction angle θ'. The solid line is obtained by assuming a perfect conductivity of the metal while the dashed line is obtained by giving to the material the complex index of silver at 0.4 μm. Obviously, the perfectly conducting model is satisfactory for TE polarization, even though numerical calculations have shown that this conclusion does not hold for very high asperities. On the other hand, the perfect conducting model completely fails for TM polarization, a conclusion already observed for gratings [10].

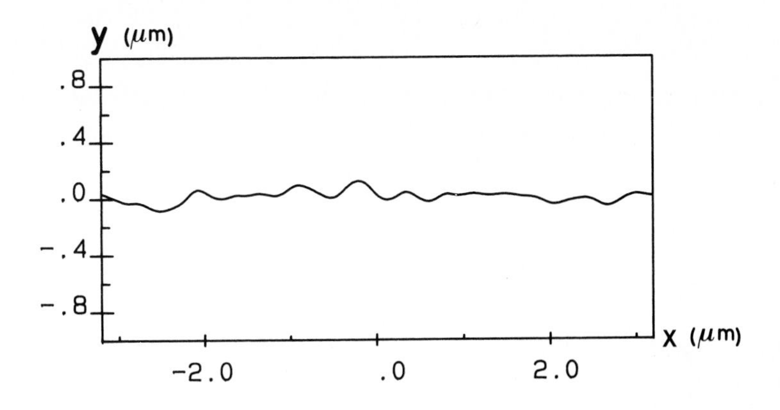

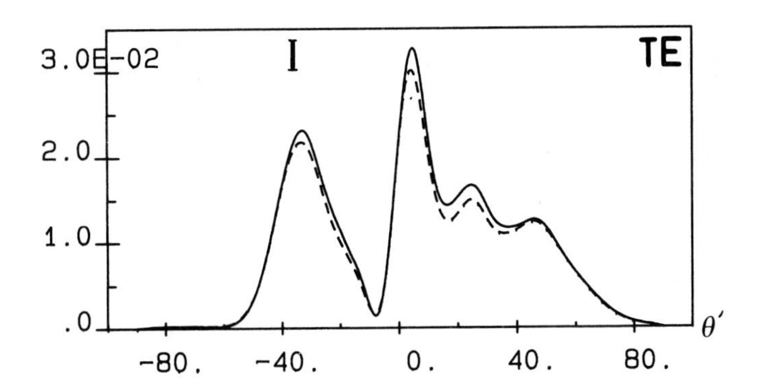

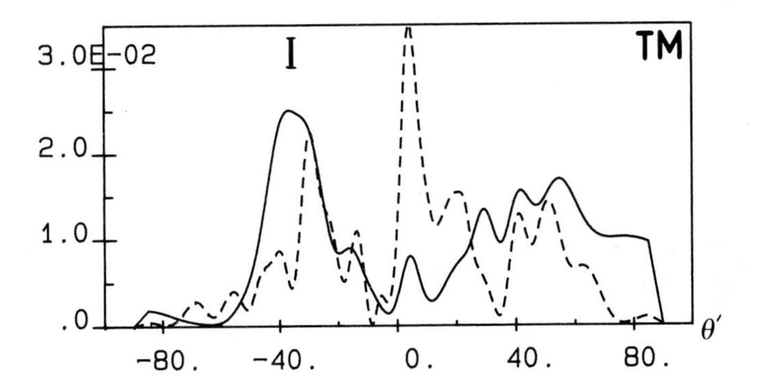

Figure 9 : Inadequacy of the perfectly conducting model in the visible region for metallic random rough surfaces.
Solid line : infinite conductivity.
Dashed line : finite conductivity.

4.4. Absorption phenomena

Figure 10 shows the part of incident energy absorbed by Joule effect by a random rough surface illuminated with a TM gaussian beam of wavelength $\lambda = 0.4$ μm in normal incidence, versus the RMS of the surface, for various correlation lengths. The main conclusion is that for a r.m.s. $\sigma < 0.06$ μm, the absorbed energy is the same whatever the correlation distance T, while for $\sigma > 0.06$ μm the maximum of absorption is obtained for T = 0.23 μm. At T = 0.23 μm and $\sigma = 0.075$ μm, we obtain a significative absorption of 22 %. The study of greater RMS requires more matching points and longer computation times, when each calculation of fig.10 needs about an hour on a CRAY2 computer!.

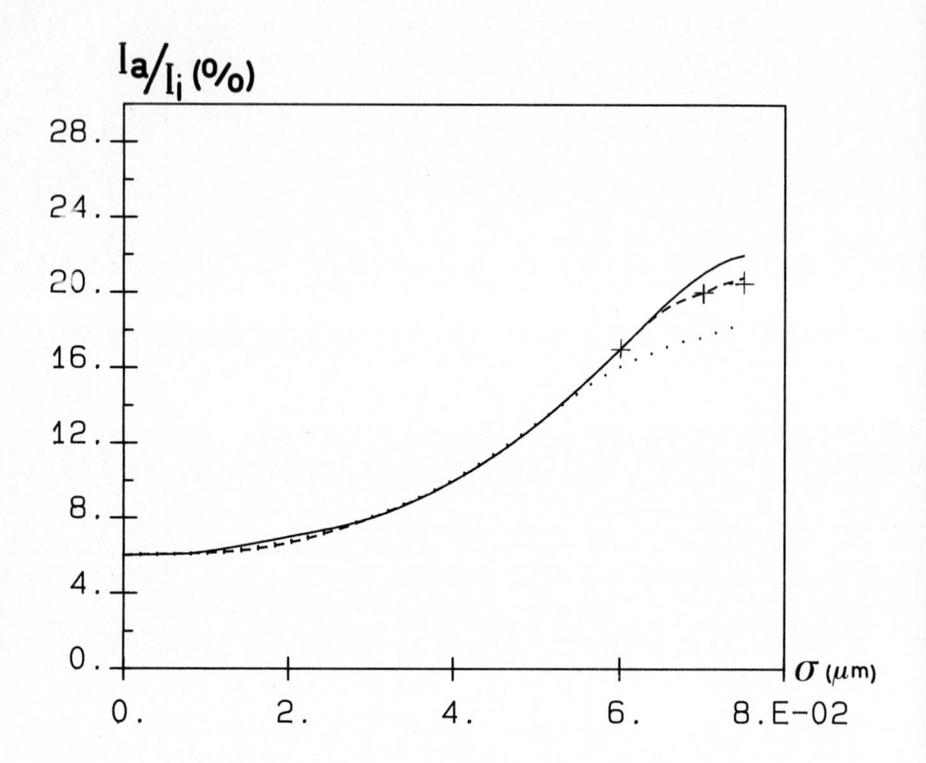

Figure 10 : Absorption by silver random rough surfaces at $\lambda = 0.4$ μm

crosses : T = 0.22 μm

solid line : T = 0.23 μm

dashed line : T = 0.24 μm

dotted line : T = 0.25 μm

4.5. Enhanced backscattering

This phenomenon has been observed experimentally by O. Donnell and Mendez [11] and, for perfectly conducting surfaces, numerically by Nieto-Vesperinas et al [7]. Figure 11 shows the mean intensity <I> of the speckle generated by an aluminum random rough surface of correlation length $T = \lambda = 0.65$ μm and RMS $\sigma = \lambda = 0.65$ μm illuminated under the incidence $\theta = 40°$ by a TE polarized gaussian beam of width 9 μm . The mean intensity has been obtained by averaging the speckles of 10 samples. The diffraction angle $\theta' = -40°$ (corresponding to the same direction as the incident wave) is particularly developped, even though a better description of the phenomenon would require an averaging process on more samples.

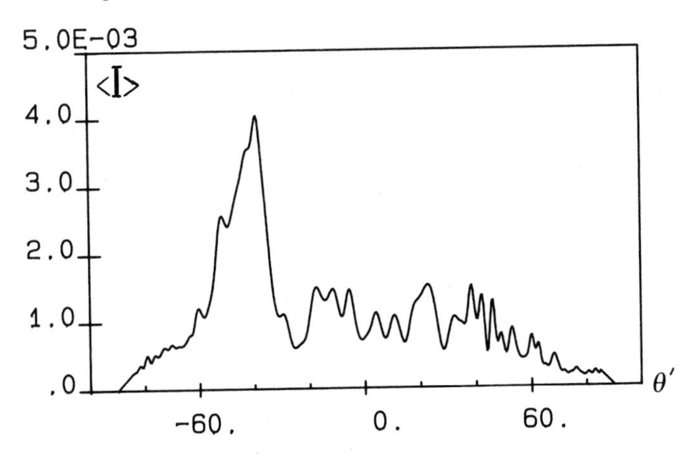

Figure 11 : Enhanced backscattering phenomenon by an aluminum random rough surface

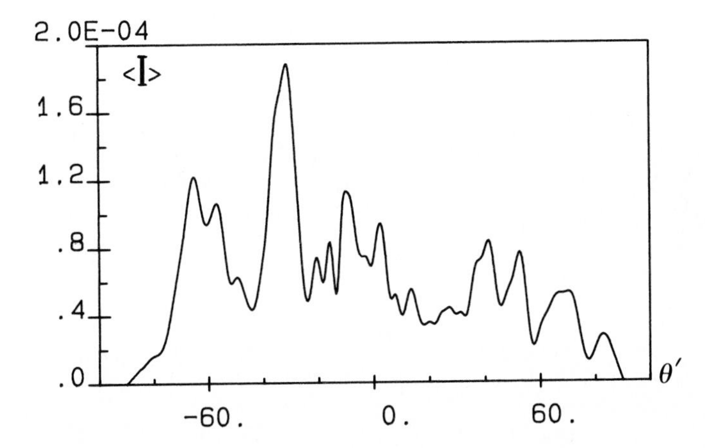

Figure 12 : Enhanced backscattering phenomenon by a dielectric random rough surface

Figure 12 shows the same curve, obtained by replacing the metal by a lossless dielectric medium of index $\nu = 1.5$. For this kind of structures, the energy balance criterion is observed with a relative precision of 0.5 %. Even though we can guess the enhanced backscattering, it is to be noticed that this phenomenon is not so obvious as in figure 11.

5. CONCLUSION

It has been shown that the rigorous integral theory can be used to solve precisely problems of scattering from large size objects. For radar cross-sections and echelle gratings, the phase regularization method provides a very simple and efficient tool, provided high computation times can be used. Even though the problem of scattering from random rough surfaces illuminated by beams of arbitrary width is much more difficult to handle, the beam simulation method is able to provide a precise tool for computations.

The first calculations on absorption and enhanced backscattering have been given. Further investigations in our laboratory will show in detail the influence of the statistical parameters of the surface and of the material on these phenomena.

REFERENCES

1) J. Van Bladel, Electromagnetic fields (Mc Graw Hill Books, New-York, 1964).
2) D. Maystre and P. Vincent, Optics Commun. 5 (1972) 327.
3) D. Maystre, IEEE Trans. Ant. Propag., Vol. AP-31, n° 6, Nov. 1983.
4) D. Maystre and J.P. Rossi, J. Opt. Soc. Am. A, Vol. 3, p. 1276, August 1986.
5) M. Saillard and D. Maystre, J. of Modern Optics, submitted.
6) M. Nieto-Vesperinas and J.M. Soto-Crespo, Optics Letters, Vol. 12, p. 979, Dec. 1987.
7) J.M. Soto-Crespo and M. Nieto-Vesperinas, J. Opt. Soc. Am. A, Vol. 6, n° 3, p. 367, march 1989.
8) D. Maystre, Progress in Optics, Vol. 21, p. 1, 1984.
9) M. Saillard and D. Maystre, J. Opt. Soc. Am., submitted.
10) D. Maystre and R. Petit, Topics in Current Physics, Vol. 22, p. 159, 1980.
11) K.A. O'Donnell and E.R. Mendez, J. Opt. Soc. Am. A, Vol. 4, n° 7, p. 1194, July 1987.

attering in Volumes and Surfaces
. Nieto-Vesperinas and J.C. Dainty (Editors)
Elsevier Science Publishers B.V. (North-Holland), 1990

The Method of Smoothing Applied To Random Volume and Surface Scattering

Gary S. Brown

Bradley Department of Electrical Engineering Virginia Polytechnic Institute and State University Blacksburg, Virginia 24061—0111 USA

ɪtroduction

The method of smoothing is a particular type of projection technique which was ɘveloped primarily for use in nonlinear and random problems; see[1] for a review of ːs history. For the random application, the unknown quantity is split into a sum ꞉ a mean or average value and a zero mean fluctuating term. If the unknown ɪantity satisfies an integral equation of the second kind then integral equations f the second kind for its mean and fluctuating parts can also be obtained. Quite requently, the kernels of these integral equations are very complicated; however, ɪere is usually some asymptotic set of parameters for which the kernels simplify ɪd an approximate solution can be obtained. Sometimes, the method of smoothing is ɪe only technique which can yield such results. The method of smoothing has also een called a modern approach[2] because it introduces averages to simplify the roblem at the start of the analysis rather than "solving" the problem as much as ossible and then averaging; this latter methodology is referred to as the lassical approach[2].

Recent uses of smoothing in random wave propagation and scattering comprise eller's work with the average scalar fields propagating in a discrete random edia[3], Lang's extension of this analysis to the electromagnetic case for an nfinite medium[4] and a half space[5], Watson and Keller's analysis of the scalar urface scattering problem for small heights and slopes[6], and Brown's formal levelopment for the electromagnetic surface scattering problem[7]. With the ɘxception of Watson and Keller's work, these efforts have been primarily concerned ɪith the determination of the average propagating field (in the volume scattering ɔroblem) or the average scattered field (in the surface scattering problem).

In this chapter, the focus will be shifted back to the fluctuating field for ɔoth the volume scattering problem and the rough surface problem. The purpose of ːhis work is to show that smoothing is capable of yielding insight into these fields also. For example, in the case of volume scattering it will be shown that smoothing leads to a result which very clearly contradicts a prediction which comes from an application of the distorted Born approximation. Furthermore, th smoothing result has an easily understood physical interpretation. In the case o surface scattering, it will be shown that smoothing is capable of both yielding new result for small surface height perturbations and, almost as important providing clear bounds on when this result is valid.

Volume Scattering

The material contained in this section is a compaction of a much large discussion presented in[8] and the interested reader should consult this referenc or more details. The total field $\vec{E}_t(\vec{r})$ propagating in a medium comprising collection of N discrete scatterers is given by

$$\vec{E}_t(\vec{r}) = \vec{E}_i(\vec{r}) + L\,\bar{\bar{K}}_\Sigma\,\vec{E}_t(\vec{r}_n + \vec{r}_o) \tag{1}$$

where $\vec{E}_i(\vec{r})$ is the incident field (in the absence of the scatterers) at the point and $\bar{\bar{K}}_\Sigma$ is the dyadic operator given by

$$\bar{\bar{K}}_\Sigma = -k_o^2 \sum_{n=1}^{N} (\epsilon_{rn}-1)\, S_n(\vec{r}_o;\, \Omega_n)\, \bar{\bar{\Gamma}}\, (\vec{r} - \vec{r}_n - r_o) \tag{2}$$

and L is the volume integral operator given by

$$L = \int_{\substack{\text{all}\\ \text{space}}} (\,\cdot\,)\, dv_n \tag{3}$$

In the above, $k_o = 2\pi/\lambda_o$ is the wavenumber for free space, ϵ_{rn} is the relativ dielectric constant of the nth scatterer and $S_n(\vec{r}_o;\, \Omega_n)$ is a support function fo the nth scatterer, e.g.

$$S_n(\vec{r}_o, \Omega_n) = \begin{matrix} 1 & \vec{r}_o \in V_n \\ 0 & \vec{r}_o \notin V_n \end{matrix}$$

which depends on the volume of the nth scatterer V_n and its orientation Ω_n. The dyadic part of $\bar{\bar{K}}_\Sigma$ is due to the dyadic Green's function $\bar{\bar{\Gamma}}$ which depends on the way one excludes the singular point $\vec{r}_o = \vec{r}_n + \vec{r}$ in volume integration in (1). For a ıall spherical exclusion volume $\bar{\bar{\Gamma}}$ becomes[9]

$$\bar{\bar{\Gamma}} = -\text{P.V.}\left[\bar{\bar{I}} + k_o^{-2}\nabla_o\nabla_o\right] g(\vec{r}-\vec{r}_n-\vec{r}_o) + \bar{\bar{I}}\delta(\vec{r}-\vec{r}_n-\vec{r}_o)/3k_o^2 \tag{4}$$

ıere

$$g(\vec{r}-\vec{r}_n-\vec{r}_o) = \frac{\exp\left[-jk\ |\vec{r}-\vec{r}_o-\vec{r}_n|\right]}{4\pi|\vec{r}-\vec{r}_n-\vec{r}_o|} \tag{5}$$

ıd P.V. denotes Principal Value. What (1) formalizes is how the fields inside ach of the scatterers are scattered to the point $\vec{r}$ were they add with the incident ield to form the total field at $\vec{r}$.

All of the potentially random quantities are contained in (2); the positions of he scatterers $\left[\vec{r}_n\right]$, their orientations $\left[\Omega_n\right]$, their volumes $\left[V_n\right]$, and their elative dielectric constants $\left[\epsilon_{rn}\right]$. Within the framework of smoothing, the ields of interest are usually the average field in the medium $<\vec{E}_t>$ and the luctuating field $\delta\vec{E}_t = \vec{E}_t - <\vec{E}_t>$ or one of its statistical moments. An pplication of the method of smoothing proceeds as follows. Since

$$\vec{E}_t = <\vec{E}_t> + \delta\vec{E}_t \tag{6}$$

1) can be written as follows;

$$<\vec{E}_t> + \delta\vec{E}_t = \vec{E}_i + L<\bar{\bar{K}}_\Sigma>(<\vec{E}_t> + \delta\vec{E}_t) + L\,\delta\bar{\bar{K}}_\Sigma(<\vec{E}_t> + \delta\vec{E}_t) \tag{7}$$

aking the fluctuating part of this equation yields

$$\delta\vec{E}_t = L<\bar{\bar{K}}_\Sigma>\delta\vec{E}_t + L\delta\bar{\bar{K}}_\Sigma<\vec{E}_t> + (1-P)\,L\delta\bar{\bar{K}}_\Sigma\,\delta\vec{E}_t \tag{8}$$

here $P(\cdot) = <\cdot>$ operates on everything to its right. Usually, but not necessarily lways, the average propagator $<\bar{\bar{K}}_\Sigma>$ is small so (8) can be simplified to

$$\delta\vec{E}_t = L\delta\bar{\bar{K}}_\Sigma<\vec{E}_t> + (1-P)\,L\,\delta\bar{\bar{K}}_\Sigma\,\delta\vec{E}_t \tag{9}$$

he equation for $<\vec{E}_t>$, involving $\delta\vec{E}_t$, can be obtained by averaging (7) and further formal manipulations can use this equation and (9) to obtain decoupled results fc both $\langle\vec{E}_t\rangle$ and $\delta\vec{E}_t$[7]. For the purposes of this chapter, (9) is sufficient.

Equation (9) clearly shows that the "source" for the scattered field is th term $L\ \delta\bar{\bar{K}}_\Sigma\langle\vec{E}_t\rangle$. Since $\delta\bar{\bar{K}}_\Sigma \approx \bar{\bar{K}}_\Sigma$, this result shows that the source of th fluctuating field is the average field in each scatterer which, in turn, i scattered to the point $\vec{r}$ by the propagator $\delta\bar{\bar{K}}_\Sigma$. An obvious approximate solution o (9) is

$$\delta\vec{E}_t \approx L\delta\bar{\bar{K}}_\Sigma \langle\vec{E}_t\rangle \qquad (10$$

It is important to note that, *insofar as the fluctuating field is concerned, th approximation in (10) is a single scattering result*. That is, it involves n interaction between the fluctuating field and the propagator as in the integra term in (9). It is tempting to say that an approximation such as (10) does includ some degree of multiple scattering because of the dependence on $\langle\vec{E}_t\rangle$ and *it* subsequent inclusion of multiple scattering. However, this interpretation i incorrect because it fails to recognize that multiple scattering *relative to* $\delta\vec{E}$ implies some degree of interaction between $\delta\vec{E}_t$ and the propagator such as in th second term on the right side of (9).

Another interesting aspect of (9) is the implied behavior as $\langle\vec{E}_t\rangle$ becomes ver small due either to the point of observation being deep in the medium containin the scatterers or every strong scatterers. In either case, the average field i small but the fluctuating field, or more precisely the incoherent scattered power may be significant especially in the forward scattering direction. What is th source of this fluctuating field if the source term in (9) is very small? Th answer may be found by solving (9) by iteration to write

$$\delta\vec{E}_t = \sum_{n=o}^{\infty} \left[(1-P)\, L\delta\bar{\bar{K}}_\Sigma\right]^n L\delta\bar{\bar{K}}_\Sigma\langle\vec{E}_t\rangle \qquad (11$$

which can be rewritten as follows;

$$\delta\vec{E}_t = \sum_{n=o}^{N} \left[(1-P)\, L\delta\bar{\bar{K}}_\Sigma\right]^n L\delta\bar{\bar{K}}_\Sigma \langle\vec{E}_t\rangle$$

$$+ \left[(1-P)\, L\delta\bar{\bar{K}}_\Sigma\right]^{N+1} \sum_{n=o}^{\infty} \left[(1-P)\, L\delta\bar{\bar{K}}_\Sigma\right]^n L\delta\bar{\bar{K}}_\Sigma\langle\vec{E}_t\rangle \qquad (12)$$

r

$$\delta\vec{E}_t = \delta\vec{S}_N + \left[(1-P)\, L\delta\bar{\bar{K}}_\Sigma\right]^{N+1} \delta\vec{E}_t \qquad (13)$$

here $\delta\vec{S}_N$ is the sum of N + 1 iterations in (12). Eqn. (13) shows that the "source" f fluctuating field can be considered to be the N scatterings of the $L\delta\bar{\bar{K}}_\Sigma\langle\vec{E}_t\rangle$ ield. Thus, even though $\langle\vec{E}_t\rangle$ may be small, there is nothing to say that $\delta\vec{S}_N$ must lso be small. In fact, this is exactly the case where multiple scattering in the edium becomes most important; the multiple scattering (N times) is represented by he partial sum $\delta\vec{S}_N$.

Before leaving the volume scattering situation it is interesting to compare 10) with the so called distorted wave Born approximation (DWBA)[4,5,10]. In this pproximation, the scatterers are assumed to be imbedded in a medium having the ame constitutive parameters as the effective medium. (The effective medium has onstitutive parameters which are derived from the propagation constant of the verage field.) These scatterers are assumed to be illuminated by the average or ean field in the medium. Ignoring, any reflections at the surface of the catterers gives the total field inside the scatterers as equal to the total utside field . Since each scatterer is assumed to be buried in the effective edium, it must also scatter into the effective medium. Thus, the DWBA for $\delta\vec{E}_t$ is s follows;

$$\left[\delta\vec{E}_t\right]_B = L\delta\bar{\bar{K}}_\Sigma(k)\,\langle\vec{E}_t\rangle \qquad (14)$$

here $\delta\bar{\bar{K}}_\Sigma(k)$ is given by (2) with k_o replaced by k which, in turn, is equal to the avenumber for the average field, i.e.

$$\langle\vec{E}_t\rangle = E_o\hat{e}_i \exp(-jkr\,\hat{k}_i\cdot\hat{r}) \qquad (15)$$

nd $\hat{k}_i$ is the direction of propagation of $\langle\vec{E}_t\rangle$. Comparing (10) with the DWBA in 14) shows that the latter has double counted for the average medium. That is, moothing says that the average field is scattered back into free space while the WBA scatterers it back into the effective or average medium. A little thought ill show that the smoothing result is correct. Scattering is the only process hereby the coherent or average field can be converted into the fluctuating field. hus, after one scattering the coherent field that is scattered becomes a luctuating field and, because this fluctuating field is only scattered once in 10), it must propagate in free space. The average or effective medium has meaning nly in so far as the average field is concerned; it has no relevance to the luctuating field. This observation is further confirmed by (13) in which even

after N scatterings, the field $\delta\vec{S}_N$ also scatters into free space and not into t
average medium.

It should be noted that the use of the DWBA can lead to a rather significa
underestimation of the fluctuating scattered field because of the double counti
for the average medium. Finally, the results from smoothing are very convenie
for estimating when the random volume situation can be modeled as an effecti
homogeneous medium having the constitutive parameters from the wavenumber for t
average field. That is, one compares the power carried by the average field wi
the mean power carried by the fluctuating field to determine when the latter can
ignored; when this is the case, the random medium can be replaced by an effecti
homogeneous medium.

Surface Scattering

For the surface scattering application, the problem of scattering of
incident magnetic field $\vec{H}_i$ by a perfectly conducting interface defined by z
$\zeta(x,y)$ will be considered. For $z > \zeta$ there is free space while for $z < \zeta$ the medi
is perfectly conducting. The incident magnetic field induces a current $\vec{J}_s$ on t
surface which reradiates into free space to (a) yield the correct scattered fie
above $z = \zeta(x,y)$ and (b) to cancel the incident field beneath the interface.

The current obeys the magnetic field integral equation (MFIE) given by

$$\vec{J}_s(\vec{r}) = \vec{J}_s^i(\vec{r}) + 2\hat{n}(\vec{r})\times \int \vec{J}_s(\vec{r}_o) \times \nabla_o g\left[|\vec{r}-\vec{r}_o|\right] ds. \qquad (1$$

where

$$\vec{J}_s^i = 2\hat{n} \times \vec{H}_i$$

$\hat{n}$ is the unit normal to the rough surface, and g is the free space scalar Green
function

$$g(|\vec{r}-\vec{r}_o|) = \exp\left[-jk_o|\vec{r}-\vec{r}_o|\right]/4\pi|\vec{r}-\vec{r}_o|$$

Equation (16) can be manipulated into the following matrix equation for J_x and
(see[7] for details);

$$\bar{J} = \bar{J}_o + L\bar{\bar{G}}\bar{J} \qquad (17$$

where

$$\bar{J} = \begin{bmatrix} J_x \\ J_y \end{bmatrix} \qquad \bar{J}_o = \begin{bmatrix} J_x^{\,i} \\ J_y^{\,i} \end{bmatrix}$$

$$\begin{bmatrix} J_x \\ J_y \end{bmatrix} = \begin{bmatrix} J_{sx} \\ J_{sy} \end{bmatrix} \exp(jk_{sz}\zeta) \sqrt{1 + (\nabla_t \zeta)^2}$$

$$\vec{k}_s = \vec{k}_{s_t} + k_{sz}\hat{z} = k_o \hat{k}_s$$

$$\vec{r}_t = x\hat{x} + y\hat{y}$$

$$\nabla_t \zeta = \frac{\partial \zeta}{\partial x}\hat{x} + \frac{\partial \zeta}{\partial y}\hat{y} = \zeta_x \hat{x} + \zeta_y \hat{y}$$

$$L = 2\int (\;)\, d\vec{r}_{t_o} = 2\int (\;)\, dx_o dy.$$

$$\bar{G} = \begin{bmatrix} G_{xx} & G_{xy} \\ G_{yx} & G_{yy} \end{bmatrix} \exp\left[jk_{sz}(\zeta - \zeta_o)\right]$$

$$G_{xx} = \zeta_{xo}\frac{\partial g}{\partial x} + \zeta_y \frac{\partial g}{\partial y}\frac{\partial g}{\partial \zeta} \qquad G_{xy} = -(\zeta_y - \zeta_{yo})\frac{\partial g}{\partial x}$$

$$G_{yy} = \zeta_x \frac{\partial g}{\partial x} + \zeta_{yo}\frac{\partial g}{\partial y} - \frac{\partial g}{\partial \zeta} \qquad G_{yx} = -(\zeta_x - \zeta_{xo})\frac{\partial g}{\partial y}$$

In addition to the use of vector and geometrical manipulations, the fact that $\hat{n} \cdot \vec{J}_s = 0$ was also employed; furthermore, (16) was multiplied by $\exp(jk_{sz}\zeta)$ because the Fourier transform of this product determines the far zone scattered field. That is, the far zone scattered field at the point $\vec{R} = R\hat{k}_s$ is given by

$$\vec{E}_s(\vec{R}) = -jk_o\eta_o g(R)\left[\tilde{\vec{J}} - (\hat{k}_s \cdot \tilde{\vec{J}})\,\hat{k}_s\right] \tag{18}$$

where

$$\tilde{\vec{J}} = \int \vec{J} \exp(j\vec{k}_{s_t} \cdot \vec{r}_t)\, d\vec{r}_t \tag{19}$$

and

$$\vec{J} = J_x\,\hat{x} + J_y\hat{y} + (\zeta_x J_x + \zeta_y J_y)\hat{z} \quad (20)$$

Thus, if (17) can be solved the Fourier transform of the z-component can be determined by

$$\mathfrak{J}_z = \mathfrak{J}_x \otimes \zeta_x + \mathfrak{J}_y \otimes \zeta_y \quad (21)$$

where $\otimes$ denotes convolution.

With the standard smoothing decomposition,

$$\overline{J} = \langle\overline{J}\rangle + \delta\overline{J}, \quad (22)$$

and proceeding exactly as from (7) to (8) yields

$$\delta\overline{J} = \delta\overline{J}_o + L\delta\overline{\overline{G}}\langle\overline{J}\rangle + L\langle\overline{\overline{G}}\rangle\delta\overline{J} + (1-P)L\,\delta\overline{\overline{G}}\delta\overline{J} \quad (23)$$

The next step in the process is to take the Fourier transform of (23) as in (19). If $F[\cdot]$ denotes the Fourier transform, the transformed terms in (23) become

$$F[\delta\overline{J}(\vec{r}_t)] = \delta\mathfrak{J}(\vec{k}_{s_t})$$

$$F[\delta\overline{J}_o(\vec{r}_t)] = \delta\mathfrak{J}_o(\vec{k}_{s_t})$$

$$F[L\delta\overline{\overline{G}}\,\langle\overline{J}\rangle] = \delta\tilde{\tilde{\overline{\overline{G}}}}(\vec{k}_{s_t}, -\vec{k}_{i_t})\overline{e}$$

$$F[L\langle\overline{\overline{G}}\rangle\delta\overline{J}] = \langle\tilde{\overline{\overline{G}}}(\vec{k}_{s_t})\rangle\delta\mathfrak{J}(\vec{k}_{s_t})$$

$$F[L\delta\overline{\overline{G}}\delta\overline{J}] = (2\pi)^{-2}\,L_{\vec{k}_t}\;\delta\tilde{\tilde{\overline{\overline{G}}}}(\vec{k}_{s_t}, -\vec{k}_t)\;\delta\mathfrak{J}(\vec{k}_t)$$

where the above results follow from the specularity of the average scattered field and the dependence of $\langle\overline{\overline{G}}\rangle$ on $\Delta\vec{r}_t = \vec{r}_t - \vec{r}_{t_o}$ only, i.e.

$$\langle\overline{J}\rangle = \overline{e}\exp(-j\vec{k}_{i_t}\cdot\vec{r}_t)$$

$$F[\langle\bar{\bar{G}}\rangle] = F[\langle\bar{\bar{G}}(\Delta\vec{r}_t)\rangle] = \langle\tilde{\bar{\bar{G}}}(k_{s_t})\rangle,$$

d $\vec{k}_{i_t}$ is the transverse (to z) part of the incident wave vector. Note that $\bar{e}$ is e complex vector amplitude of the average scattered field. Substituting the ove results into the transform of (23) yields

$$\delta\tilde{\bar{J}} = [1 - \langle\tilde{\bar{\bar{G}}}\rangle]^{-1} (\delta\tilde{\bar{J}}_o + \bar{e}\delta\tilde{\tilde{\bar{\bar{G}}}}) + [1 - \langle\tilde{\bar{\bar{G}}}\rangle]^{-1} \frac{(1-P)}{(2\pi)^2} L_{k_t} \delta\tilde{\tilde{\bar{\bar{G}}}}\, \delta\tilde{\bar{J}} \tag{24}$$

ich is an integral equation for the fluctuating parts of $\mathfrak{J}_x$ and $\mathfrak{J}_y$; the rresponding value for $\mathfrak{J}_z$ is obtained from the fluctuating part of (21), i.e.

$$\delta\mathfrak{J}_z = (1 - P)\zeta_x \otimes \delta\mathfrak{J}_x + (1 - P)\zeta_y \otimes \delta\mathfrak{J}_y \tag{25}$$

ere, as before, $P = \langle\cdot\rangle$ is the average operator.

The term $[1 - \langle\tilde{\bar{\bar{G}}}\rangle]^{-1}$ comes from summing all of the interactions on the surface tween the average propagator $\langle\bar{\bar{G}}\rangle$ and the fluctuating current $\delta\bar{J}$. Interestingly ough, this renormalization affects not only the effective source term in (24) t also the interaction or last term in (24). By carrying out an analysis of the erage scattered field, it is possible to provide the estimate that for small rface heights and slopes the renormalization term is probably not important. wever, when the surface slopes become large, it is possible that this term may come very significant.

The term $\delta\tilde{\bar{J}}_o$ is the standard Kirchhoff approximation and it has been augmented y the addition of the factor $\bar{e}\delta\tilde{\tilde{\bar{\bar{G}}}}$ to provide a new "source" or Born term. The irchhoff term treats the rough surface essentially like a randomly elevated and ilted planar surface. As such, it provides almost no accurate accounting for the orizontal surface structure. The new term $\bar{e}\delta\tilde{\tilde{\bar{\bar{G}}}}$ accounts for this horizontal urface structure to a degree and it is this degree that must be determined if the orn term in (24) is to be a useful approximation, i.e.

$$\delta\tilde{\bar{J}} \approx [1 - \langle\tilde{\bar{\bar{G}}}\rangle]^{-1} (\delta\tilde{\bar{J}}_o + \bar{e}\delta\tilde{\tilde{\bar{\bar{G}}}}) \tag{26}$$

If the average scattered field is small then $\bar{e}$ must also be small and (26) becomes essentially the Kirchhoff approximation. If the surface is such that $\bar{e}$ not negligible then $\bar{e}\delta\tilde{\bar{\bar{G}}}$ in (26) must be retained. The important question is und what surface conditions is (26) a valid approximation? To answer this question, is necessary to go back to (23) and rewrite it as follows;

$$\delta\bar{J} = \delta\bar{J}_o + L\langle\bar{\bar{G}}\rangle\delta\bar{J} + (1-P)L\delta\bar{\bar{G}}\,[\langle\bar{J}\rangle + \delta\bar{J}] \quad (2$$

where we have added the superfluous factor $(1-P)$ in front of $L\delta\bar{\bar{G}}\langle\bar{J}\rangle$. Now, order to obtain (26) from (27) one must satisfy

$$F[L\delta\tilde{\bar{\bar{G}}}\langle\bar{J}\rangle] > (1-P)F[L\delta\bar{\bar{G}}\delta\bar{J}]$$

or

$$\delta\tilde{\bar{\bar{G}}}(k_{s_t}, -\vec{k}_{i_t})\,\vec{e} > (1-P)\,L_{k_t}\,\delta\tilde{\bar{\bar{G}}}(\vec{k}_{s_t}, -\vec{k}_t)\,\delta\tilde{\bar{J}}(\vec{k}_t)/(2\pi)^2 \quad (2$$

The condition in (28) is difficult to understand in its general form, so it will considered under "worst case" conditions; that is, when is (28) most difficult satisfy? It turns out that (28) is most difficult to satisfy when the scattering from a <u>randomly</u> <u>elevated</u> <u>planar</u> <u>surface</u>. In this case, the fluctuating scatter field is given by

$$\delta\tilde{\bar{J}}(\vec{k}_t) = \{\exp[j(k_{sz} - k_{iz})\zeta] - |\vec{e}|\}\,\hat{e}\,\delta(\vec{k}_t + \vec{k}_{i_t})\,(2\pi)^2 \quad (2$$

where $\hat{e}$ is the polarization of the reflected field and ζ is now independent of $\vec{r}_t$ The normalization is such that the incident field has unit amplitude and $|\vec{e}|$ is th amplitude of the average reflected field. Substituting (29) into (28) yields

$$\vec{e} > \{\exp[-j2k_{iz}\zeta] - |\vec{e}|\}\,\hat{e} \quad (30$$

Squaring and averaging both sides of (30) given rise to the following inequality

$$|\vec{e}|^2 > 1 - |\vec{e}|^2$$

or

$$|\vec{e}|^2 > 0.5$$

ich says that the square of the magnitude of the normalized average scattered eld must be greater than 0.5. If the average scattered field drops below this lue, the fluctuating current can no longer be ignored in(27).

This represents a worst case because of the following reasoning. For a ndomly elevated planar surface, all of the incoherent or fluctuating field is opagating in the specular direction — the same direction as the average attered field. For a fixed mean square height, the fluctuating scattered field reads out in angle as horizontal structure is added to the surface but the total wer carried by it remains fixed because the average scattered field or power does t change. As the surface slopes, curvature, etc becomes large enough (but $\langle\zeta^2\rangle$ mains fixed), the average scattered field increases in amplitude and the uctuating power integrated over all angles must decrease. Continuing this ocess leads to all the power going back into the specularly scattered direction d being carried by the average scattered field. The fluctuating field carries no al power in this limit and acts simply to store energy in the vicinity of the rface. The important point of this discussion is that horizontal structure in e surface roughness will lead to a larger mean or average scattered field than is edicted by the undulating planar analysis[2,11]. Hence (31) represents a worst se situation because it is exactly this latter case.

The above discussion is important for another reason. If (31) is satisfied d, thus, (26) is a good approximation then it will continue to be a good proximation regardless of how large the surface slopes become. This is because the surface slopes become larger, (31) will be satisfied even more strongly and 26) will increase in accuracy. Thus, (26) is an approximation whose accuracy epends only on the surface height and not on the surface slopes. In fact, one can asily show that if the small surface slope approximation is introduced into (26) nat results is the classical Rice boundary perturbation result which has recently een obtained by Holliday[12] from a first iteration of (16) along with the small eight and slope approximations. To the author's knowledge, (26) is the first nalytical approximation which does not simultaneously require small height and lopes.

Clearly, much numerical work is needed to fully understand (26). However, ince it is not restricted to small surface height derivatives it should be capable f predicting the class of backscattering enhancement associated with large slopes nd small heights. It may also be capable of explaining enhancement from surfaces aving large roughness provided we can treat the surface by composite roughness echniques.

Summary

The purpose of this chapter is to illustrate the elegance and power of t method of smoothing when applied to the fluctuating part of the field scattered a volume distribution of discrete scatters and rough surfaces. In both cases was found that smoothing was capable of providing new results and added insight.

Smoothing shows that the distorted wave Born approximation overestimates t attenuation of the fluctuating scattered field because it assumes that a partic scatters into the average medium rather than free space. Furthermore, smoothi also shows the importance of multiple scattering when the point of observation deep in a random medium.

For the surface scattering problem, smoothing has led to a new analytic result for the fluctuating scattered field. This result is valid for small heig but arbitrary surface height derivatives. Consequently, it has the potential provide insight into enhanced backscattering from surfaces with large slopes.

References

1) Frisch, U., "Wave propagation in random media," in Probabilistic Metho in Applied Mathematics, A.T. Barucha-Reid, Ed., New York; Academ Press, Chapter 2, 1968.

2) DeSanto, J.A. and G.S. Brown, "Analytical techniques for multip scattering from rough surfaces," in Progress in Optics, Vol. XXIII, Wolf, Ed., New York; Elsevier, Chapter 1, 1986.

3) Keller, J.B., "Wave propagation in random media," Proc. Symposium Applied Math., Vol. 13, Am. Math. Soc., Providence, RI, pp. 227–246, 1962.

4) Lang, R.H., "Electromagnetic backscattering from a sparse distribution lossy dielectric scatterers," Radio Science, Vol. 16, pp. 15–30, 1981.

5) Lang, R.H. and J.S. Sidhu, "Electromagnetic backscattering form a lay of vegetation: discrete approach," IEEE Trans. Geoscience & Remot Sensing, Vol. GE–21, pp. 62–70, 1983.

6) Watson, J.G. and J.B. Keller, "Rough surface scattering via the smoothi method," J. Acoust. Soc. Am., Vol. 75, pp. 1705–1708, 1984.

7) Brown, G.S., "Application of the integral equation method of smoothing t random surface scattering," IEEE Trans. Antennas & Propag., Vol. AP–32 pp. 1308–1312, 1984.

8) Brown, G.S. "A theoretical study of the effects of vegetation on terrai scattering," Final Report RADC–TR–88–64, Rome Air Development Center AFSC, Griffiss AFB, NY, March, 1988.

9) Van Bladel, J., "Some remarks on Green's dyadic for infinite space, IRE Trans. Antennas & Propag., Vol. AP–9, pp. 563–566, 1961.

10) Tsang, L., J.A. Kong, and R.T. Shin, Theory of Microwave Remote Sensing New York; Wiley–Interscience, 1985.

) Brown, G.S., "New results on coherent scattering from randomly rough conducting surfaces," IEEE Trans. Antennas & Propagation, Vol. AP—31, pp. 5—11, 1983.

) Holliday, D., "Resolution of a controversy surrounding the Kirchhoff approach and the small perturbation method in rough surface scattering," IEEE Trans. Antennas & Propaga., Vol. AP—35, pp. 120—122, 1987.

Scattering in Volumes and Surfaces
M. Nieto-Vesperinas and J.C. Dainty (Editors)

SCATTERING OF EM WAVES FROM PARTICLES WITH RANDOM ROUGH SURFACES

Ralf SCHIFFER

Institut für Reine und Angewandte Kernphysik der Universität Kiel, Olshausenstr. 40, 2300 Kiel, F. R. Germany

The statistical ideas of the Rayleigh–Rice–method for scattering from rough planar surfaces are applied to scattering from slightly aspherical particles. The method represents a statistical version of a well-known perturbation approach and leads, up to the second perturbation order, to rather simple analytical results, in which the expansion coefficients of the surface autocorrelation function in Legendre polynomials enter as descriptive roughness parameters. Finally plots are shown which illustrate some typical features of aspherical scattering.

1. INTRODUCTION

There are several important fields of physical research that deal with the interaction of EM waves with non–planar surfaces: two typical examples are diffraction by gratings in the optical domain and the reflection of radiation by "rough" surfaces for radar as well as optical wavelengths. Both fields have much in common, the decisive difference between their theoretical descriptions is that the former considers periodic deterministic surfaces, whereas the latter employs a statistical description of the surface irregularities; the reason for this "randomization" is on the one hand that the theoretical treatment of scattering from a very irregular deterministic surface is difficult, on the other hand the measured data, e. g. in radar physics, generally are space- and time–averages, and it makes sense for the theoreticians to concentrate on those quantities that can in fact be measured.

In some respect, the situation for scattering from particles is similar: whether one considers atmospheric aerosols, interplanetary or interstellar dust — one is always dealing with scattering or extinction measurements for an (e. g. randomly oriented) ensemble of particles. In most cases one is confronted with an inverse scattering problem: properties of the particles are to be inferred from the measurements. The usual way to attack this problem is to model the particles as spheres, for Mie theory[1] provides an easily applicable and numerically fast algorithm for treating scattering of light by spheres of arbitrary size. However, particles occurring in nature are not spherical in general, and so one has to ask in how far the inversion results found by Mie theory are reliable. Other questions are: From which measurements can the deviations of the scattering bodies from sphericity be determined? Which particle properties can at all be inferred from the available average data?

So there is a considerable interest in scattering theories for nonspherical particles, and the most successful, mainly numerical, methods appear to be the spheroid solution by Asano and Yamamoto[2] and the extended boundary condition method[3,4] (EBCM). But in regard of the remarks made above, the scattering properties of a deterministic irregular particle in a specific orientation is not of interest for applications; so extensive parameter studies for a randomly oriented particle ensemble have been carried out by Asano and Sato[5] for the spheroid solution and by Mugnai and Wiscombe[6–8] for the EBCM. These studies considerably improved our knowledge of aspherical scattering.

The numerical procedures involved in the above–mentioned methods are, however, rather time–consuming, and quite generally it seems to be uneconomic to solve the scattering problem for a deterministic particle and to average afterwards over orientation, shape, size, or refractive index, because much information is lost in the averaging process. In analogy with methods for scattering from rough planar areas mentioned in the first paragraph of this section, one should instead replace the particle ensemble by a single particle with stochastic properties and take the average at an early stage of the calculations, if possible in the analytical part: this saves computer time, and moreover one deals only with practically measurable quantities like average intensities etc.

A possible realization of this idea is to apply results for rough surface scattering directly to the surface of a model particle[9,10], but this method accounts only for the reflected component and is applicable only for very large particles. Instead, in this paper results are presented which are obtained by applying the statistical approach to a perturbation method for scattering from slightly aspherical particles devised by Yeh[11,12] and Erma[13–15]. Many of the basic concepts of this method are quite analogous to those of the perturbational Rayleigh–Rice–approach for scattering from rough planar areas. In a recent paper[16] the procedure is described in detail for perfectly conducting particles, here we only sketch the method, give results for particles of arbitrary material, and show some typical plots.

2. STATISTICAL DESCRIPTION

For the statistical approach to be applicable the cloud of scattering particles under consideration is required to be optically thin, and the positions of the particles are assumed to be random: then multiple scattering can be neglected, and the total scattering intensity is the incoherent sum of the intensities scattered from the individual particles. If these conditions are met we may replace the cloud by a single scattering body with stochastic properties: the set of deterministic objects which it represents consists of the members of the particle cloud, and from the remarks made above we may conclude that the average scattering cross section of the stochastic particle is equivalent to the mean cross section of the cloud.

Our model particle is assumed to be homogeneous with a deterministic refractive index M, but with a stochastic shape: choosing the origin in the interior of the particle we define its "rough" surface as

$$r(\vartheta,\varphi) = R[1 + f(\vartheta,\varphi)] \quad , \tag{1}$$

where f is a stochastic function of the polar angles ϑ and φ; in the following we shall restrict ourselves to randomly oriented particle ensembles. The "mean radius" R is chosen such that

$$\langle f(\vartheta, \varphi)\rangle = 0 \quad , \tag{2}$$

where the angle bracket denotes the average.

For a perturbation calculation up to the second order, a complete statistical description only requires, as in the Rayleigh–Rice–approach, the two–point autocorrelation function $\hat{\rho}(\kappa)$:

$$\langle f(\vartheta_1, \varphi_1) f(\vartheta_2, \varphi_2)\rangle \equiv \hat{\rho}(\kappa) \quad , \tag{3}$$

where $\hat{\rho}$ only depends on the angular distance κ between (ϑ_1, φ_1) and (ϑ_2, φ_2):

$$\cos\kappa = \cos\vartheta_1 \cos\vartheta_2 + \sin\vartheta_1 \sin\vartheta_2 \cos(\varphi_1 - \varphi_2) \quad . \tag{4}$$

An expansion of $\hat{\rho}$ in Legendre polynomials will turn out useful:

$$\hat{\rho}(\kappa) = \sum_{L=0}^{\infty} \rho_L \mathrm{P}_L(\cos\kappa) \quad ; \tag{5}$$

one can easily prove that $\rho_L \geq 0$ for all L. The "roughness coefficients" ρ_L correspond to the surface power spectrum in the Rayleigh–Rice–method.

If only ρ_0 is non–vanishing we have an ensemble of spheres with a certain size distribution, for $\rho_2 \neq 0$ we are dealing with ellipsoidal deformations, and with increasing index the complexity of the particle shapes increases. R, M, and $\{\rho_L\}$ will be the only parameters that enter the final results, and it is important to note that there is in general a great amount of different particle ensembles belonging to the same parameter set, and by means of scattering measurements alone one cannot distinguish between them. The fact that randomly oriented prolate and oblate spheroids of the same volumes and axial ratios correspond to nearly identical roughness coefficients ρ_2 shows without any calculation that both particle types will produce very similar scattering patterns, at least within the range of validity of the perturbation method. This conclusion is strongly supported by theoretical as well as experimental work[5–8,17].

There is, however, still one degree of freedom: the position of the origin of our coordinate system. It can be shown that this position can be chosen such as to make ρ_1 vanish: this appears to be the most reasonable choice, because it minimizes the deviation from the spherical shape.

As we are dealing with a statistical scattering problem, the scattered light too must be described statistically, namely by means of Stokes parameters $\{I, Q, U, V\}$, as defined by van de Hulst[18]. All information on the scattering process is then contained in the so–called Mueller matrix, which for the far–field region connects the Stokes parameters of the scattered wave (subscript s) with those of the incident beam (subscript i). For point– and mirror–symmetry in the average, the Mueller matrix has only six independent (real) elements, which are of course functions of the scattering angle θ:

$$\begin{pmatrix} I_s(\theta) \\ Q_s(\theta) \\ U_s(\theta) \\ V_s(\theta) \end{pmatrix} \sim \frac{1}{k^2r^2} \begin{pmatrix} A_1(\theta) & B_1(\theta) & 0 & 0 \\ B_1(\theta) & A_2(\theta) & 0 & 0 \\ 0 & 0 & A_3(\theta) & B_2(\theta) \\ 0 & 0 & -B_2(\theta) & A_4(\theta) \end{pmatrix} \cdot \begin{pmatrix} I_i \\ Q_i \\ U_i \\ V_i \end{pmatrix} \quad ; \tag{6}$$

k is the wavenumber of the monochromatic incoming radiation. Scattering cross sections for any polarization states of the incident and scattered waves are simply linear combinations of these six functions, which can be calculated analytically in a perturbative manner, as indicated in the next section.

3. PERTURBATION RESULTS

Only a brief description of the perturbation method is given here, because a detailed derivation for the case of perfectly conducting material can be found in Ref. 16.

In the Rayleigh–Rice–approach for scattering from rough planar areas the scattered fields are expanded in plane and evanescent waves; in contrast to this, for scattering from particles the appropriate functional system is furnished, as in Mie theory, by the vector spherical harmonics $\mathbf{X}_{n,m}$ defined as

$$\mathbf{X}_{n,m}(\theta,\phi) \equiv \frac{-i}{\sqrt{n(n+1)}} \mathbf{r} \times \nabla \mathrm{Y}_{n,m}(\theta,\phi) \tag{7}$$

($\mathrm{Y}_{n,m}$: spherical harmonics; $n \geq 1, |m| \geq n$).

The electric field of an incident monochromatic plane wave, linearly polarized in x–direction and propagating in z–direction, can e. g. be written as

$$\begin{aligned} \mathbf{E}_i &= \mathbf{e}_x \exp(ikr\cos\theta) \\ &= \sum_{n=1}^{\infty} i^n \sqrt{(2n+1)\pi} \Big([\mathbf{X}_{n,1}(\theta,\phi) + \mathbf{X}_{n,-1}(\theta,\phi)]\, \mathrm{j}_n(kr) \\ &\qquad + \frac{1}{k} \nabla \times \{[\mathbf{X}_{n,1}(\theta,\phi) - \mathbf{X}_{n,-1}(\theta,\phi)]\, \mathrm{j}_n(kr)\} \Big) \quad , \end{aligned} \tag{8}$$

where j_n is the spherical Bessel function of order n; for the scattered field we write

$$\begin{aligned} \mathbf{E}_s &= -\sum_{n=1}^{\infty} i^n \sqrt{(2n+1)\pi} \sum_{m=-n}^{n} \Big[b_{n,m} \mathbf{X}_{n,m}(\theta,\phi) \mathrm{h}_n(kr) \\ &\qquad + a_{n,m} \frac{1}{k} \nabla \times \mathbf{X}_{n,m}(\theta,\phi) \mathrm{h}_n(kr) \Big] \end{aligned} \tag{9}$$

(h_n: spherical Hankel function of order n), and similarly for the field inside the particle.

The yet unknown coefficients $a_{n,m}$ and $b_{n,m}$ are to be determined by the perturbed boundary conditions at the particle's surface given by Eq. (1); this can be done analytically only in a perturbative manner, i. e. by expanding in powers of αf, where $\alpha \equiv kR$ is the mean size parameter. At this point validity of the Rayleigh hypothesis is required, in other words, the expansions in spherical harmonics are assumed to converge at any point of the surface. This problem is more thoroughly discussed in Ref. 16, but some remarks are in order: For treating scattering from rough surfaces one can circumvent the Rayleigh hypothesis e. g. by employing the extinction theorem[19], which leads to rigorous integral equations. If one, however, tries to solve the resulting equations in an iterative way by expansion in powers of αf, one is of course bound to the same restrictions as in the Rayleigh–Rice–approach — it is quite obvious that both procedures yield identical perturbation terms, though this has explicitly been proved only through fifth order[20]. It appears to me that theoretical methods making no explicit use of the Rayleigh hypothesis nevertheless run into numerical difficulties when this hypothesis is violated; this is also true for scattering from highly aspherical particles[21].

In any case, carrying out the perturbation expansion only up to the second order imposes severe restrictions on the range of validity of the resulting expressions: not only have the relative deformations from the spherical shape to be small compared to unity, also the absolute deviations are required to be much smaller than wavelength: this condition obviously becomes more and more restrictive with increasing size parameter.

The perturbation ansatz leads to series expansions in powers of αf like

$$a_{n,m} = a^{(0)}_{n,m} + a^{(1)}_{n,m} + a^{(2)}_{n,m} + \dots \tag{10}$$

for the scattering coefficients, and the average scattering intensities (and consequently the Mueller matrix elements) are built up by the moduli squared of these expressions; up to the second order we have

$$\left\langle |a_{n,m}|^2 \right\rangle \approx \left|a^{(0)}_{n,m}\right|^2 + \left\langle \left|a^{(1)}_{n,m}\right|^2 \right\rangle + 2\mathrm{Re}\left(a^{(0)*}_{n,m} \left\langle a^{(2)}_{n,m} \right\rangle\right) \quad , \tag{11}$$

and this partition in a zero–order term, an incoherent and a coherent contribution will also result for the Mueller matrix. For scattering from randomly rough planar surfaces we obtain the same structure, but in that case the coherent term is of minor importance, because it does not contribute to the diffuse scattering: up to the second order it only affects the specular intensity.

3.1. Zero–Order Contributions

The zero–order terms are those resulting from Mie theory, the only non–vanishing coefficients are

$$a^{(0)}_{n,1} = -a^{(0)}_{n,-1} = \frac{1}{\Delta^{(1)}_n}\left[M\psi'_n(\alpha)\psi_n(\beta) - \psi_n(\alpha)\psi'_n(\beta)\right] \tag{12}$$

$$b^{(0)}_{n,1} = b^{(0)}_{n,-1} = \frac{1}{\Delta^{(2)}_n}\left[\psi'_n(\alpha)\psi_n(\beta) - M\psi_n(\alpha)\psi'_n(\beta)\right] \tag{13}$$

with

$$\Delta^{(1)}_n \equiv M\xi'_n(\alpha)\psi_n(\beta) - \xi_n(\alpha)\psi'_n(\beta) \tag{14}$$

$$\Delta^{(2)}_n \equiv \xi'_n(\alpha)\psi_n(\beta) - M\xi_n(\alpha)\psi'_n(\beta) \quad , \tag{15}$$

where

$$\beta \equiv M\alpha \equiv MkR \tag{16}$$

and

$$\psi_n(\alpha) \equiv \alpha\, \mathrm{j}_n(\alpha) \tag{17}$$

$$\xi_n(\alpha) \equiv \alpha\, \mathrm{h}_n(\alpha) \quad . \tag{18}$$

The explicit form of the zero–order Mueller matrix elements has been derived by de Rooij and van der Stap[22] and can also be found in Ref. 16.

3.2. Incoherent Contributions

The resulting incoherent Mueller matrix elements are

$$\begin{aligned} A_1^{\mathrm{INC}}(\theta) \pm A_4^{\mathrm{INC}}(\theta) = \\ \frac{1}{2} \sum_{\substack{n_1 n_2 L \\ l_1 l_2 \mathcal{L}}} \rho_L (2n_1+1)(2n_2+1)(2l_1+1)(2l_2+2)(2\mathcal{L}+1)(-)^{L+\mathcal{L}}\, i^{n_1-n_2-l_1+l_2} \\ \times \begin{Bmatrix} n_2 & n_1 & \mathcal{L} \\ l_1 & l_2 & L \end{Bmatrix} \begin{pmatrix} n_1 & l_1 & L \\ -1 & 1 & 0 \end{pmatrix} \begin{pmatrix} n_2 & l_2 & L \\ -1 & 1 & 0 \end{pmatrix} \begin{pmatrix} n_1 & n_2 & \mathcal{L} \\ -1 & 1 & 0 \end{pmatrix} \begin{pmatrix} l_1 & l_2 & \mathcal{L} \\ -1 & 1 & 0 \end{pmatrix} \\ \times \left(\pm\right)^{l_1+l_2+\mathcal{L}} \mathrm{H}_{\substack{2\\1}}(n_1, l_1, L) \mathrm{H}^*_{\substack{2\\1}}(n_2, l_2, L) \mathrm{P}^{\mathcal{L}}_{0,0}(\cos\theta) \end{aligned} \tag{19}$$

$$\begin{aligned} A_2^{\mathrm{INC}}(\theta) \pm A_3^{\mathrm{INC}}(\theta) = \\ \frac{1}{2} \sum_{\substack{n_1 n_2 L \\ l_1 l_2 \mathcal{L}}} \rho_L (2n_1+1)(2n_2+1)(2l_1+1)(2l_2+1)(2\mathcal{L}+1)\, i^{n_1+n_2+l_1+l_2} \\ \times \begin{Bmatrix} n_2 & n_1 & \mathcal{L} \\ l_1 & l_2 & L \end{Bmatrix} \begin{pmatrix} n_1 & l_1 & L \\ -1 & 1 & 0 \end{pmatrix} \begin{pmatrix} n_2 & l_2 & L \\ -1 & 1 & 0 \end{pmatrix} \begin{pmatrix} n_1 & n_2 & \mathcal{L} \\ -1 & -1 & 2 \end{pmatrix} \begin{pmatrix} l_1 & l_2 & \mathcal{L} \\ -1 & -1 & 2 \end{pmatrix} \\ \times \left(\mp\right)^{l_1+l_2+\mathcal{L}} \mathrm{H}_{\substack{2\\1}}(n_1, l_1, L) \mathrm{H}^*_{\substack{2\\1}}(n_2, l_2, L) \mathrm{P}^{\mathcal{L}}_{2,\pm 2}(\cos\theta) \end{aligned} \tag{20}$$

$$\begin{aligned} B_1^{\mathrm{INC}}(\theta) + iB_2^{\mathrm{INC}}(\theta) = \\ \frac{1}{2} \sum_{\substack{n_1 n_2 L \\ l_1 l_2 \mathcal{L}}} \rho_L (2n_1+1)(2n_2+1)(2l_1+1)(2l_2+1)(2\mathcal{L}+1)(-)^{L}\, i^{n_1-n_2+l_1-l_2} \\ \times \begin{Bmatrix} n_2 & n_1 & \mathcal{L} \\ l_1 & l_2 & L \end{Bmatrix} \begin{pmatrix} n_1 & l_1 & L \\ -1 & 1 & 0 \end{pmatrix} \begin{pmatrix} n_2 & l_2 & L \\ -1 & 1 & 0 \end{pmatrix} \begin{pmatrix} n_1 & n_2 & \mathcal{L} \\ -1 & 1 & 0 \end{pmatrix} \begin{pmatrix} l_1 & l_2 & \mathcal{L} \\ -1 & -1 & 2 \end{pmatrix} \\ \times \mathrm{H}_1(n_1, l_1, L) \mathrm{H}^*_2(n_2, l_2, L) \mathrm{P}^{\mathcal{L}}_{2,0}(\cos\theta) \quad , \end{aligned} \tag{21}$$

where

$$\begin{aligned} \mathrm{H}_{\substack{1\\2}}(n,l,L) \equiv\ & \alpha(M^2-1) \left\{ \frac{1+(-)^{n+l+L}}{2} \frac{\psi_n(\beta)\psi_l(\beta)}{\Delta_n^{(1)}\Delta_l^{(1)}} Z \right. \\ & \pm \frac{1}{2} \left[\frac{\psi_n(\beta)}{\Delta_n^{(2)}} - i\frac{\psi_n'(\beta)}{\Delta_n^{(1)}} \right] \left[\frac{\psi_l(\beta)}{\Delta_l^{(2)}} \mp i\frac{\psi_l'(\beta)}{\Delta_l^{(1)}} \right] \\ & \left. \pm \frac{(-)^{n+l+L}}{2} \left[\frac{\psi_n(\beta)}{\Delta_n^{(2)}} + i\frac{\psi_n'(\beta)}{\Delta_n^{(1)}} \right] \left[\frac{\psi_l(\beta)}{\Delta_l^{(2)}} \pm i\frac{\psi_l'(\beta)}{\Delta_l^{(1)}} \right] \right\} \end{aligned} \tag{22}$$

with

$$Z \equiv \frac{2n(n+1)l(l+1)\alpha^{-2}}{L(L+1)-n(n+1)-l(l+1)} \tag{23}$$

In Eqs. (19)–(21) the objects in round and curly brackets are, respectively, Wigner's 3j– and 6j–symbols[23], and the functions $P^{l}_{m,j}$ represent generalized spherical functions discussed in detail e. g. by Gel'fand et al.[24]

It is interesting to note that the expansion of the Mueller matrix elements in generalized spherical functions, which arises quite naturally in the calculations[16], is just the kind of expansion used in certain analytical approaches to radiative transfer problems for polarized light propagating in a horizontally layered isotropic medium[25–28].

3.3. Coherent Contributions

The second order coherent scattering coefficients are

$$\left\langle a^{(2)}_{l,1}\right\rangle = -\left\langle a^{(2)}_{l,-1}\right\rangle = \sum_{nL}(2n+1)\rho_L \begin{pmatrix} n & l & L \\ -1 & 1 & 0 \end{pmatrix}^2 \mathrm{J}_1(n,l,L) \tag{24}$$

$$\left\langle b^{(2)}_{l,1}\right\rangle = \left\langle b^{(2)}_{l,-1}\right\rangle = \sum_{nL}(2n+1)\rho_L \begin{pmatrix} n & l & L \\ -1 & 1 & 0 \end{pmatrix}^2 \mathrm{J}_2(n,l,L) \tag{25}$$

with

$$\begin{aligned} \mathrm{J}_1(n,l,L) \equiv & -\frac{i}{\left(\Delta^{(1)}_l\right)^2}\alpha^2(M^2-1)\Big(\psi_l(\beta)\left[M\psi'_l(\beta)+\frac{l(l+1)}{\alpha^3}\psi_l(\beta)\right] \\ & \frac{1+(-)^{n+l+L}}{2\Delta^{(1)}_n}\Big\{[-\psi'_n(\beta)\psi'_l(\beta)+Z\psi_n(\beta)\psi_l(\beta)]\,[\xi'_n(\alpha)\psi'_l(\beta)+MZ\xi_n(\alpha)\psi_l(\beta)] \\ & \quad +[M\psi'_n(\beta)\psi'_l(\beta)+\frac{Z}{M}\psi_n(\beta)\psi_l(\beta)]\,[M\xi'_n(\alpha)\psi'_l(\beta)-Z\xi_n(\alpha)\psi_l(\beta)]\Big\} \\ & +\frac{1-(-)^{n+l+L}}{2\Delta^{(2)}_n}(M^2-1)\xi_n(\alpha)\psi_n(\beta)\,[\psi'_l(\beta)]^2\Big) \end{aligned} \tag{26}$$

$$\begin{aligned} \mathrm{J}_2(n,l,L) \equiv & -\frac{i}{\left(\Delta^{(2)}_l\right)^2}\alpha^2(M^2-1)\Big\{-M\psi_l(\beta)\psi'_l(\beta) \\ & +(M^2-1)\left[\frac{1+(-)^{n+l+L}}{2\Delta^{(2)}_n}\xi_n(\alpha)\psi_n(\beta)+\frac{1-(-)^{n+l+L}}{2\Delta^{(1)}_n}\xi'_n(\alpha)\psi'_n(\beta)\right][\psi_l(\beta)]^2\Big\}. \end{aligned} \tag{27}$$

The scattering coefficients in Eqs. (12),(13),(24), and(25) have to be inserted in the following expressions for the second–order coherent Mueller matrix elements:

$$\begin{aligned} A^{\mathrm{COH}}_{\nu}(\theta) = & \,\mathrm{Re}\sum_{nl\mathcal{L}}(2n+1)(2l+1)(2\mathcal{L}+1) \\ & \times\Bigg\{\left[\frac{1\pm(-)^{n+l+\mathcal{L}}}{2}a^{(0)*}_{n,1}+\frac{1\mp(-)^{n+l+\mathcal{L}}}{2}b^{(0)*}_{n,1}\right]\left\langle a^{(2)}_{l,1}\right\rangle \\ & +\left[\frac{1\mp(-)^{n+l+\mathcal{L}}}{2}a^{(0)*}_{n,1}+\frac{1\pm(-)^{n+l+\mathcal{L}}}{2}b^{(0)*}_{n,1}\right]\left\langle b^{(2)}_{l,1}\right\rangle\Bigg\} \\ & \times\begin{pmatrix} n & l & \mathcal{L} \\ -1 & 1 & 0 \end{pmatrix}^2 \mathrm{P}^{\mathcal{L}}_{0,0}(\cos\theta) \quad , \end{aligned} \tag{28}$$

where the upper signs are to be taken for $\nu = 1,2$, the lower ones for $\nu = 3,4$;

$$B^{\rm COH}_{\frac{1}{2}}(\theta) = \begin{matrix}\mathrm{Re}\\ \mathrm{Im}\end{matrix} \sum_{nl\mathcal{L}} (2n+1)(2l+1)(2\mathcal{L}+1)$$
$$\times \left\{ \left[\frac{1 \pm (-)^{n+l+\mathcal{L}}}{2} a^{(0)*}_{n,1} + \frac{1 \mp (-)^{n+l+\mathcal{L}}}{2} b^{(0)*}_{n,1} \right] \left\langle a^{(2)}_{l,1} \right\rangle \right.$$
$$\left. - \left[\frac{1 \mp (-)^{n+l+\mathcal{L}}}{2} a^{(0)*}_{n,1} + \frac{1 \pm (-)^{n+l+\mathcal{L}}}{2} b^{(0)*}_{n,1} \right] \left\langle b^{(2)}_{l,1} \right\rangle \right\} \tag{29}$$
$$\times \begin{pmatrix} n & l & \mathcal{L} \\ -1 & 1 & 0 \end{pmatrix} \begin{pmatrix} n & l & \mathcal{L} \\ -1 & -1 & 2 \end{pmatrix} \mathrm{P}^{\mathcal{L}}_{2,0}(\cos\theta) \quad ;$$

it can be shown that the cross sections resulting from Eqs. (19)–(29) satisfy the Optical Theorem[16].

The sums occurring the final results formally run to infinity, but the behaviour of the Hankel functions in the denominators of the H– and J–functions as well as the properties of the 3j– and 6j–symbols lead to a cut–off of the summations — with one exception: for the coherent terms the infinite sum over L is left, which can lead to convergence problems. This becomes more transparent in the next subsection.

3.4. Long–Wavelength Limit

The Rayleigh limit $\alpha \to 0$ is easily carried out, and as is well known, the only relevant scattering coefficient in this limit is $a_{1,1}$ (except for perfectly conducting material). In the following we just quote the perturbation results for the matrix element $A_1(\theta)$, i. e. the scattering efficiency for unpolarized incident light, because the other quantities show up the same structure;

$$A^{\rm INC}_1 \underset{\alpha \ll 1}{\longrightarrow} \frac{9}{2}\alpha^6 \left| \frac{M^2-1}{M^2+2} \right|^2 \left[\rho_0 (1+\cos^2\theta) + \frac{\rho_2}{25} \left| \frac{M^2-1}{M^2+2} \right|^2 (13+\cos^2\theta) \right] \tag{30}$$

$$A^{\rm COH}_1 \underset{\alpha \ll 1}{\longrightarrow} 3\alpha^6 \left| \frac{M^2-1}{M^2+2} \right|^2 \mathrm{Re} \left[\frac{1}{M^2+2} \sum_L \rho_L F(L) \right] (1+\cos^2\theta) \tag{31}$$

with

$$F(L) \equiv \frac{(L-1)L[L(M^4-6M^2+1)+3M^2-1]}{[L(M^2+1)-M^2](2L+1)} + \frac{(M^2+1)(L+1)(L+2)}{2L+1} - M^2 \quad . \tag{32}$$

We see that in the incoherent cross sections only the roughness coefficients ρ_0 and ρ_2 enter, so that more complicated than ellipsoidal deformations do not affect the incoherent intensities. In contrast to this, the coherent expressions contain all roughness coefficients, and the fact that

$$F(L) \underset{L\to\infty}{\longrightarrow} \frac{(M^2-1)^2}{M^2+1} L \tag{33}$$

shows that the sum in Eq. (31) is not necessarily convergent. In fact one can prove that for convergence one has to require that the correlation function $\hat{\rho}(\kappa)$ defined in Eq. (3) has vanishing slope for zero argument[16]. So e. g. an exponential correlation function leads to divergent second–order results, which is, by the way, also the case in the Rayleigh–Rice–approach[29]. This is in accordance with the fact that for stochastic gaussian surfaces with an exponential correlation function almost all realizations are nowhere differentiable[30]: for such non–analytic surfaces the Rayleigh hypothesis is obviously violated.

Consideration of the long–wavelength limit is useful also for another reason: scattering from ellipsoids can be treated rigorously in this limit, and for small eccentricities the resulting expressions can be compared with our formulas with only ρ_2 being non–vanishing. Carrying out this procedure one obtains perfect agreement; this is explicitly shown for perfectly conducting spheroids in Ref. 16.

4. NUMERICAL RESULTS

The analytical formulas of the preceding section can easily be evaluated numerically, because one can exploit certain recurrence relations for the generalized spherical functions as well as for the 3j– and 6j–symbols. In the following some plots for the three simplest cases are presented, namely the cases that only one of the coefficients ρ_0, ρ_2, or ρ_4 is different from zero; that corresponds to an ensemble of spheres, "convex", and "concave" particles in the sense of Wiscombe and Mugnai[8]; more complex deformations apparently to not lead to essentially new features in the scattering patterns, at least not in the range of size parameters considered.

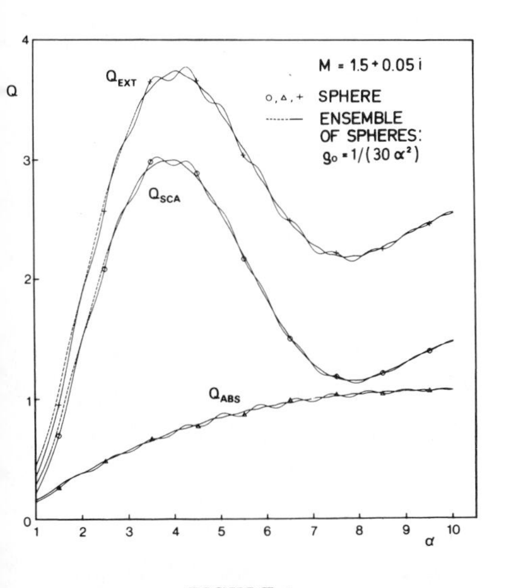

FIGURE 1
Scattering, extinction, and absorption efficiencies as functions of size parameter for a single sphere and an ensemble of spheres

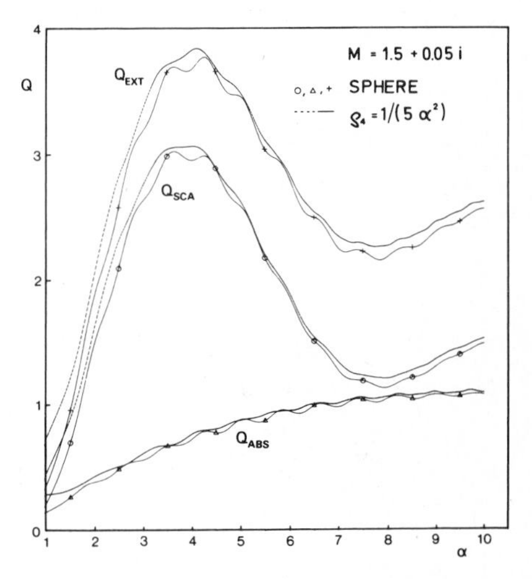

FIGURE 2
Scattering, extinction, and absorption efficiencies as functions of size parameter for a single sphere and a ρ_4–particle

Figs. 1 and 2 show scattering, extinction, and absorption efficiencies (cross sections divided by geometrical cross sections) as functions of the size parameter for spheres, ρ_0– and ρ_4–particles of slightly absorbing material. The roughness coefficients ρ_0 and ρ_4 have been chosen proportional to α^{-2} so that the absolute surface deviations from the reference sphere are independent of α: as mentioned above, the smallness of these deviations relative to the wavelength λ is a necessary condition for the validity of the perturbation results values of $1/(30\alpha^2)$ and $1/(5\alpha^2)$ for the roughness coefficients correspond to r. m. s. surface deviations of $0.03\,\lambda$ and $0.07\,\lambda$, respectively.

As could be expected, the ρ_0–curve is a smoothed version of the single–sphere curve (Fig. 1); in contrast to this, slightly aspherical particles cause an extra enhancement of the scattering as well as the absorption cross section (Fig. 2): this is in agreement with the generally accepted phenomenon that nonspherical particles show up larger extinction cross sections than equal–volume spheres[31].

Further numerical studies have shown, however, that difficulties with the perturbation results arise near sharp resonances, which occur for larger size parameters, $\alpha > 10$, and for non–absorbing material. Here the perturbation terms tend to overcompensate the peaks of the spherical curves, so that the restrictions on the magnitude of the deviations have to be tightened up. Similar conclusions are drawn by Kiehl et al. [32], who consider first–order perturbation results for deterministic particles. This situation can partly be remedied either by introducing a small imaginary part of the refractive index or by considering size–averages, because in both cases the resonances are damped.

No problems arise for the parameters chosen in Figs. 3–12: here the Mueller matrix elements (all elements except A_1 are normalized by A_1) as functions of the scattering angle θ are shown, for $\alpha = 3$ and non–absorbing material in Figs. 3–7 and for $\alpha = 5$ and strong absorption in Figs. 8–12. Appropriate values for the roughness coefficients have been chosen in the following way: scattering efficiencies for an ensemble of spheres have been calculated in two different ways, rigorously by averaging over single–sphere curves, and on the basis of our perturbation results with only ρ_0 being non–vanishing; that value of ρ_0 which yielded errors smaller than 5% has been taken and adopted for ρ_2 and ρ_4 too.

The contents of these and many other plots can be summarized as follows: Figs. 3 and 8 show that deviations from the spherical shape lower the backscattered intensity (in contrast to an ensemble of spheres) and slightly enhance forward scattering. Quite generally, near-forward scattering is clearly less affected by shape variations than large–angle scattering. From the degree of polarization for unpolarized incident light (Figs. 4 and 9) probably not much information can be extracted, because this quantity is sensitive to shape as well as to size and refractive index. The deviations from sphericity can be estimated by the quantities shown in Figs. 7 and 12, which vanish for spherical scatterers. For most angles, the cross polarization, $(A_1 - A_2)/A_1$, is larger than $(A_4 - A_3)/A_1$ so that the former quantity should favourably be used to determine the asphericity of particles. The ratio least sensitive to

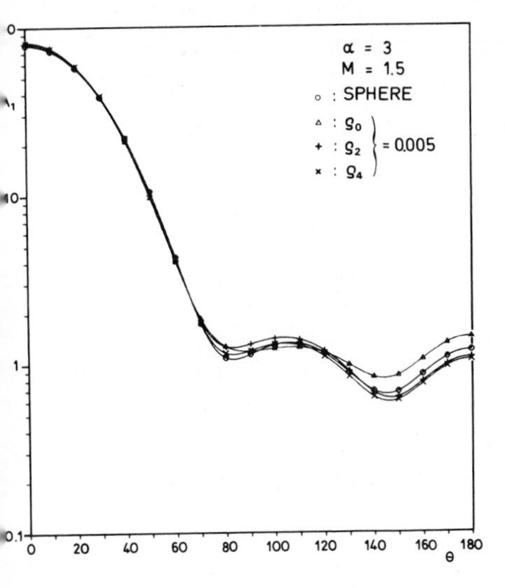

FIGURE 3

Mueller matrix element A_1 for several particle types of non–absorbing material

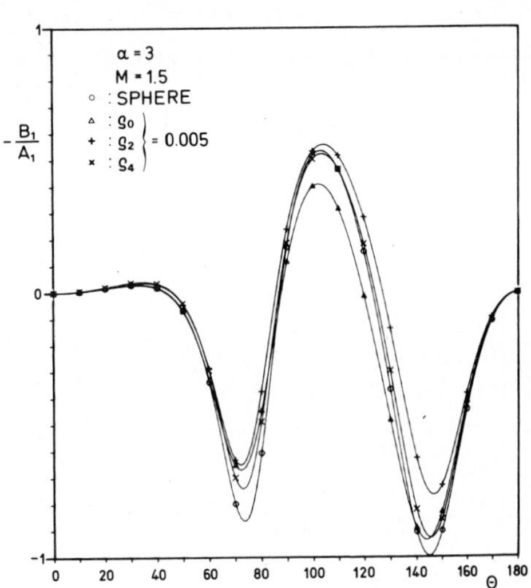

FIGURE 4

Polarization for several particle types of non–absorbing material

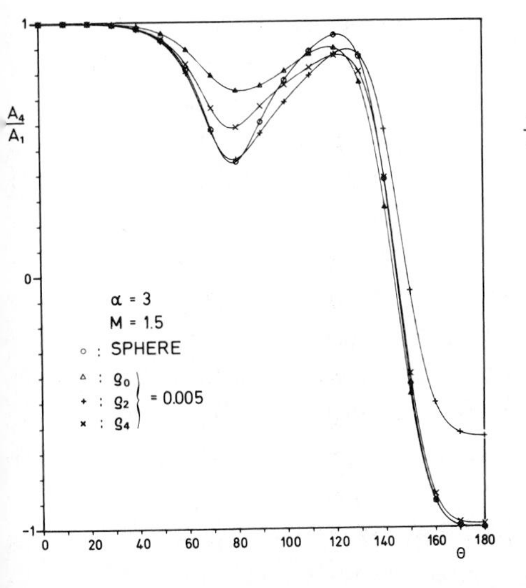

FIGURE 5

Ratio A_4/A_1 for several particle types of non–absorbing material

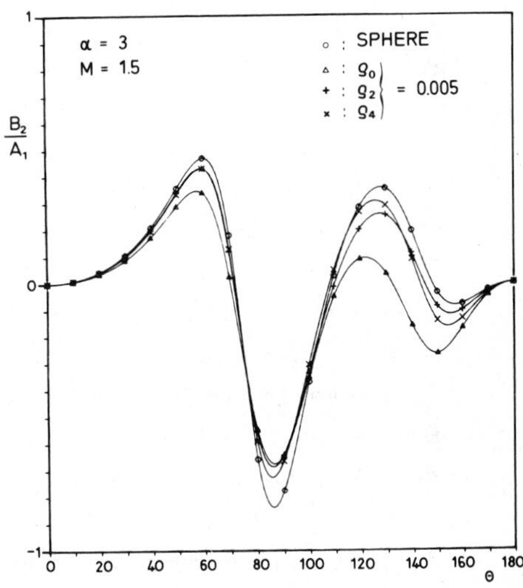

FIGURE 6

Ratio B_2/A_1 for several particle types of non–absorbing material

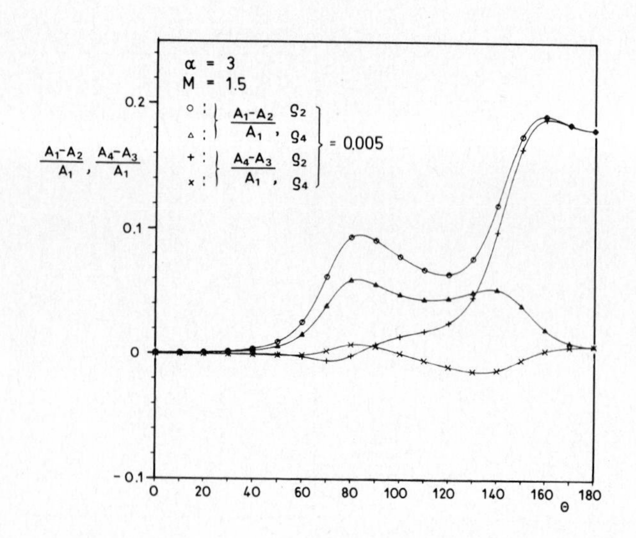

FIGURE 7
Ratios $(A_1 - A_2)/A_1$ and $(A_4 - A_3)/A_1$ for ρ_2– and ρ_4–particles of non-absorbing material

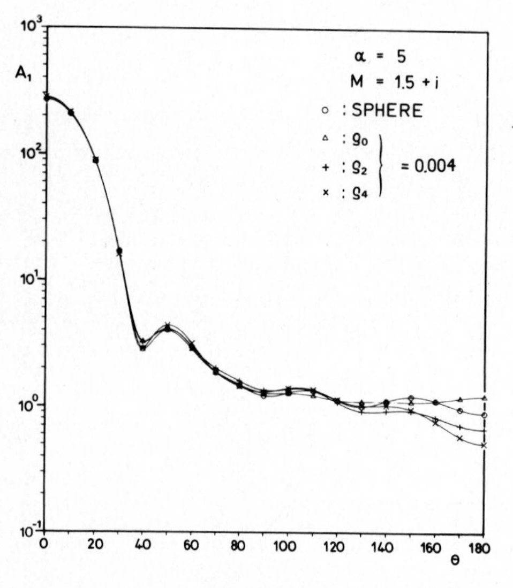

FIGURE 8
Mueller matrix element A_1 for several particle types of strongly absorbing material

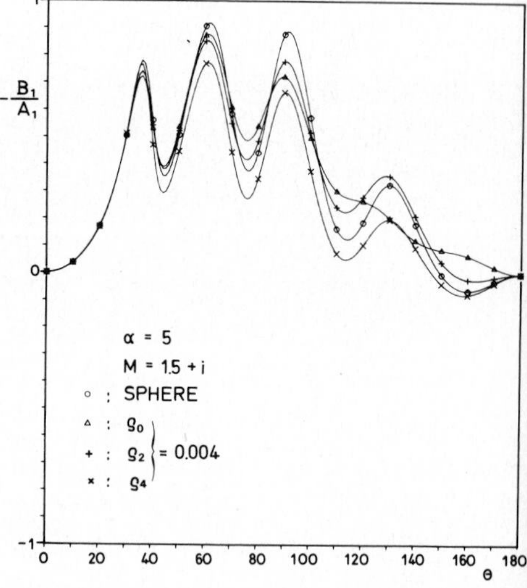

FIGURE 9
Polarization for several particle types of strongly absorbing material

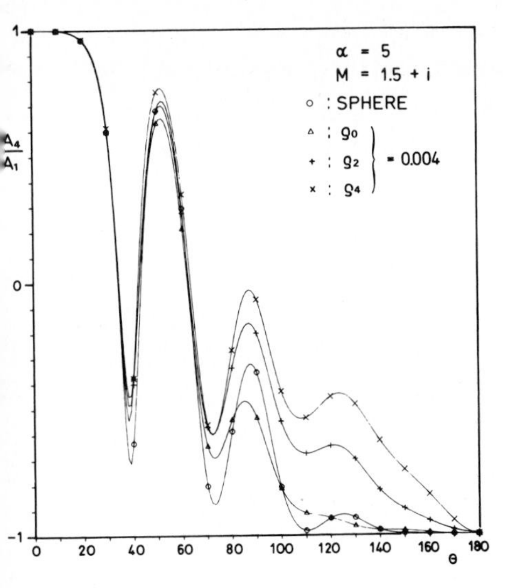

FIGURE 10
Ratio A_4/A_1 for several particle types of strongly absorbing material

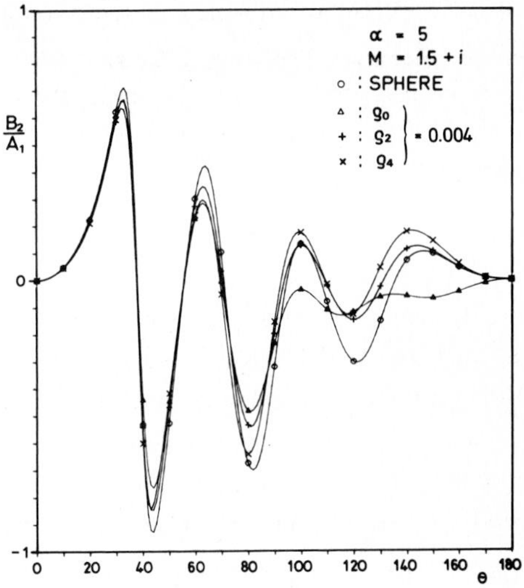

FIGURE 11
Ratio B_2/A_1 for several particle types of strongly absorbing material

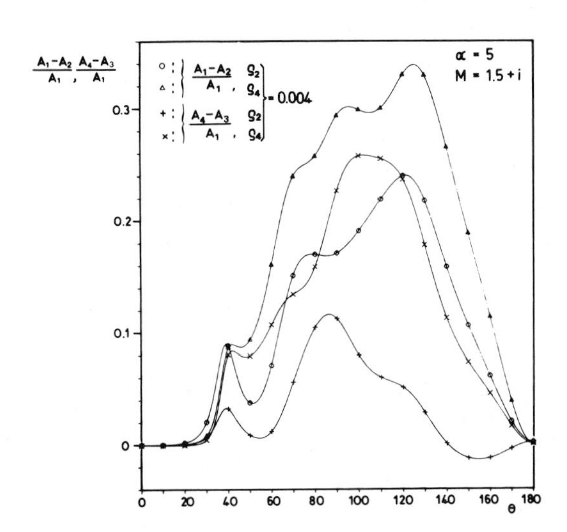

FIGURE 12
Ratios $(A_1 - A_2)/A_1$ and $(A_4 - A_3)/A_1$ for ρ_2– and ρ_4–particles of strongly absorbing material

shape effects appears to be B_2/A_1 (Figs. 6 and 11), especially for larger scattering angles: here the ρ_0–curves (ensemble of spheres) deviate more from the single–sphere curves than the ρ_2– and ρ_4–curves do. So B_2/A_1 seems to be suitable for particle sizing if only scattering data for large scattering angles are available.

All these findings are in accordance with other theoretical as well as experimental work[5–8,17,33,34].

5. CONCLUDING REMARKS

By means of a statistical description the vast amount of possible particle shapes has been reduced to the (still infinite) parameter set R, M, and $\{\rho_L\}$, which contains all information relevant to the scattering properties of a randomly oriented particle ensemble; at least up to the second order of the perturbation approach presented in this paper these parameters represent a complete description, for a surface function $f(\vartheta, \varphi)$ [cf. Eq. (1)] obeying gaussian statistics this is even true through all perturbation orders. The analytical results for the Mueller matrix elements appeal by their symmetrical structure and are well amenable to numerical evaluation.

The range of validity of the perturbation results is of course limited by the condition that the surface deviations from the reference sphere be much smaller than wavelength. On the other hand these results offer an easy access to a theoretical treatment of scattering from arbitrarily shaped particles and yield correct trends of aspherical scattering even outside the permitted parameter range. The problems arising near sharp resonances for larger non–absorbing particles are due to the rather erratic behaviour of the single–sphere curves, which therefore do not represent suitable zero–order terms for a perturbation expansion: instead a size–average should be carried out from the very beginning, e. g. in the manner of Wiscombe and Mugnai[8].

Finally, I want to emphasize that the statistical approach to scattering from particles is of course not confined to the perturbation method and can e. g. also be applied to inhomogeneous scatterers.

REFERENCES

1) G. Mie, Ann. Phys. 25 (1908) 377.

2) S. Asano and G.Yamamoto, Appl. Opt. 14 (1975) 29.

3) P. C. Waterman, Phys. Rev. D3 (1971) 825.

4) P. Barber and C. Yeh, Appl. Opt. 24 (1975) 2864.

5) S. Asano and M. Sato, Appl. Opt. 19 (1980) 962.

6) A. Mugnai and W. J. Wiscombe, J. Atmos. Sci. 37 (1980) 1291.

7) A. Mugnai and W. J. Wiscombe, Appl. Opt. 25 (1986) 1235.

8) W. J. Wiscombe and A. Mugnai, Appl. Opt. 27 (1988) 2405.

9) R. Schiffer and K. O. Thielheim, J. Appl. Phys. 53 (1982) 2825.

10) R. Schiffer and K. O. Thielheim, J. Appl. Phys. 57 (1985) 2437.

11) C. Yeh, Phys. Rev. 135 (1964) A1193.

12) C. Yeh, J. Math. Phys. 6 (1965) 2008.

13) V. A. Erma, Phys. Rev. 173 (1968) 1243.

14) V. A. Erma, Phys. Rev. 176 (1968) 1544.

15) V. A. Erma, Phys. Rev. 179 (1969) 1238.

16) R. Schiffer, J. Opt. Soc. Am. A6 (1989) 385.

17) D. W. Schuerman, R. Wang, B. Gustafson, and R. Schaefer, Appl. Opt. 20 (1981) 4093.

18) H. C. van de Hulst, Light Scattering from Small Particles (Wiley, New York, 1957) Ch. 5.1.

19) M. Nieto–Vesperinas and N. Garcia, Opt. Acta 28 (1981) 1651.

20) S. L. Broschat, L. Tsang, A. Ishimaru, and E. I. Thorsos, J. Electrom. Waves Appl. 2 (1987) 85.

21) M. F. Iskander and A. Lakhtakia, Appl. Opt. 23 (1984) 948.

22) W. A. de Rooij and C. C. A. H. van der Stap, Astron. Astrophys. 131 (1984) 237.

23) A. R. Edmonds, Angular Momentum in Quantum Mechanics (Princeton Univ. Press, 1957).

24) I. M. Gel'fand, R. A. Minlos, and Z. Ya. Shapiro, Representations of the Rotation and Lorentz Groups and Their Applications (Pergamon, Oxford, 1963) Ch. II.7.

25) I. Kuščer and M. Ribarič, Opt. Acta 6 (1959) 42.

26) C. E. Siewert, Astrophys. J. 245 (1981) 1080.

27) J. W. Hovenier and C. V. M. van der Mee, Astron. Astrophys. 128 (1983) 1.

28) S. Ito and T. Oguchi, Radio Sci. 22 (1987) 873.

29) R. Schiffer, Appl. Opt. 26 (1987) 704.

30) H. P. McKean, Stochastic Integrals (Acad. Press, New York, 1969).

31) P. Chýlek, J. Opt. Soc. Am. 67 (1977) 1348.

32) J. T. Kiehl, M. W. Ko, A. Mugnai, and P. Chýlek, Perturbation approach to light scattering by non-spherical particles, in: Light Scattering by Irregularly Shaped Particles, ed. D. W. Schuerman (Plenum, New York, 1980) pp. 135–140.

33) R. J. Perry, A. J. Hunt, and D. R. Huffman, Appl. Opt. 17 (1978) 2700.

34) P. E. Geller, T. G. Tsuei, and P. Barber, Appl. Opt. 24 (1985) 2391.

Scattering in Volumes and Surfaces
M. Nieto-Vesperinas and J.C. Dainty (Editors)

SURFACE IMPEDANCE BOUNDARY CONDITIONS USED TO STUDY LIGHT SCATTERING FROM METALLIC SURFACES

Ricardo A. DEPINE

Universidad de Buenos Aires, Departamento de Física,
Ciudad Universitaria, Pabellón I, 1428 Buenos Aires, Argentina

1. INTRODUCTION

The usual methods of solving the scattering of a wave at a boundary separating two media require the solution of two wave equations, one for each medium. Both solutions are then matched across the interface through a pair of boundary conditions to express the continuity of the tangential components of the field. Problems treated using this approach usually require large amounts of numerical work and the use of sophisticated computer codes[1,2]. In the particular case of metallic boundaries one is also faced with numerical difficulties specially for those metals used in the optical and infrared range in order to achieve high reflectivities (for example, aluminium, silver and gold).

On the other hand, the mathematical complexity of many problems in the theory of wave scattering from a metallic obstacle is appreciably reduced if the metal can be considered as being perfectly conducting. In this case, the electromagnetic fields cannot penetrate inside the scatterer and then the boundary conditions take the simple form of Neumann's or Dirichlet's boundary conditions. Only the fields outside the scatterer must be calculated, the fields inside it being zero. Unfortunately, the assumption of perfect conductor does not allways hold for the metals used in optics. Aluminium, silver and gold have very high reflectivity[3] but the perfectly conducting model does not properly describe the observed behaviour of the scattered fields in a quantitative manner[1]. Up to this point, two kinds of boundary conditions can be used in the electromagnetic treatment of metals: the first (Maxwell boundary condition) leads to rather complex methods that give good results but face numerical difficulties and the second and very simple one (perfect conductor) does not always give good results. The subject of this article is concerned with the difficulties and advantages involved in using an approximate boundary condition as a third possibility for solving light scattering from metallic surfaces.

2. THE SURFACE IMPEDANCE

From a mathematical point of view the wave equation has unique solution if the ratio between the unknown funtion and its normal derivative along a closed surface is known. So, the scattering problem (see Fig. 1) can also be solved if the relation between the tangential components of the fields $E_{//}$ and $H_{//}$ along some closed surface Σ is known. Σ may coincide or not with the actual diffracting boundary. In this third approach, Maxwell boundary conditions are replaced by a boundary condition in the form

$$\vec{E}_{//} = Z\ \hat{n} \times \vec{H}_{//} \qquad \text{at } \Sigma \tag{1}$$

$\hat{n}$ is a normal unit vector pointing towards the exterior of Σ and Z is a second rank tensor called the surface impedance tensor.

If Z is known the fields outside Σ can be found without solving Maxwell equations inside it. Thus the concept of surface impedance looks very advantageous because it enables us to recover the simplicity inherent to perfectly conducting models (in the sense that only a single wave equation must be solved). But since Z depends on the shape and nature of the scatterer and on the

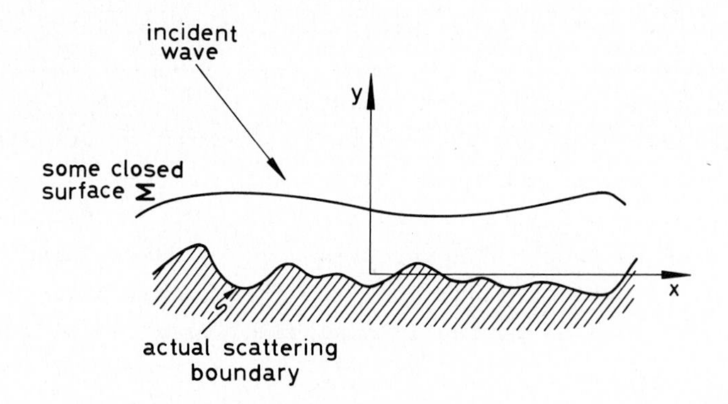

FIGURE 1
The scattering problem. Σ is a closed surface which may coincide or not with the actual scattering boundary.

external configuration of the fields, its determination generally involves the whole solution of the problem, a fact that makes eq. (1) a poor candidate to be postulated as a boundary condition. Only when the dependence of Z on the scatterer or on the external field parameters is known, can it be useful as a boundary condition. For this reason, it is of interest to derive fictitious surface impedances which can account in an approximate way for the effects of the metallic boundary.

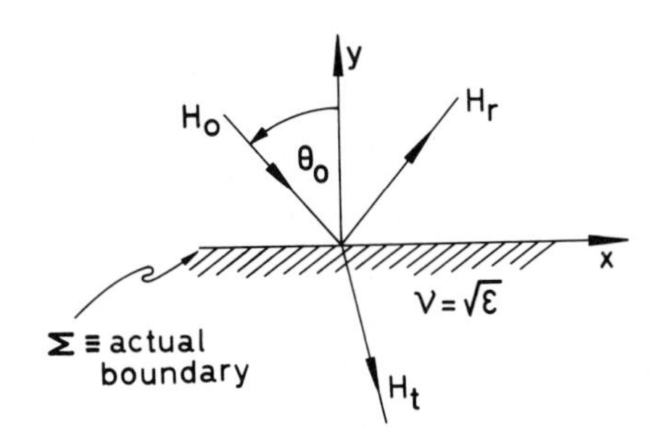

FIGURE 2
Reflection of a plane wave at a smooth boundary between air and a metal having refractive index ν.

A hint on the behaviour of Z can be obtained by considering the reflection of a plane wave at a smooth boundary (Fig. 2). When the incident magnetic field is polarized along the z-axis (S polarization), the tangential components of the fields at the boundary are

$$\vec{H}_{//} = (H_o + H_r)\ \exp(ikx \sin \vartheta_o)\ \exp(-i\omega t)\ \hat{z} \quad , \tag{2}$$

$$\vec{E}_{//} = (H_o - H_r)\ \exp(ikx \sin \vartheta_o)\ \exp(-i\omega t)\ \hat{x} \quad , \tag{3}$$

and using the very well known Fresnel coefficient

$$\frac{H_r}{H_o} = \frac{\nu^2 \cos \vartheta_o \quad (\nu^2 - \sin^2\vartheta_o)^{1/2}}{\nu^2 \cos \vartheta_o + (\nu^2 - \sin^2\vartheta_o)^{1/2}} \tag{4}$$

we can calculate Z in a rigorous manner as

$$Z = \nu^{-2}(\nu^2 - \sin^2\vartheta_o)^{1/2} \quad . \tag{5}$$

For metals like Al, Ag and Au in the visible and infrared range, the following relation holds approximately for the refractive index ν

$$\text{Re } \nu^2 >> 1 \quad , \tag{6}$$

and then it is valid to approximate Z in (5) by Z_o, the inverse of the refractive index

$$Z \cong Z_o = \nu^{-1} \quad . \tag{7}$$

The same result is obtained when P polarization is considered.

So, in this problem we have found a surface Σ which coincides with the diffracting boundary and on which the surface impedance does not depend on the external field parameters (it is a constant).

3. Z_o AS A BOUNDARY CONDITION

The approximated value Z_o, obtained in the case of a plane boundary, was used as a boundary condition to solve scattering problems involving non-plane surfaces[4-9]. For the particular case of diffraction gratings it was possible to handle highly conducting metals by using a differential method and a conformal mapping technique[8,9]. The validity of the results was studied by comparing the constant-impedance results to those obtained by means of rigorous electromagnetic theories[10]. In Figs. 3 and 4 we observe the P efficiency curves as a function of

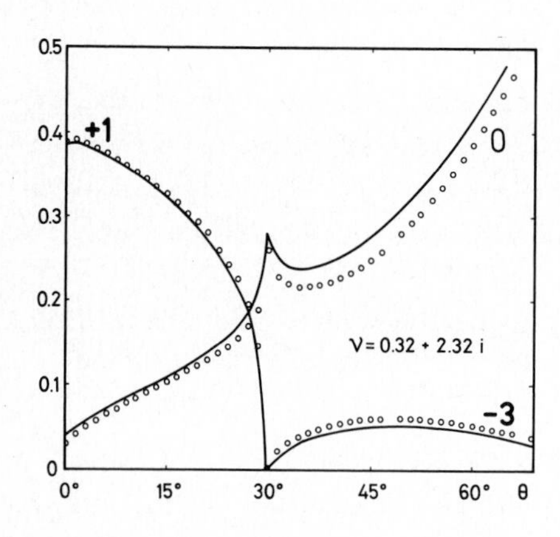

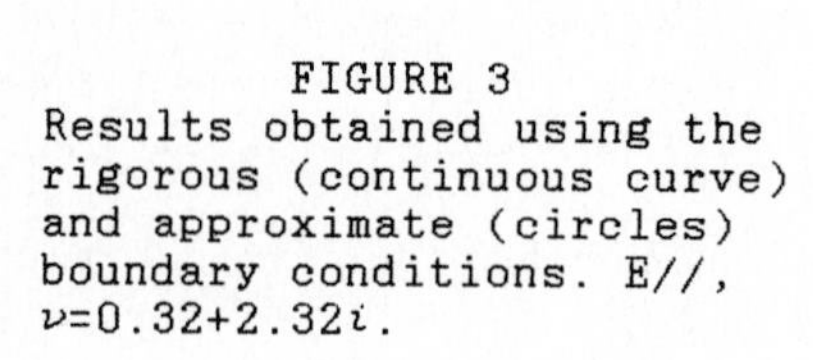
FIGURE 3
Results obtained using the rigorous (continuous curve) and approximate (circles) boundary conditions. E//, ν=0.32+2.32i.

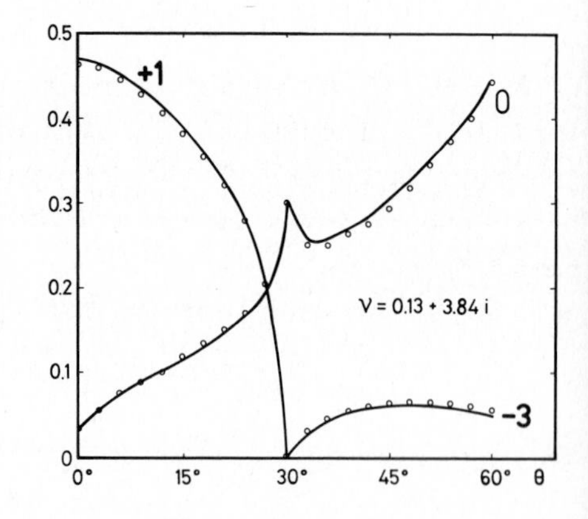

FIGURE 4
Results obtained using the rigorous (continuous curve) and approximate (circles) boundary conditions. E//, ν=0.13+3.84i.

angle of incidence for a grating with groove-height to period ratio h/d=0.2 for two different values of the refractive index. When ν=0.32+2.32i (gold, 550 nm), a not too favorable value for the Z_o-approximation to be valid, the rigorous and approximated theories give very similar curves. The agreement is better when we approach the high conductivity region as can be seen from Fig. 4, ν=0.13+3.84i (gold, 700nm). The same behaviour is observed in S-polarization (Figs. 5 and 6). In Fig. 6, ν=0.15+10.4i (gold, 1.47 μm) the agreement between both methods is excellent and it is interesting to note that in spite of the high value of ν^2 considered here, the infinite conductivity model does not give good results.

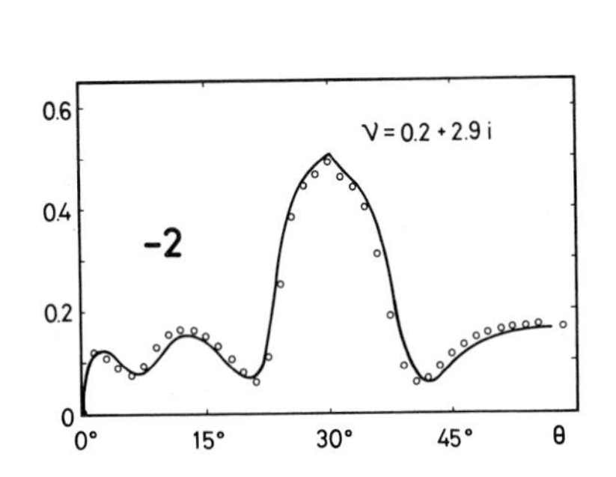

FIGURE 5
Results obtained using the rigorous (continuous curve) and approximate (circles) boundary conditions. H//, ν=0.32+2.32i.

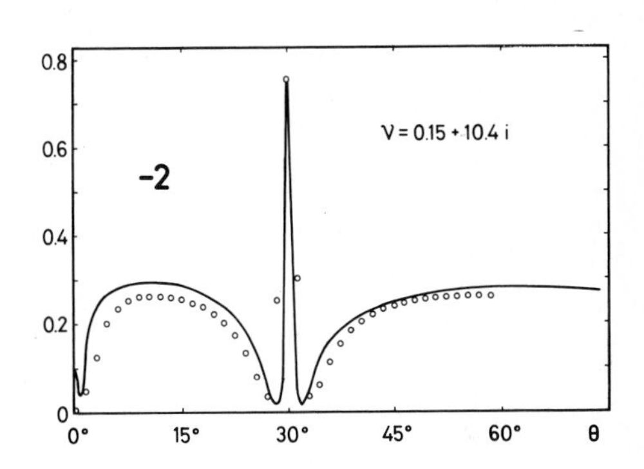

FIGURE 6
The differences between the approximate and rigorous boundary conditions (circles) cannot be represented at this value of ν. The continuous curve was obtained assuming perfect conductor.

4. CHECKS ON THE Z_o- APPROXIMATION

The comparisons between the efficiency curves obtained using the Z_o- approximation and those obtained using rigorous methods[8,9,11] can be regarded as an indirect check on the validity of the constant impedance approximation. The obvious direct check can be performed by using a rigorous method to calculate the ratio of the tangential fields along the surface. In Figs. 7-9 we can observe plots of the true value of the surface impedance as a function of angle of incidence and position along the boundary for different cycloidal gratings[12]. This kind of profile was chosen because it permits strong variations of the minimum local radius of curvature (ρ_{min}) with relatively small variations of the groove-depth to period ratio. The left side corresponds to P

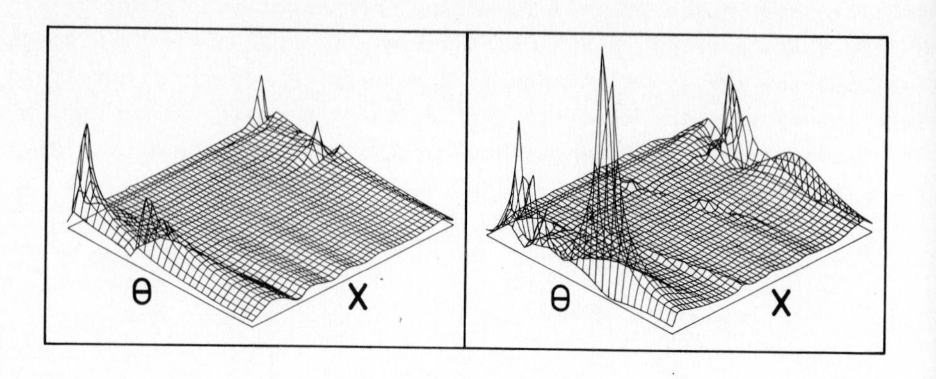

FIGURE 7
Absolute value of the surface impedance corresponding to a cycloidal diffraction grating plotted as a function of angle of incidence and position along a period of the grating surface. The left column corresponds to P polarization and the right column to S polarization. $\nu = 0.32 + 2.32\iota$, $\rho_{min}/d=0.0018$.

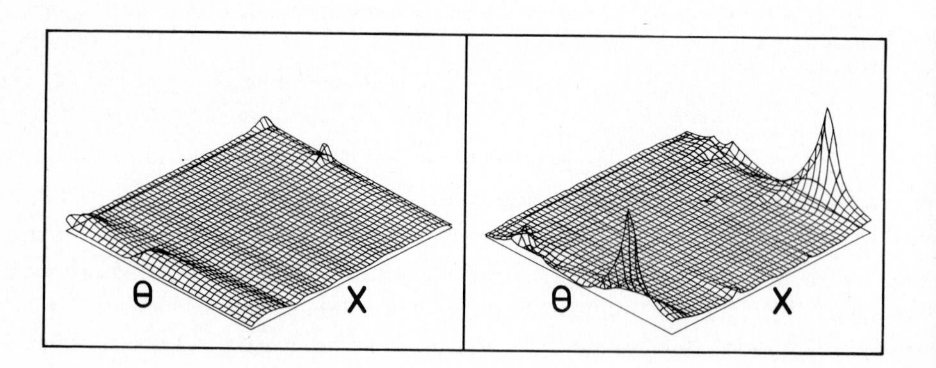

FIGURE 8
The same as in Fig. 7 but increasing the value of the refractive index. $\nu=0.15+4.65\iota$.

polarization and the right side to S polarization. In Fig. 7 we see the plots obtained for a grating with highly curved regions (ρ_{min}/d=0.0018) and a low value of the refractive index (ν=0.32+2.32i) where we do not expect the constant impedance approximation to be valid. The true surface impedance exhibits peaks located near the highly curved portion of the grating, that is at the vertex-like points of the groove profile. In Fig. 8 the groove shape is the same as in Fig. 7 and the refractive index is increased (ν=0.15+4.65i). The plots are smoother and for P polarization the surface impedance is almost a constant. Finally, for a grating with smaller local curvature (ρ_{min}/d=0.035) the plots are flat but a small peak remains in S polarization at grazing incidences (Fig. 9).

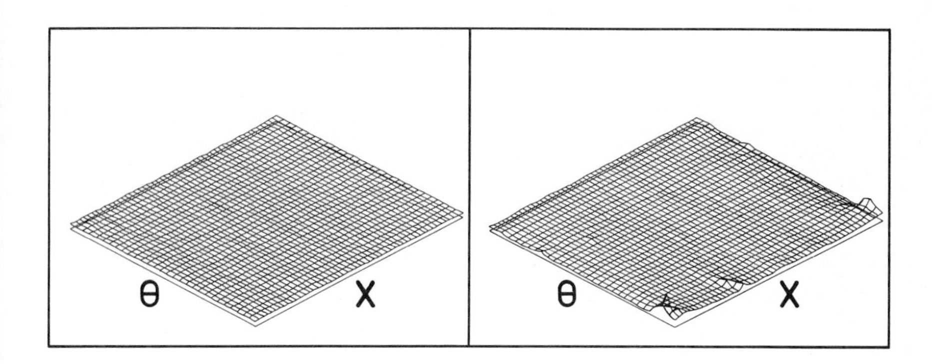

FIGURE 9
The same as in Fig. 8 but increasing the value of the refractive index. ν=0.15+4.65i.

5. ANALYTICAL EXPRESSION FOR THE SURFACE IMPEDANCE

Can we know the dependence of Z on the shape of the surface and on the angle of incidence without solving numerically the whole scattering problem? For highly conducting metals the answer is yes. It can be shown[13] that this task can be accomplished for the case of periodical cylindrical structures by means of the integral method developed by Maystre[14]. Omitting the mathematics, the final result is a coordinate dependent surface impedance which depends on the external fields only through their polarization

$$Z^{p}(x) = Z_{o}\{1 + i/ [2k\nu\ \rho(x)]\}^{-1} \tag{8}$$

$$Z^{s}(x) = Z_{o}\{1 + i/ [2k\nu\ \rho(x)]\} \tag{9}$$

$\rho(x)$ is the radius of curvature of the cylindrical periodic surface $y=f(x)$

$$\rho(x) = \{1 + f'^{2}(x)\}^{3/2} / f''(x) \tag{10}$$

$k=\omega/c$ and Zo is given by eq. (5). In the case of highly conducting materials and low surface curvatures the following relation holds

$$|\ 2k\nu\ \rho(x)\ | >> 1 \tag{11}$$

and both Z^{p} and Z^{s} tend to the value Z_{o} obtained for a flat boundary. Then Z_{o} might be considered as the zero order term in the expansion of $Z(x)$ in terms of the curvature (the curvature is defined as the inverse of the radius of curvature). These analytical expressions for the surface impedance explain the results shown in Section 3, when the Z_{o}-approximation was tested against exact methods. For fixed curvatures, better agreement is achieved when higher values of $|\nu|$ are considered because the greater $|2k\nu\ \rho(x)|$ is with respect to unity, then the better $Z(x)$ is approximated by Z_{o}. The relation $|2k\nu\ \rho_{min}|>>1$ (where ρ_{min} is the lowest value of $\rho(x)$) can be regarded as a sufficient condition for the validity of the Z_{o}-approximation. The same kind of approach used to obtain the dependence of Z on the local curvature could also be used to obtain the dependence of Z on other parameters like the angle of incidence.

Due to the fact that expressions (8) and (9) contain nothing referring to the periodicity of the diffracting surface, we could think that its validity exceeds the particular case of periodic surfaces. It has been demonstrated[15] that for a two dimensional rough surface, the analytical expressions (8) and (9) still hold.

6. EQUIVALENT SURFACE IMPEDANCE

In the previous sections the surface impedance was evaluated at the actual scattering boundary but as suggested in Fig. 1, Z can be evaluated at any closed surface Σ. For example, Hessel and Oliner[16] replaced a diffraction grating by an equivalent plane boundary having a periodic surface impedance and in that way they were able to predict in a very simple theoretical manner all the known features of the fields diffracted at a periodically modulated structure. The main advantage of this kind of approach is that the fields can be represented in terms of Rayleigh expansions everywhere above Σ.

Let $Z_g(x)$ be the surface impedance of the flat surface replacing the actual grating (period d). Under plane wave incidence the fields outside Σ can be written as

$$f(x,y) = f_i(x,y) + f_d(x,y) \quad , \tag{12}$$

$$f_i(x,y) = \exp\,[ik(\alpha_o x - \beta_o y)] \quad , \tag{13}$$

$$f_d(x,y) = \sum_n A_n \exp\,[ik(\alpha_n - \beta_n y)] \quad . \tag{14}$$

$\alpha_n=\alpha_o+n\lambda/d$, $\beta_n=(1-\alpha_n^2)^{1/2}$, $\alpha_o=\sin\vartheta_o$. f(x,y) represents either the electric field for the E// case (electric field parallel to the grooves, $\hat{z}$ direction) or the magnetic field for the H// case (magnetic field parallel to the grooves). A time dependence of the form $\exp[-i\omega t]$ is assumed. Imposing boundary condition (1) at y=0 we obtain

$$f_i+f_d = i\,[\partial_y f_i + \partial_y f_d]\,Z_g(x)/k \qquad (E//) \quad , \tag{15}$$

$$[f_i+f_d]Z_g(x) = i\,[\partial_y f_i + \partial_y f_d]/k \qquad (H//) \quad , \tag{16}$$

$Z_g(x)$ can be expanded into a Fourier series

$$Z_g(x) = \sum_{n=-\infty}^{\infty} \tilde{Z}_g \exp\,(inKx), \quad K=2\pi/d \quad , \tag{17}$$

and introducing expansion (17) into eqs. (15) and (16) we obtain a set of linear equations for the amplitudes A_n

$$\sum_n A_n\,(\beta_n \tilde{Z}_{gl-n} + \delta_{ln}) = -\delta_{lo} + \beta_o \tilde{Z}_{gl} \qquad (E//) \quad , \tag{18}$$

$$\sum_n A_n\,(\tilde{Z}_{gl-n} + \beta_l \delta_{ln}) = \beta_o \delta_{lo} - \tilde{Z}_{gl} \qquad (H//) \quad . \tag{19}$$

7. DIFFUSE LIGHT BANDS IN A REAL GRATING

The approach used by Hessel and Oliner is very simple and although it is not clear how to construct $Z_g(x)$ from the knowledge of the grating structure, it proved to be so fruitfull in the insight it provided to the problem of the diffraction by a periodic structure that recently it has been also applied to solve the problem of a rough diffraction grating[17].

When light impinges on a diffraction grating it is scattered not only into discrete directions (diffracted orders) but it also forms a diffuse background that under certain circumstances can

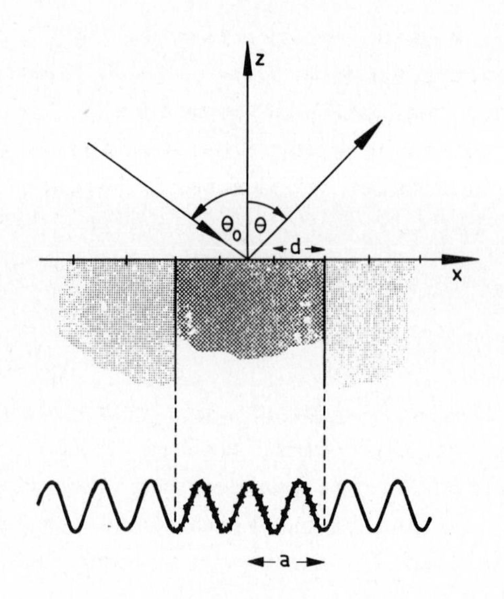

FIGURE 10
A rough diffraction grating (bottom) is replaced by a plane having surface impedance Z(x) (top). The rough portion of the grating is 2a.

present intensity maxima called diffuse light bands. Their main experimental features are[18]: i) they are only observed when the incident beam is polarized with its electric vector in the plane of incidence, ii) the pattern of the bands can be detected at all angles of incidence of the beam, iii) there are some angles of incidence for which the intensity of the bands is significantly stronger and iv) as the grating is rotated about an axis parallel to the grooves the position of the bands with respect to the grating does not change (i.e. their relation with the diffracted orders changes with the angle of incidence).

An explanation for these effects was sought in the interaction of surface waves travelling on the grating with the statistical surface roughness which exists on a real grating. Some phenomenological models[19,20] were developed using this concept but they can only account for the geometrical aspects of the process and not for the energetic ones. On the other hand, exact electromagnetic methods developed to predict the efficiency of diffraction gratings do not take into account the statistical surface roughness, and so they cannot predict the occurrence of the bands. Let us see a simple surface impedance approach to the treatment of this problem, which predicts the existence of the bands from an electromagnetic point of view. As in Section 6 we shall suppose that the real grating can be replaced by a flat surface possesing a surface impedance Z(x) varying only along a direction perpendicular to the grooves as shown in Fig. 10

$$Z(x) = Z_g(x) + Z_r(x) \tag{20}$$

but now apart from the deterministic modulation $Z_g(x)$ there is also a contribution $Z_r(x)$ due to the random roughness. The roughness is supposed to obey a gaussian distribution law characterized by its mean square height δ and its correlation length σ. Bearing in mind numerical applications we shall consider that it is spread over a finite extension of the grating, i.e. $Z_r(x)=0$ when $|x| > a$.

The total field in the region $y \geq 0$ can be written as

$$f = f_i + f_d + f_s \quad , \tag{21}$$

f_i and f_d are given by eqs. (13) and (14) and f_s is defined as the diffuse scattered field

$$f_s(x,y) = \int_{-\infty}^{\infty} r(\alpha) \exp[ik(\alpha x+\beta y)]\, d\alpha \quad , \tag{22}$$

$\beta=(1-\alpha^2)^{1/2}$. The imposition of boundary condition (1) at y=0 yields eqs. (15) and (16) representing the interaction of the fields with the periodic deterministic structure and the following equations representing the interaction of the fields with the periodic and rough surface

$$f_s = i\left[Z_g\, \partial_y f_s + Z_r\, (\partial_y f_i + \partial_y f_d + \partial_y f_s)\right] / k \qquad (E//) \quad , \tag{23}$$

$$Z_g\, f_s + Z_r\, (f_i + f_d + f_s) = i\, \partial_y f_s / k \qquad (H//). \tag{24}$$

Introducing expansions (13), (14), (17) and (22) into eqs. (23), and (24) it can be shown that the amplitude distribution of the diffuse fields $r(\alpha)$, is a solution of the following integral equations

$$\sum_n [\beta_n \tilde{Z}_{gn} + \delta_{no}]\, r(\alpha - n\lambda/d) + \int_{-\infty}^{\infty} \beta' r(\alpha')\, \tilde{Z}_r(\alpha-\alpha')\, d\alpha' = \sum_n \tilde{Z}_r(\alpha-\alpha_n)\, \beta_n [\delta_{no} - A_n] \quad (E//), \tag{25}$$

$$\sum_n [\tilde{Z}_{gn} + \beta_n \delta_{no} / k]\, r(\alpha - \lambda n/d) + \int_{-\infty}^{\infty} r(\alpha')\, \tilde{Z}_r(\alpha-\alpha')\, d\alpha' = -\sum_n \tilde{Z}_r(\alpha-\alpha_n)\, [\delta_{no} + A_n] \quad (H//), \tag{26}$$

$\tilde{Z}_r(\alpha)$ is the Fourier transform of $Z_r(x)$.

The method was implemented for the particular case of a periodic impedance

$$Z_g(x) = Z_o \left(1 + \frac{h}{d} \cos Kx\right) \quad ,$$

$h/d=0.4$, $Z_o=-1.9i$, $\lambda/d=1.5$. Solving eqs. (18) and (19) for A_n it can be shown that the amplitude of the zero order presents two strong peaks for 41° and 57° approximately in H// (Fig. 11) while for E// the zero order amplitude is a smooth function of the angle

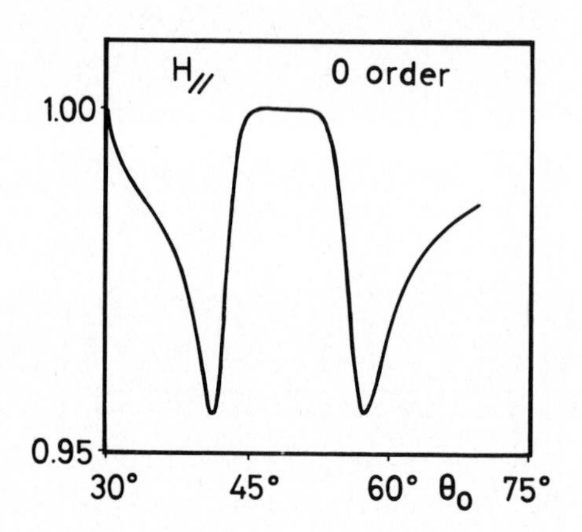

FIGURE 11
Efficiency of the order zero as a function of angle of incidence. $Z_r(x)=0$ (grating without random roughness).

of incidence. The peaks in H// correspond to the excitation of a surface wave and it is for angles of incidence near these peaks that the diffuse light bands could be expected to be stronger. The random component Z_r was chosen as a sum of rectangles of width $\sigma/d=0.25$ and mean square height $\delta/d=0.004$ spread over 6 periods

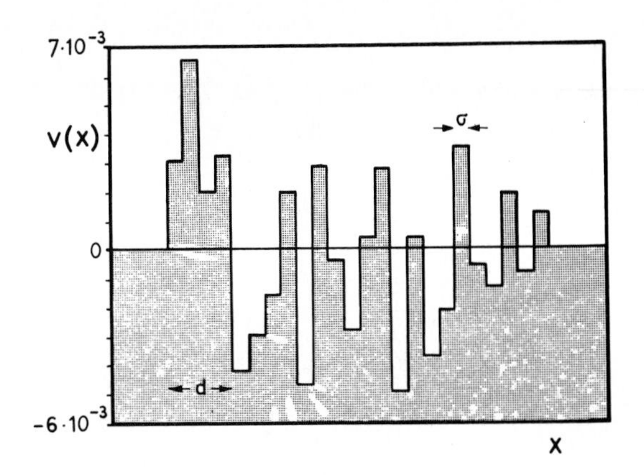

FIGURE 12
The random surface impedance is $Z_g(x) = Z_o\ v(x)$

(Fig. 12).The intensity distribution of the scattered field for several angles of incidence for the H// case is shown in Figs. 13-15. In all these plots the intensity presents two maxima located near 36° and -36°; these are the diffuse light bands. The maximum intensity increases in about two orders of magnitude when the angle of incidence changes from 30° to 41° as expected from the observation made above concerning the excitation of a surface wave. Although the intensity of the peaks is much lower for other

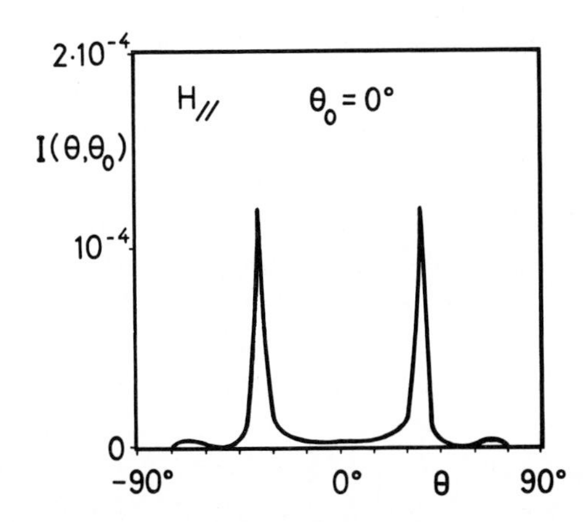

FIGURE 13
Intensity distribution of the scattered field. Normal incidence, H// polarization.

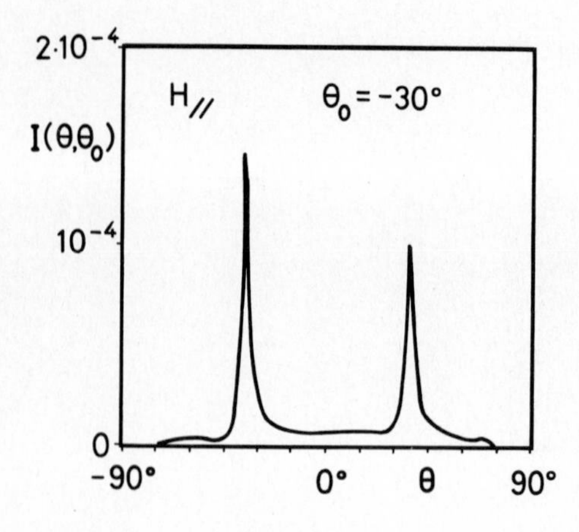

FIGURE 14
Intensity distribution of the scattered field for $\vartheta_o = -30^\circ$.

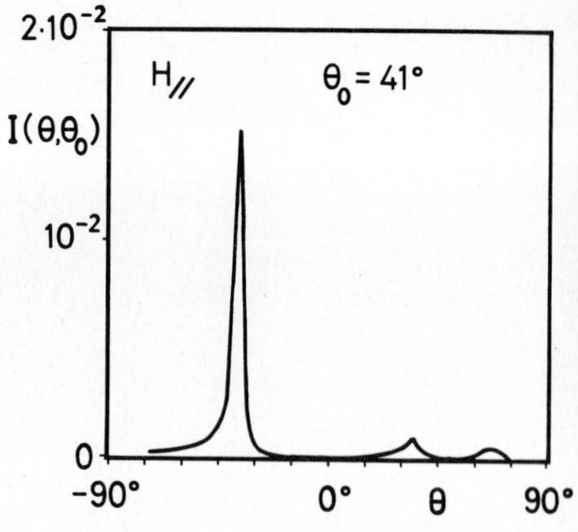

FIGURE 15
Intensity distribution of the scattered field for $\vartheta_o = 41^\circ$.

angles of incidence, the bands can always be identified and their position does not vary with the angle of incidence. In the E// case no bands can be recognized for any angle of incidence. Thus, this very simple model reproduces the main experimental features of the diffuse light background reported for real diffraction gratings.

8 CONCLUSION

The main difficulties and advantages involved in using different kinds of surface impedance boundary conditions for solving light scattering from metallic surfaces have been presented. Although not rigorous, this kind of boundary condition deserves more attention because its use:

i) simplifies the boundary value problem,

ii) gives good results for finitely conducting materials in regions where the infinite conductivity model does not hold and

iii) conserves the simplicity inherent to infinitely conducting methods.

ACKNOWLEDGMENT

Part of the results reported here were obtained in collaboration with V. L. Brudny and J. M. Simon. The author thanks Marta Pedernera for the drawing of the figures.

REFERENCES

1) R. Petit ed., Electromagnetic theory of gratings (Springer, Berlin, 1982).

2) D. Maystre, Rigorous solution of problems of scattering by large size surfaces, Workshop on Recent Progress in Surface and volume scattering, Madrid, Spain, 14-16 September 1988.

3) G. Hass, Mirror coatings, in: Applied Optics and Optical Engeneering, Vol. 3, ed. R.Kingslake (Academic Press, New York 1966) p. 309.

4) T. Senior, Appl. Sci. Res. B8 (1960) 418.

5) N. G. Alexopoulos and G. A. Tadler, J. Appl. Phys. 46 (1975) 3326.

6) D. S. Wang, IEEE Trans. Antennas Propagat. AP35 (1987) 453.

7) F. Grasso, F. Musumeci, A. Scordino, A. Triglia and L. Ronchi, Optica Acta 33 (1986) 1415.

8) R. A. Depine and J. M. Simon, Optica Acta 29 (1982) 1459.

9) R. A. Depine and J. M. Simon, Optica Acta 30 (1983) 313.

10) R. A. Depine and J. M. Simon, Optica Acta 30 (1983) 1273.

11) V. L. Brudny, R. A. Depine and J. Simon, Optik 76 (1987) 157.

12) R. A. Depine and V. L. Brudny, IEEE Antennas Prop., in print.

13) R. A. Depine, J. Opt. Soc. Am. A5 (1988) 507.

14) D. Maystre, Integral methods, in: Ref. 1 pp. 63-100.

15) R. A. Depine, Optik 79 (1988) 75.

16) A. Hessel and A. A. Oliner, Applied Optics 4 (1965) 1275.

17) R. A. Depine and V. Brudny, Journal modern Optics, in print.

18) M. C. Hutley, Diffraction gratings (Academic Press, London, 1982)

Scattering in Volumes and Surfaces
M. Nieto-Vesperinas and J.C. Dainty (Editors)

THE USE OF THE CALDERON PROJECTORS AND THE CAPACITY OPERATORS IN SCATTERING

Michel CESSENAT

CEA, CEL/V BP 27, 94190 Villeneuve Saint Georges, France.

The usual physicist concepts of surface impedance and of capacity are taken into account and generalized by the Calderon projectors and the capacity operators in the case of scalar waves which satisfy the Helmholtz equation and in the case of electromagnetic waves which satisfy the stationary Maxwell equations. We show how the scattering problem of an incident wave on a bounded obstacle can be solved by their use.
The trace spaces, which are the usual mathematical framework of these integral singular Calderon operators are given.
Many uses may be put forward : for instance the outside Calderon projector allows to transform any scattering problem with an inhomogeneous bounded obstacle Ω into a differential problem in Ω with an integral boundary condition.
The use of the Calderon projectors and the capacity operators is a very powerful means to solve numerous problems (diffraction, interference, scattering, ...) in very different situations.

1. INTRODUCTION

The notion of the Calderon projectors (and of the capacity operators) which comes from the theory of partial differential equations is a leading idea in the treatment of many problems such that scattering by surface integral method. The aim of my talk is to popularize these notions.

The applications will be limited here for a question of time to the Helmholtz and the Maxwell equations with a bounded obstacle, but it would be possible to have similar treatment for an infinite obstacle, and for other equations like the Cauchy-Rieman equation, or the elasticity equations and so on...
First I consider the

2. MODEL PROBLEM FOR HELMHOLTZ

Let Ω be a regular open bounded set of $\mathbb{R}^N$ $(N = 2,3)$, $\Omega' = \mathbb{R}^N/\bar{\Omega}$ its complement, $\Gamma = \partial\Omega$ its boundary[1]. Let k be a positive number (or more generally a complex number with $0 \leqslant \arg k < \pi$) be given.

---

[1] With Ω and Ω' connected

Problem : Find a function u ou $\mathbb{R}^N$ satisfying :

i) the *Helmholtz equation* in Ω and Ω' (with the *same* k) :

$$\Delta u + k^2 u = 0 \qquad \text{in } \mathcal{D}'(\Omega) \text{ and } \mathcal{D}'(\Omega')$$

ii) the *radiation Sommerfeld condition at infinity* ("outgoing wave" for an evolution $u(t) = e^{-i\omega t} u_0$) :

$$u(r) = 0(\frac{1}{r}), \quad \frac{\partial u}{\partial r} - iku = 0(\frac{1}{r^2})$$

for $r = |x| \to \infty$, *if k is a real number*,

iii) the **jumps** of **the boundary values** of u and its normal derivative $\frac{\partial u}{\partial n}$ across the surface Γ *are given by* ρ, ρ' :

$$\begin{cases} [u]_\Gamma = u|_{\Gamma_i} - u|_{\Gamma_e} = \rho \\ \left[\frac{\partial u}{\partial n}\right]_\Gamma = \frac{\partial u}{\partial n}|_{\Gamma_i} - \frac{\partial u}{\partial n}|_{\Gamma_e} = \rho' \end{cases}$$

iv) u is of *finite local energy* in Ω and Ω'.
Using the notations $H^1(\Omega)$, $H^1(\Gamma)$ of the Sobolev spaces, iv) may be written :
iv)' $u|_\Omega \in H^1(\Omega)$, $u|_{\Omega'} \in H^1_{loc}(\bar{\Omega}')$ and the given functions ρ and ρ' must be in the trace spaces $H^{½}(\Gamma)$, $H^{-½}(\Gamma)$ respectively :

$$\rho \in H^{½}(\Gamma), \ \rho' \in H^{-½}(\Gamma).$$

Problem : Find the boundary values of u and its normal derivative on each side of Γ :

$$\left(u|_{\Gamma_i}, \frac{\partial u}{\partial n}|_{\Gamma_i}\right), \ \left(u|_{\Gamma_e}, \frac{\partial u}{\partial n}|_{\Gamma_e}\right).$$

2.1. Resolution of the model problem

i) We write the Helmholtz equation in the sense of distributions in $\mathbb{R}^N$, using the Dirac δ_Γ distribution on Γ :

$$\Delta u + k^2 u = -(\rho' \delta_\Gamma + \text{div}(\rho\, n\, \delta_\Gamma)).$$

ii) Let Φ be the elementary (Green) outgoing solution of the Helmholtz equation :

$$\Delta \Phi + k^2 \Phi = -\delta$$

(for $N = 3$, $\Phi(x) = \dfrac{e^{ikr}}{4\pi r}$, $r = |x|$).

iii) Then there exists an unique solution of the Model Problem, and it is given by the convolution product :

$$u = \Phi * [\rho' \delta_\Gamma + \operatorname{div} (\rho\, n\, \delta_\Gamma)],$$

which may be written by (for $x \notin \Gamma$) :

$$u(x) = \int_\Gamma \left[\rho'(y)\, \Phi\, (x - y) - \rho(y) \frac{\partial \Phi}{\partial n_y} (x - y)\right] d\Gamma_y$$

u is the sum of a single-layer and of a double-layer potential :

$$u = \mathcal{L}\, \rho' + \mathcal{D}\rho.$$

iv) Using the properties of the single-layer and the double-layer potential, we can give the boundary values of u and $\dfrac{\partial u}{\partial n}$ on each side of Γ by the formula

$$\begin{pmatrix} u|_{\Gamma_i} \\ \dfrac{\partial u}{\partial n}|_{\Gamma_i} \end{pmatrix} = \underbrace{\frac{1}{2} (I + S)}_{P_i} \begin{pmatrix} \rho \\ \rho' \end{pmatrix} ; \quad \begin{pmatrix} u|_{\Gamma_e} \\ \dfrac{\partial u}{\partial n}|_{\Gamma_e} \end{pmatrix} = \underbrace{-\frac{1}{2} (I - S)}_{P_e} \begin{pmatrix} \rho \\ \rho' \end{pmatrix}$$

with S a matrix operator :

$$S = \begin{pmatrix} K & 2L \\ 2R & J \end{pmatrix},$$

where K, L, J, R are singular integral operators *on* Γ, defined by :

$$\begin{cases} L\, \rho'(z) = \displaystyle\int_\Gamma \rho'(y)\, \Phi\, (z - y)\, d\Gamma_y \\ J\, \rho'(z) = 2 \displaystyle\int_\Gamma \rho'(y) \frac{\partial \Phi}{\partial n_z} (z - y)\, d\Gamma_y \end{cases} \qquad z \in \Gamma$$

(given by the single-layer potential) and :

$$\begin{cases} K\rho(z) = -2\int_\Gamma \rho(y)\,\dfrac{\partial\Phi}{\partial n_y}(z-y)\,d\Gamma_y \\ \\ R\rho(z) = -2\int_\Gamma \rho(y)\,\dfrac{\partial^2\Phi}{\partial n_z\,\partial n_y}(z-y)\,d\Gamma_y \end{cases} \quad , \quad z \in \Gamma,$$

(given by the double layer potential).

Properties of K, L, J, R :

i) Transposition:

$${}^tL = L,\quad {}^tJ = -K,\ {}^tK = -J,\ {}^tR = R$$

ii) "Regularity properties" : $\forall s \in \mathbb{R}$

$$\begin{cases} L \text{ is an isomorphism of } H^s(\Gamma) \text{ onto } H^{s+1}(\Gamma) \\ I \pm J,\ I \pm K \text{ are isomorphisms of } H^s(\Gamma) \\ R \text{ is an isomorphism of } H^s(\Gamma) \text{ onto } H^{s-1}(\Gamma) \end{cases}$$

<u>if</u> k^2 is not an eigenvalue of $-\Delta_D$ or of $-\Delta_N$ in Ω.

$$H^s(\Gamma) \underset{R}{\overset{L,\ J,\ K}{\rightleftarrows}} H^{s+1}(\Gamma)$$

1) $k^2 \notin \sigma(-\Delta_D) \Leftrightarrow L,\ I - J,\ I + K$ are isomorphisms

$k^2 \in \sigma(-\Delta_D) \Leftrightarrow 0 \in \sigma(C) \Leftrightarrow 1 \in \sigma(J) \Leftrightarrow -1 \in \sigma(K)$

2) $k^2 \notin \sigma(-\Delta_N) \Leftrightarrow K,\ I + J,\ I - K$ are isomorphisms

$k^2 \in \sigma(-\Delta_N) \Leftrightarrow 0 \in \sigma(R) \Leftrightarrow -1 \in \sigma(J) \Leftrightarrow 1 \in \sigma(K)$

2.2. Calderon projectors :

The operators $P_i = \frac{1}{2}(I + S)$, $P_e = \frac{1}{2}(I - S)$ are projectors, because if you do again the same trick with ρ and ρ' equal to the interior or the exterior boundary values of u and its normal derivative $\frac{\partial u}{\partial n}$, you will find the same values

$$P_i^2 = P_i \quad , \quad P_e^2 = P_e$$

but they are not *hermitian operators*. P_i is a projector from the space of the jumps (ρ, ρ'): $X = H^{1/2}(\Gamma) \times H^{-1/2}(\Gamma)$ onto the *space of the boundary values* of u and its normal derivative on Γ when u satisfies the Helmholtz equation in the *interior* domain Ω ; P_e is a projector from the space of the jumps (ρ, ρ') $X = H^{1/2}(\Gamma) \times H^{-1/2}(\Gamma)$ onto the space of the *boundary values* of u and its normal derivative on Γ, when u is an outgoing wave in the *exterior* domain Ω'.

First consequences of the properties of P_i and P_e :

$$P_i + P_e = I(\text{identity}),\ P_i - P_e = S,$$

with $S^2 = I$ which is equivalent to the following relations :

$$\begin{cases} KL + LJ = 0 & RK + JR = 0 \\ K^2 + 4L\,R = I & J^2 + 4\,RL = I \end{cases}.$$

Second consequences :

Let G_e and G_i be the image spaces of P_e and P_i :

$$G_e = \operatorname{Im} P_e\ ,\quad G_i = \operatorname{Im} P_i$$

then

$$G_e = \operatorname{Ker} P_i\ ,\quad G_i = \operatorname{Ker} P_e$$

G_e and G_i are complementary spaces, but not *orthogonal spaces*

Diagram of decomposition :

$$(\rho,\rho') \in X = H^{1/2}(\Gamma) \times H^{1/2}(G) = G_e \oplus G_i \quad \begin{array}{l} \xrightarrow{P_e} - \left(u|_{\Gamma_e}, \dfrac{\partial u}{\partial n}|_{\Gamma_e}\right) \in G_e \\ \xrightarrow{P_i} \left(u|_{\Gamma_i}, \dfrac{\partial u}{\partial n}|_{\Gamma_i}\right) \in G_i \end{array}$$

2.3. Capacity operators :

i) G_e is the graph of an operator, the capacity operator for the exterior domain :

$$C_e\ :\ u|_{\Gamma_e} \in H^{1/2}(\Gamma) \to \frac{\partial u}{\partial n}|_{\Gamma_e} \in H^{-1/2}(\Gamma)$$

C_e is an *isomorphism* from $H^s(\Gamma)$ onto $H^{s-1}(\Gamma)$.

ii) If $-k^2$ is not *an eigenvalue of the Laplacien* with the *Dirichlet condition*, G_i is the graph of an operator, the capacity operator for the interior domain:

$$C_i\ :\ u|_{\Gamma_i} \in H^{1/2}(\Gamma) \to \frac{\partial u}{\partial n}|_{\Gamma_i} \in H^{-1/2}(\Gamma).$$

if Γ is of C^∞ regularity.

If $-k^2 \notin \sigma(\Delta_D) \cup \sigma(\Delta_N)$, C_i is an isomorphism from $H^s(\Gamma)$ onto $H^{s-1}(\Gamma)$, if Γ is of $\mathcal{C}^\infty$ regularity. These operators may be define directly, by considering limit problems : exterior or interior Dirichlet problem. By considering exterior or interior Neumann problem, we show that C_i and C_e are isomorphisms. If $-k^2$ is not an eigenvalue of the Laplacien the capacity operators are given with the integral operators L, J, K, R, by :

$$\begin{cases} C^e = -\dfrac{1}{2} L^{-1}\ (I + K) = -\ 2\ (I + J)^{-1}\ R \\ C^i = -\dfrac{1}{2} L^{-1}\ (-\ I + K) = -\ 2\ (-\ I + J)^{-1}\ R; \end{cases}$$

C^i and C^e are pseudo-differential operators. The capacity operators occur naturally by the Green formula. They may be thought as surface impedance. They have natural positivity properties in the following sense :

i) Interior capacity :

$$\text{Im}\ (-\,C^i u,\ u) = \text{Re}\ (i\ C^i u,\ u) = 2\ k_R\ k_I \int_\Omega |u|^2\ dx$$

$$\text{if } k = k_R + ik_I$$

ii) Exterior capacity :

$$\text{Im}\ (C^e u,\ u) = \text{Re}\ (-\ i\ C^e u,\ u) = k \int_{\Gamma^2} |\mathcal{F}(\alpha)|^2\ d\alpha;$$

it is the total radar cross section ($\mathcal{F}(\alpha)$ is the radiation pattern). These operators are "symmetric" i.e.: ${}^tC^e = C^e$, ${}^tC^i = C^i$ if k is real $(C^i)^* = C^i$

3. AN EXAMPLE OF THE USE OF THE CALDERON PROJECTORS: A SCATTERING PROBLEM

Let an incident wave u_I be given (for instance produced by a finite source, or a plane wave) propagating in an homogeneous isotropic medium (the vacuum) with a wavenumber k, falling on a bounded penetrable obstacle (an other homogeneous isotropic medium) in which the wave number will be k_i.

Scattering problem :

Find the "total" wave $u_t = u_I + u$ in Ω' (u being the scattered wave), and $u_t = u$ in Ω such that :

i) u is a solution of the Helmholtz equations

$$1)\quad \begin{cases} \Delta u + k^2 u = 0 & \text{in } \Omega' \quad k \text{ real} \geqslant 0 \\ \Delta u + k_i^2 u = 0 & \text{in } \Omega \quad k_i \in \mathbb{C} \quad 0 \leqslant \arg k_i < \pi \end{cases}$$

ii) u is an outgoing wave in Ω',

iii) the total wave u_t and its normal derivation have no jump across the boundary Γ of Ω :

$$[u_t]_\Gamma = 0 \quad , \quad \left[\frac{\partial u_t}{\partial n}\right]_\Gamma = 0$$

which is equivalent to : the jumps of u and its normal derivative are given by :

$$2)\quad [u]_\Gamma = u_I|_\Gamma \quad , \quad \left[\frac{\partial u}{\partial n}\right]_\Gamma = \frac{\partial u_I}{\partial n}|_\Gamma .$$

The problem is equivalent to find the boundary values of u and $\frac{\partial u}{\partial n}$ on each side of Γ : $\left(u|_{\Gamma_i}, \frac{\partial u}{\partial n}|_{\Gamma_i}\right)$, $\left(u|_{\Gamma_e}, \frac{\partial u}{\partial n}|_{\Gamma_e}\right)$.

Using the Calderon projectors, we have (by *definition*)

$$3)\quad P_i^{k_i}\left(u|_{\Gamma_i}, \frac{\partial u}{\partial n}|_{\Gamma_i}\right) = \left(u|_{\Gamma_i}, \frac{\partial u}{\partial n}|_{\Gamma_i}\right) \Leftrightarrow P_e^{k_i}\left(u|_{\Gamma_i}, \frac{\partial u}{\partial n}|_{\Gamma_i}\right) = 0$$

$$4)\quad P_e^{k}\left(u|_{\Gamma_e}, \frac{\partial u}{\partial n}|_{\Gamma_e}\right) = \left(u|_{\Gamma_e}, \frac{\partial u}{\partial n}|_{\Gamma_e}\right) \Leftrightarrow P_i^{k}\left(u|_{\Gamma_e}, \frac{\partial u}{\partial n}|_{\Gamma_e}\right) = 0$$

Then writing the transmission relation (2) by :

$$5)\quad \left(u|_{\Gamma_e}, \frac{\partial u}{\partial n}|_{\Gamma_e}\right) + \left(u_I|_{\Gamma}, \frac{\partial u_I}{\partial n}|_{\Gamma}\right) = \left(u|_{\Gamma_i}, \frac{\partial u}{\partial n}|_{\Gamma_i}\right)$$

and applying the projectors P_i^k or $P_e^{k_i}$ to it, we obtain the equations :

$$6)\quad P_i^k\left(u|_{\Gamma_i}, \frac{\partial u}{\partial n}|_{\Gamma_i}\right) = \left(u_I|_{\Gamma}, \frac{\partial u_I}{\partial n}|_{\Gamma}\right)$$

$$7)\quad P_e^{k_i}\left(u|_{\Gamma_e}, \frac{\partial u}{\partial n}|_{\Gamma_e}\right) = P_e^{k_i}\left(u_I|_{\Gamma}, \frac{\partial u_I}{\partial n}|_{\Gamma}\right)$$

Then the interior boundary values $\left(u|_{\Gamma_i}, \frac{\partial u}{\partial n}|_{\Gamma_i}\right)$ are determined by the equations (3) and (6), and the exterior boundary values $\left(u|_{\Gamma_e}, \frac{\partial u}{\partial n}|_{\Gamma_e}\right)$ are determined by the equations (4) and (7). There are 4 equations for 2 unknowns.

[The problems is equivalent to do the decomposition :

$$X = H^{1/2}(\Gamma) \times H^{-1/2}(\Gamma) = G_i^{k_i} \oplus G_e^{k}$$

from the two decompositions :

$$X = H^{1/2}(\Gamma) \times H^{-1/2}(\Gamma) = G_i^{k} \oplus G_e^{k} = G_i^{k_i} \oplus G_e^{k_i}]$$

- It is possible to reduce the number of equations and unknowns to one by surching (for example) the exterior boundary values in the following form (with the new unknowns ρ') :

$$8) \quad \left(u|_{\Gamma_e}, \frac{\partial u}{\partial n}|_{\Gamma_e}\right) = -P_e^k(0,\rho'), \; \rho' \in H^{-1/2}(\Gamma)$$

equivalent to surching u as a single-layer potential in Ω', like Maystre-Vincent[2], or in the form :

$$9) \quad \left(u|_{\Gamma_e}, \frac{\partial u}{\partial n}|_{\Gamma_e}\right) = -P_e^k(\rho, 0), \; \rho \in H^{1/2}(\Gamma),$$

with the new unknown ρ, which is equivalent to surching u as a double-layer potential in Ω' [because the projector P_e^k is an isomorphism from $\{0\} \times H^{-1/2}(\Gamma)$ and from $H^{1/2}(\Gamma) \times \{0\}$ onto G_e^k if k^2 is not an eigenvalue for the interior problem (irregular frequencies)]. This method may be called the method of equivalent current (M.E.C). By (8) or (9), the relation (4) is satisfied, and the relation (7) give the equations :

$$10) \quad P_e^{k\,i} P_e^k(0, \rho') = -P_e^{k\,i}\left(u_I|_\Gamma, \frac{\partial u_I}{\partial n}|_\Gamma\right)$$

or

$$11) \quad P_e^{k\,i} P_e^k(\rho, 0) = -P_e^{k\,i}\left(u_I|_\Gamma, \frac{\partial u_I}{\partial n}|_\Gamma\right)$$

Thus there are 2 equations for one unknown ρ' (or ρ), but the two equations are equivalent [because the projectors P_1 and P_2 on the first (resp second) component (of $X = H^{1/2}(\Gamma) \times H^{-1/2}(\Gamma)$) are isomorphisms when reduced to G_e $\forall k$, (or to G_i (if k^2 is not an eigenvalue)).

To eliminate the irregular frequencies, it seems better, like Kleinman-Martin[1], to surch the exterior boundary values $\left(u|_{\Gamma_e}, \frac{\partial u}{\partial n}|_{\Gamma_e}\right)$ in the following form (with the new unknown μ)

$$12) \quad \left(u|_{\Gamma_e}, \frac{\partial u}{\partial n}|_{\Gamma_e}\right) = -P_e^k(a\,\mu, b\,\mu)$$

with $a, b \in \mathbb{C}$, $\mu \in H^{1/2}(\Gamma)$.

Let us define the space V by :

$$V = V_{a,b} = \{(a\,\mu, b\,\mu), \; \mu \in H^{1/2}(\Gamma)\}$$

(V is contained in the space $X = H^{1/2}(\Gamma) \times H^{-1/2}(\Gamma)$).

Then the restriction of P_e^k to V is an isomorphism onto G_e^k if k^2 is not an eigenvalue of the problem :

$$\begin{cases} \Delta u + k^2 u = 0 \\ a \dfrac{\partial u}{\partial n} - bu|_\Gamma = 0 \end{cases}$$

If the imaginary part of the ratio $\lambda = b/a$ is not equal to zero, there are *no real eigenvalue*, so μ is determined by the two (equivalent) equations :

$$13) \quad P_e^{k\,i}\ P_e^k\ (a\ \mu,\ b\ \mu) = -\ P_e^{k\,i} \left(u_I\,|_\Gamma,\ \frac{\partial u_I}{\partial n}|_\Gamma\right).$$

For other choices see for example Kleinman-Martin[1].

4. ELECTROMAGNETISM ; MODEL PROBLEM

Find the electromagnetic field (E, H) satisfying:

i) the Maxwell equations in Ω and Ω', with $\Omega \subset \mathbb{R}^3$

$$\begin{cases} \text{curl } H + i\ \omega\ \varepsilon\ E = 0 \\ \text{curl } E - i\ \omega\ \mu\ H = 0 \end{cases}$$

with ω, ε, μ constants with the same value in Ω and Ω', with the conditions :

$$0 \leqslant \arg\ (\omega\ \varepsilon) < \pi \qquad 0 \leqslant \arg\ (\omega\ \mu) < \pi.$$

- Then we define the wave number k, and the admittance Z by :

$$\begin{cases} k^2 = \omega^2\ \varepsilon\ \mu \qquad 0 \leqslant \arg k < \pi \\ Z = \dfrac{\omega\ \varepsilon}{k} = \dfrac{k}{\omega\ \mu},\ Z^2 = \dfrac{\varepsilon}{\mu}\ ;\ -\dfrac{\pi}{2} < \arg Z < \dfrac{\pi}{2}\ - \end{cases}$$

ii) the Silver-Müller conditions at infinity

$$\begin{cases} E(x) = O\left(\dfrac{1}{r}\right),\ H(x) = O\left(\dfrac{1}{r}\right) \\ \omega\ \varepsilon\,(\alpha \wedge E) - k\ H = o\left(\dfrac{1}{r}\right),\ \omega\ \mu(\alpha \wedge H) + kE = o\left(\dfrac{1}{r}\right) \end{cases}$$

(uniformly in α) $r = |x|$, $\alpha = \dfrac{x}{r}$;

iii) the jumps of the tangential components of the electromagnetic field are given by electric and magnetic currents on the surface Γ

$$[n \wedge E]_\Gamma = j' \quad , \quad [n \wedge H]_\Gamma = - j$$

iv) the electromagnetic field is of locally finite energy in Ω and Ω':

E, H curl E, curl $H \in L^2\ (\Omega)^3$ and $L^2_{loc}\ (\bar{\Omega}')^3$

4.1. Resolution of the Model problem

i) From the above conditions we have the mathematical framework for the trace of the tangential components of (E, H) or of the exterior product of the normal with (E, H) on Γ :

$$n \wedge E|_\Gamma,\ n \wedge H|_\Gamma \in H^{-1/2}(\text{div},\ \Gamma)\ :$$

$$H^{-1/2}(\text{div},\Gamma) = \left\{ j_\Gamma \in H^{-1/2}(\Gamma)^3,\ n\ .\ j_\Gamma = 0,\ \text{div}_\Gamma\ j_\Gamma \in H^{-1/2}(\Gamma) \right\}$$

$$\text{or } E_t\,|_\Gamma,\ H_t\,|_\Gamma \in H^{-1/2}\ (\text{curl},\ \Gamma)$$

$$H^{-1/2}(\text{curl},\Gamma) = \left\{ \tilde{j}_\Gamma \in H^{-1/2}(\Gamma)^3,\ n.\tilde{j}_\Gamma = 0,\ \text{curl}_\Gamma\ \tilde{j}_\Gamma \in H^{-1/2}(\Gamma) \right\}$$

ii) Remark: The jumps of n.E, n.H are determined by :

$$\begin{cases} i\ \omega\ \varepsilon\ [n.E]_\Gamma = \text{div}_\Gamma([n \wedge H]_\Gamma) \\ -i\ \omega\ \mu\ [n.H]_\Gamma = \text{div}_\Gamma([n \wedge E]_\Gamma) \end{cases}$$

iii) Write the Maxwell equations in the sense of distributions on $\mathbb{R}^3$:

$$\begin{cases} \text{curl } H + i\ \omega\ \varepsilon\ E = -\ [\ n \wedge H]_\Gamma\ \delta_\Gamma = j_\Gamma\ \delta_\Gamma \\ -\text{curl } E + i\ \omega\ \mu\ H = [n \wedge E]_\Gamma\ \delta_\Gamma = j'_\Gamma\ \delta_\Gamma \end{cases}$$

The solution is obtained by the convolution :

$$\begin{cases} E = (i\ \omega\ \mu\ j\ \delta_\Gamma - \text{curl}\ (j'\ \delta_\Gamma) - \dfrac{1}{\varepsilon}\ \text{grad}\ \rho) * \Phi \\ H = (i\ \omega\ \varepsilon\ j'\ \delta_\Gamma + \text{curl}\ (j\ \delta_\Gamma) - \dfrac{1}{\mu}\ \text{grad}\ \rho') * \Phi \end{cases}$$

with ρ and ρ' defined by :

$$i\ \omega\ \rho = \text{div}_\Gamma j\ , \quad i\ \omega\ \rho' = \text{div}_\Gamma\ j'$$

These relations can be written in the Chu-Stratton form :

$$\begin{cases} E = \int_\Gamma \left[i\,\omega\,\mu\,\Phi\,j - \text{grad}\,\Phi \wedge j' - \frac{1}{\varepsilon}\,\text{grad}\,\Phi\,\rho \right] d\Gamma \\ H = \int_\Gamma \left[i\,\omega\,\varepsilon\,\Phi\,j' + \text{grad}\,\Phi \wedge j - \frac{1}{\mu}\,\text{grad}\,\Gamma\,\rho' \right] d\Gamma. \end{cases}$$

4.2. The Calderon propectors

Thus the boundary values of the exterior product of the normal n with (E, H) are given by the operators P_i and P_e (the Calderon projectors) :

$$\begin{pmatrix} n \wedge E|_{\Gamma_i} \\ - n \wedge H|_{\Gamma_i} \end{pmatrix} = P_i \begin{pmatrix} j' \\ j \end{pmatrix}, \quad \begin{pmatrix} - n \wedge E|_{\Gamma_e} \\ n \wedge H|_{\Gamma_e} \end{pmatrix} = P_e \begin{pmatrix} j' \\ j \end{pmatrix}$$

with a matrix operator S :

$$P_i = \frac{1}{2}(I + S), \quad P_e = \frac{1}{2}(I - S)$$

which is of the following form :

$$S = \begin{pmatrix} - M & \frac{1}{Z} R \\ - Z\,R & - M \end{pmatrix}$$

with the singular integral operators R, M on Γ:

$$\begin{cases} Rj = 2\,ik\,n \wedge [Lj + \frac{1}{k^2}\,\text{grad}_\Gamma\,L\,\text{div}_\Gamma\,j] \\ (\text{with } Lj = \int_y j\,\Phi\,d\Gamma) \\ \text{and} \\ Mj(z) = 2\int_\Gamma n_z \wedge (j(y) \wedge \text{grad}_z\,\Phi(z - y))\,d\Gamma_y \\ \text{or} \\ Mj(z) = -2\int_\Gamma [j(y)\,\frac{\partial\Phi}{\partial n_z}(z - y) - (n_z \cdot j(y))\,\text{grad}\,\Phi\,(z - y)]\,d\Gamma_y \end{cases}$$

First consequences of the properties of P_i and P_e :

$$P_i + P_e = I, \quad P_i - P_e = S$$

with $S^2 = I$ (identity in $H^{-½}(\text{div}, \Gamma)^2$). Thus:

$$M^2 - R^2 = I \ , \ MR + RM = 0.$$

Second consequences :

Let G_e and G_i be the images of the Calderon projectors P_e and P_i

$$G_e = \text{Im } P_e \quad , \quad G_i = \text{Im } P_i$$

$$\text{then } G_e = \text{Ker } P_i \quad , \quad G_i = \text{Ker } P_e$$

G_e and G_i are complementary spaces but not orthogonal spaces

Diagram of decomposition

$$(j', j) \in H = H^{\frac{1}{2}}(\text{div}, \Gamma)^2 = G_i \oplus G_e \quad \xrightarrow{P_i} G_i \quad \xrightarrow{P_e} G_e$$

4.3. The capacity operators C_e and C_i

i) G_e is always the graph of an operator $- C_e$:

$$C_e \ (n \wedge E|_{\Gamma_e}) = n \wedge H|_{\Gamma_e}$$

ii) G_i is the graph of an operator $- C_i$:

$$C_i \ (n \wedge E|_{\Gamma_i}) = n \wedge H|_{\Gamma_i}$$

iff k^2 is not an eigenvalue of the operator A_0 in the space

$$\mathcal{H}_0 = \{E \in L^2(\Omega)^3, \ \text{div } E = 0\}$$

defined by :

$$A_0 \ E = - \Delta E$$

$$D(A_0) = \left\{E, \ \text{cur } E, \ \text{curl curl } E \in L^2(\Omega)^3, \ \text{div } E = 0, \ n \wedge E|_\Gamma = 0\right\}$$

$$= \left\{E \in H^1(\Omega)^3, \ \Delta E \in L^2(\Omega)^3, \ \text{div } E = 0, \ n \wedge E|_\Gamma = 0\right\} ;$$

A_0 is a self adjoint operator with compact resolvant.

The capacity operators can be defined directly by considering limit problems. They occur naturally by the Green formulas. They correspond to the physical idea of *surface impedance*. They have also *positivity properties* (given by the flux of the Poynting vector) :

i) Interior :

$$\mathrm{Re} \int_\Gamma n \cdot E \wedge \bar{H} \, d\Gamma = \int_\Omega [\mathrm{Re}\,(\overline{-\, i\,\omega\,\varepsilon})\, E \cdot \bar{E} + \mathrm{Re}\,(i\,\omega\,\mu)\, H \cdot \bar{H}]\, dx \leqslant 0.$$

Thus with $c = n \wedge E|_\Gamma$

$$\mathrm{Re} \int_\Gamma - C^i c \cdot n \wedge \bar{c} \geqslant 0$$

(= 0 is ε and μ are real).

ii) Exterior :

$$\mathrm{Re} \int_\Gamma n \cdot E \wedge \bar{H} \, d\Gamma > 0,$$

$$\mathrm{Re} \int_\Gamma n \cdot E \wedge \bar{H} \, d\Gamma = \frac{Z}{(4\pi)^2} \int_{S^2} |\mathcal{F}_E(\alpha)|^2 \, d\alpha \text{ (for \underline{real Z})}$$

$\mathcal{F}_E(\alpha)$ is the *radiation pattern*.

$$\mathrm{Re} \int_\Gamma C^e c \cdot n \wedge \bar{c} = \frac{Z}{(4\pi)^2} \int_{S^2} |\mathcal{F}_E(\alpha)|^2 d\alpha \geqslant 0$$

(Z real $\rightarrow$ $Z \geqslant 0$).

Dependence in ε, μ, ω :

C^e and C^i depend of ε, μ, ω : $C^{e,i} = C^{e,i}_{\varepsilon,\,\mu,\,\omega}$, but are the product of the admittance Z by operators C^e_k and C^i_k which depend only of k :

$$C^{e,i}_{\varepsilon,\,\mu,\,\omega} = Z\, C^{e,i}_k$$

and which have the following property :

$$\left(C^{e,i}_k\right)^2 = -\, I \rightarrow \left(C^{e,i}_k\right)^{-1} = -\, C_k$$

for the interior or the exterior.

The *relations* with the singular integral operators R and M are given by :

$$\begin{cases} C^e_k = R^{-1}(I - M) = -(I - M)^{-1}\, R = (I + M)\, R^{-1} \\ C^i_k = -R^{-1}(I + M) = (I + M)^{-1}\, R = -(I + M)\, R^{-1}, \end{cases}$$

iff k^2 is not an eigenvalue of A_0.

The scattering problems could be solved with the same formalism and the same ideas as for the Helmholtz equation.

REFERENCES

1) R.E.Kleinman, P.A. Martin. On single integral equations for the transmission problem of acoustics. Siam J. Appl. Math. Vol. 48 n° 2 April 1988.

2) D. Maystre and P. Vincent. Diffraction d'une onde électromagnétique plane par un objet cylindrique non infiniment conducteur de section arbitraire, Optics Commun 5 (1972), pp. 327-330.

Scattering in Volumes and Surfaces
M. Nieto-Vesperinas and J.C. Dainty (Editors)

GRATINGS AS ELECTROMAGNETIC FIELD AMPLIFIERS FOR SECOND HARMONIC GENERATION

R. Reinisch

Laboratoire d'Electromagnétisme, Micro-Ondes et Opto-Electronique UA CNRS n° 833,
23 Avenue des Martyrs - 38016 GRENOBLE Cédex FRANCE

M. Nevière

Laboratoire d'Optique Electromagnétisme - UA CNRS n° 843, Faculté des Sciences et Techniques, Centre de Saint Jérôme Service 262 - 13397 MARSEILLE Cédex 13 FRANCE

1. INTRODUCTION

Various nonlinear optical effects, among which are Raman Scattering and Second Harmonic Generation, can be enhanced by several orders of magnitude by surface roughness. In these surface enhanced phenomenae, electromagnetic resonances linked with the excitation of surface waves play a key role. Two kinds of surface waves will be considered here, the surface plasmon resonance which occurs at a vacuum-metal interface for TM polarization, and the guided wave resonance which can be excited inside a corrugated waveguide for both TE and TM polarizations. The excitation of one of these surface waves results in an important increase of the local field, which produces an increase of the nonlinear polarization, i.e. of the sources of a Second Harmonic or a Raman signal. Such signal field is then diffracted by the rough surface. If we idealize the rough surface by a grating, the determination of the signal field confronts us to the resolution of a grating diffraction problem in nonlinear optics. We concentrate here on Second Harmonic Generation (SHG). Two main theories have been developed depending on the fact that the nonlinear medium is a metal or a dielectric. For metals, the problem is mainly tackled by the integral method, previously developed for gratings in linear optics[1,2], although a differential technique has also been derived and implemented numerically. For dielectrics, the differential method for gratings[3] has been generalized to nonlinear optics. As a result, we have now at our disposal a new electromagnetic theory of gratings in nonlinear optics which is able to calculate and optimize the signal diffracted by metallic gratings, dielectric gratings, dielectric coated metallic gratings whatever the nonlinear medium (metal or dielectric) may be, and for any polarization (TE or TM) of pump beams and signal.

The aim of this chapter is to present the formalism of diffraction in nonlinear optics.

2. ENHANCED SECOND HARMONIC GENERATION BY GRATINGS : GENERAL CONSIDERATIONS

Figure 1 describes the main grating configurations which may be used to increase SHG through electromagnetic resonances. In figures 1.a and 1.b the pump field will have to be TM polarized in order to be able to excite surface plasmons or polaritons. In figure 1.c the incident field will be TM or TE polarized depending on the fact that surface plasmons (SP) or guided waves (GW) are excited.

Thus we have to consider the interaction $(\omega,\omega) \rightarrow 2\omega = \omega + \omega$ in modulated structures (gratings). This interaction corresponds to a multistep process : a pump beam with frequency ω, incident on a bare or coated grating, is diffracted by the grating. The electromagnetic (EM) resonance involves one of the evanescent diffracted orders. The interaction among all the diffracted orders at frequency ω gives rise to a nonlinear polarization at frequency 2ω. This polarization radiates an EM field at 2ω, which in turn is diffracted by the grating. Since the frequency of the diffracted signal (2ω) differs from that of the incident field (ω), this kind of enhanced SHG appears as a special case of diffraction in nonlinear optics. We compute the enhancement of the signal with respect to the associated smooth structure, i.e., deduce it from the modulated structure by letting the groove depth go to zero.

Our goal is :

a) to determine the directions of diffraction of the SH field at circular frequency 2ω, i.e. to derive the grating equation in nonlinear optics, equivalent to the well known Fraunhoffer equation in linear optics ;

b) to compute the diffracted efficiencies in the different spectral orders at SH frequency in the absence of any electromagnetic resonance ;

c) to do the same work as b) but in presence of different kinds of EM resonances (SP, GW,...) and compute the induced EM resonance enhancement in order to determine the most efficient resonance ;

d) to study the influence of the grating parameters (groove spacing, groove depth, groove shape, grating material...) on the SH intensity.

3. ELECTROMAGNETIC THEORY OF GRATING DIFFRACTION FOR SHG

We consider the geometry shown in figure 1. A pump beam (frequency ω) is incident at angle θ on a bare or coated metallic grating (periodicity d, groove depth δ).

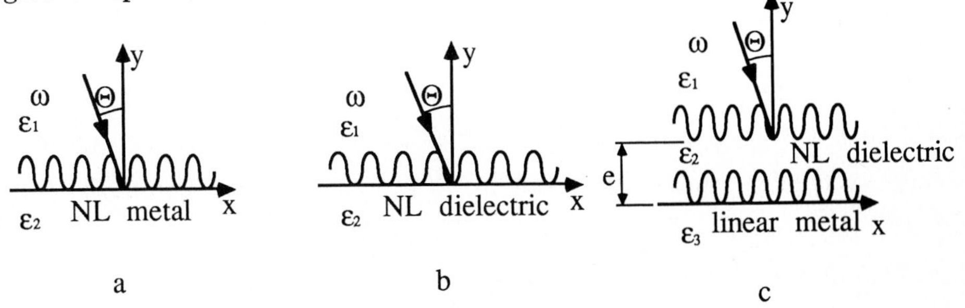

Figure 1
Main grating configurations

We assume $\partial/\partial z = 0$. Therefore, the signal at 2ω is either TM [specified by $H_z(2\omega)$] or TE [specified by $E_z(2\omega)$].

We make the usual undepleted pump approximation for the pump beam. Thus, the Maxwell equations at frequency ω read (with an $e^{-j\omega t}$ time dependance).

$$\nabla \times \mathbf{E}(\omega) = j\omega\mu_0\mathbf{H}(\omega) \qquad (1.a)$$
$$\nabla \times \mathbf{H}(\omega) = -j\omega\varepsilon_0\varepsilon(\omega)\mathbf{E}(\omega) \qquad (1.b)$$

From these two equations, we see that the pump beam is linearly diffracted by the grating and that this diffraction occurs independently from that of the signal at 2ω. Consequently, we can study the diffraction of the pump beam using standard methods : differential, integral, Yasuura and modal methods[1-4].

Now let us focus on the diffraction process at 2ω. The starting point is the Maxwell equations at 2ω (with a time dependence of $e^{-2j\omega t}$) :

<u>In the linear media</u> :

$$\nabla \times \mathbf{E}_q(2\omega) = j2\omega\mu_0\mathbf{H}_q(2\omega) \qquad (2.a)$$
$$\nabla \times \mathbf{H}_q(2\omega) = -j2\omega\varepsilon_0\varepsilon_q(2\omega)\mathbf{E}_q(2\omega) \qquad (2.b)$$

where $E_q(2\omega)$ and $H_q(2\omega)$ are, respectively the electric and magnetic fields at 2ω in media q (q = 1,3).

In the nonlinear medium :

$$\nabla \times \mathbf{E}_2(2\omega) = j2\omega\mu_0\mathbf{H}_2(2\omega) \qquad (2.c)$$

$$\nabla \times \mathbf{H}_2(2\omega) = -j2\omega\varepsilon_0\varepsilon_2(2\omega)\mathbf{E}_2(2\omega) - j2\omega\mathbf{P}^{NL}(2\omega) \qquad (2.d)$$

where $\varepsilon_2(2\omega)$ is the relative permittivity of the medium at 2ω, and $\mathbf{P}^{NL}(2\omega)$ is the nonlinear (NL) polarization which depends on the nature of the NL medium.

Consequently, we develop the following theory. First, we obtain the expression of the electric field of the diffracted pump beam at circular frequency ω in any nonlinear medium using standard linear methods. Then we determine the expression of the NL polarization at circular frequency 2ω in any nonlinear regions including the modulated one. Finally, we resolve the boundary value problem corresponding to the diffraction at circular frequency 2ω.

3.1. Expression of the nonlinear polarization

3.1.1. Dielectrics

For noncentrosymmetric dielectrics, $\mathbf{P}^{NL}(2\omega)$ is given by[5] :

$$P_j^{NL}(2\omega) = \varepsilon_0\chi_{jih}(2\omega)\, E_i(\omega)E_h(\omega)$$

$\mathbf{E}(\omega)$: electric field at ω in the NL dielectric.

3.1.2. Metals

In the case of metals, we are faced with the problem of SHG in centrosymmetric media. As is well known, SHG results from the breaking of inversion symmetry at the metal surface[6-10]. Following the work of Bloembergen et al[6], in the metal, the pump beam at frequency ω gives rise to a nonlinear polarization :

$$\mathbf{P}^{NL}(2\omega) = \gamma\nabla[\mathbf{E}(\omega)\,.\,\mathbf{E}(\omega)] + \beta\mathbf{E}(\omega)\nabla\,.\,\mathbf{E}(\omega) \qquad (3)$$

where

$$\beta = \frac{e\varepsilon_0}{2m\omega^2},$$

$$\gamma = \frac{\beta}{4}[1 - \varepsilon_2(\omega)],$$

$$\varepsilon_2(\omega) = 1 - \frac{\omega_P^2}{\omega^2},$$

e = electron charge ,

m = conduction band electron mass ,

$\varepsilon_2(\omega)$ = permittivity of the metal at the pump circular frequency ω,

ω_p = plasma frequency ,

$\mathbf{E}(\omega)$ = electric field at the pump circular frequency ω in the metal.

Inside the bulk material, $\nabla.E(\omega) = 0$. Thus the second term in equation (3) only gives a contribution at the metal surface.

We have shown that, contrary to a generally held opinion, Eq. (3) is well defined, i.e., it can be used everywhere in space, including the metal surface S. To this end, S is considered as the limit, when $\eta \to 0$, of a continuous transition region with thickness η between metal and vacuum. We have calculated $\mathbf{P}^{NL}$ [Eq. (3)] in this transition region and let $\eta \to 0$. Since Maxwell equations are valid in the sense of distributions,[4,11] the limit, when $\eta \to 0$, of the terms in Eq. (3) are calculated in the sense of distributions.

As a result $\mathbf{P}^{NL}(2\omega)$ includes not only a bulk term but also a surface term, which behaves like a Dirac δ function :

$$\mathbf{P}^{NL}(2\omega) = \mathbf{P}_V^{NL}(2\omega) + \mathbf{P}_S^{NL}(2\omega)\,\delta_S \qquad (4)$$

$\mathbf{P}_V^{NL}$ is the bulk nonlinear polarization in the metal :

$$\mathbf{P}_V^{NL}(2\omega) = \frac{\beta}{4}\,[1 - \varepsilon_2(\omega)]\,\{\nabla[\mathbf{E}(\omega)\,.\,\mathbf{E}(\omega)]\} \qquad (5.a)$$

where { } is the operator written in the usual sense of functions[4,11], $\mathbf{P}_S^{NL}(2\omega)\,\delta_S$ is the surface nonlinear polarization :

$$\mathbf{P}_S^{NL}(2\omega)\,\delta_S = E_t \beta E_v^- \,[\varepsilon_2(\omega) - 1]\,\delta_S + \hat{n}\,\frac{3}{4}\,\beta\, E_v^{-2}\,[\varepsilon_2(\omega) - 1]\left[\frac{1}{3}\,\varepsilon_2(\omega) + 1\right]\delta_S \qquad (5.b)$$

where E_v^- = normal component of the pump electric field just below the grating profile S,

$\hat{n}$ = unit vector of the normal to S oriented outwards the metal.

δ_S = distribution located on S, defined by $\langle\delta_S,\varphi\rangle = \iint_S \varphi\, dS$, with φ being an infinitely differentiable function with compact support.

Note that the surface nonlinear polarization, Eq. (5.b), is located on S and not above it[7,8]. Moreover, the expression of $\mathbf{P}^{NL}{}_S\,\delta_S$ is independent of the profile of ε in the transition region.

Due to the existence of $\mathbf{P}^{NL}{}_S\,\delta_S$, the usual boundary conditions at 2ω (continuity of the tangential components of $\mathbf{E}(2\omega)$ and $\mathbf{H}(2\omega)$) are no longer valid. New boundary conditions have to be established at frequency 2ω. This is achieved using Maxwell equations written in the sense of distributions[4,11] in a curvilinear coordinate system. The calculation is developed in Ref. 10. The results are :

$$\hat{n} \times J\left[\mathbf{H}(2\omega)\right] = -j2\omega\hat{n} \times \left[\mathbf{P}_S^{NL}(2\omega) \times \hat{n}\right] \qquad (6.a)$$

$$J[\mathbf{E}_{v,t}(2\omega)] = -\frac{\beta}{3\varepsilon_0}\left[\frac{4\varepsilon_2^2(\omega)}{3}\ln\frac{\varepsilon_2(\omega)}{\varepsilon_2(2\omega)} + [\varepsilon_2(\omega) - 1]\,[3 - \varepsilon_2(\omega)]\right]\nabla_t E_v^{-2}(\omega) \qquad (6.b)$$

where $\varepsilon_2(2\omega)$ = permittivity of the metal at frequency 2ω, J[ ] = jump of [ ] across S in the direction of $\hat{n}$, $\mathbf{E}_{v,t}(2\omega)$ = tangential component of the classical (volume) part of $\mathbf{E}(2\omega)$ [see (ref. 10), where it is shown that $\mathbf{E}(2\omega)$ also includes a distributive part], and ∇_t = two-dimensional gradient taken in the plane tangent to S. In eq. 6.b, the term $\nabla_t\, E^{-2}{}_v(\omega)$ component is due to the existence of a normal component in the surface nonlinear polarization $\mathbf{P}^{NL}{}_S$.

Eq. (6.a) is a known result of electromagnetism, but eq. (6.b) is not. Equation (6.b) shows that the tangential component of the electric field at 2ω is no longer continuous when there exists a surface nonlinear polarization having a component perpendicular to S.

We keep in mind anyway that the free-electron model may be questionned from a physical point of view[7,8,12]. Nevertheless the method used to rigorously derive the new boundary conditions is not linked to a particular model and can be reutilized as soon as a more realistic expression of $\mathbf{P}^{NL}$ (taking into account non local effects for example...) is derived.

3.2. Propagation equations at SH frequency

3.2.1. Nonlinear material : dielectric

Included in this paragraph are dielectric gratings and linear metallic gratings overcoated by a nonlinear dielectric. As recalled in paragraph 3.1.1., $\mathbf{P}^{NL}(2\omega)$ has only a bulk contribution and the classical boundary conditions in electromagnetics apply. The Maxwell equations 2.c and d used in the sense of distributions[4,11] lead to the following propagation equations :

For TE polarization of the signal $\left(\mathbf{E}(2\omega) = E(2\omega)\,\mathbf{e}_z\right.$, where $\mathbf{e}_z$ is the unit vector of the z axis$\left.\right)$

$$\Delta E(2\omega) + k^2(2\omega)\,E(2\omega) = -\frac{4\omega^2}{\varepsilon_0 c^2} P_Z^{NL}(2\omega) \qquad (7.a)$$

For TM polarization $\left(\mathbf{H}(2\omega) = H(2\omega)\,\mathbf{e}_z\right)$

$$\nabla \cdot \left[\frac{1}{k^2(2\omega)} \nabla H(2\omega)\right] + H(2\omega) = j2\omega \left[\nabla \times \frac{\mathbf{P}^{NL}(2\omega)}{k^2(2\omega)}\right]_z \qquad (7.b)$$

where $[\ \ ]_z$ is the z component of $[\ \ ]$,

$$k^2(2\omega) = k^2(2\omega,x,y) = \frac{(2\omega)^2}{c^2}\varepsilon(2\omega,x,y), \text{ and}$$

$\varepsilon(2\omega,x,y)$ is the relative permittivity at SH frequency at point of coordinates x,y.

Of course, eqs. 7.a and 7.b are valid in the sense of distributions and, as a consequence, in the entire space. They include the boundary conditions at any frontier between two media.

From eqs. 7, it is seen that the polarization of the signal is TE or TM depending on the fact that $\mathbf{P}^{NL}$ is perpendicular to the plane of incidence or belongs to this plane. The orientation of $\mathbf{P}^{NL}$ is obviously related to the cristalline class of the dielectric and the polarization of the pump beam.

Let us see what is the main difference between the diffraction problem in nonlinear and linear optics. In linear optics we are led to the same eqs. as 7.a and b except that they are homogeneous ones. On the other hand, the source of the field is an incident plane wave, which is not the case in NL optics. Thus the second members of eqs. 7 replace the incident wave of the linear problem. They give rise to sources located inside the bulk material. The consequence is that the integral method is not utilizable. The resolution of the problem is then conducted through a generalization of the Differential Formalism[13].

Taking into account the periodicity of $\varepsilon(2\omega,x,y)$ and the pseudo-periodicity of E and H with respect to x, which allows representing these functions by their Fourier series, eqs. 7 are transformed into and infinite set of coupled differential equations. This set, after truncation, is numerically integrated with respect to y in the modulated region, and the numerical solution is matched with the Rayleigh expansions of the fields outside the modulated region. We then obtain the fields everywhere and the efficiencies in the different spectral orders at SH frequency.

3.2.2 Nonlinear material : metal

The existence of the surface nonlinear polarization (Eq. 5.b), together with Maxwell equations 2.c and d, leads to the following propagation equation :
For TE polarization

$$\Delta E_v(2\omega) + k^2(2\omega,x,y)\, E_v(2\omega) = J\left[\frac{\partial E}{\partial n}\right]\delta_S \qquad (8.a)$$

For TM polarization[14]

$$\nabla . \left[\frac{1}{k^2(2\omega,x,y)}\left(\nabla H(2\omega) - \hat{n}\, J(H(2\omega))\, \delta_S\right)\right] + H(2\omega) = J\left[\frac{1}{k^2(2\omega,x,y)}\frac{\partial H(2\omega)}{\partial n}\right]\delta_S \qquad (8.b)$$

A similar differential method as in paragraph 3.2.1. can be used, except that it requires the calculation of the Fourier coefficients of distributions[14]. Anyway, the method has been developped and implemented numerically. However, it suffers from numerical instabilities when the pump wavelength belongs to the visible or infrared region and can only be used for very shallow gratings in these spectral domains (this limitation does not hold for UV and X-ray regions).

Thus we seeked for an alternative method. It follows from eqs. 8 that, contrary to eqs. 7, the only source terms are located on the grating surface, i.e. there is no bulk source terms. This may be surprising since $\mathbf{P}^{NL}$ includes a volume term $\mathbf{P}^{NL}{}_v$. But, as is deduced from Maxwell eqs, $\mathbf{P}^{NL}{}_v$ only contributes through its jump accross the grating surface. This is coherent with the well known fact that a centrosymmetrical medium does not exhibit SHG.

Accordingly, eqs. 8.a and b can be cast in the form :

$$\Delta F_q + k_q^2(2\omega)F_q = 0 \quad (q = 1,2 \text{ depending on the medium})$$

$$F = \begin{cases} H(2\omega) & \text{for } 2\omega \text{ TM } (2\omega_{TM}) \\ \\ E(2\omega) & \text{for } 2\omega \text{ TE } (2\omega_{TE}) \end{cases} \qquad (9.a)$$

$$J(F) = \begin{cases} \text{right-hand member of Eq. (6.a) for } 2\omega_{TM} \\ \\ 0 \text{ for } 2\omega_{TE} \end{cases} \qquad (9.b)$$

$$J\left(c\frac{\partial F}{\partial n}\right) = \begin{cases} -j2\omega\varepsilon_0\hat{t}\,.\,J[E_v(2\omega)] - j2\omega\hat{t}\,.\,J\left[\dfrac{\mathbf{P}_V^{NL}(2\omega)}{\varepsilon(2\omega)}\right], \; c = \dfrac{1}{\varepsilon(2\omega)} & \text{for } 2\omega_{TM} \\ \\ -4\omega^2\mu_0 P_{S,z}^{NL}(2\omega)\,, \qquad c = 1 & \text{for } 2\omega \text{ TE} \end{cases} \qquad (9.c)$$

t : unit vector in the plane of incidence, tangent to the grating surface.

Note that this boundary value problem is the same as in linear optics[1,2] except for the values of J(F) and J(c(∂F/∂n)). Thus the integral formalism for linear optics[1] can be used for metallic gratings in nonlinear optics. In NL optics, the same computer code has to be called twice, the first time to calculate the diffracted pump field, the second time for the signal. In the meantime, the jumps of F and c(∂F/∂n) have to be calculated from eqs. 9.b or 9.c.

The polarization of the signal depends on the polarization of the pump field. Let be ψ the angle between $\mathbf{E}(\omega)$ and the plane of incidence. We consider the three situations $\psi = \pi/2$(TE), $\psi = 0$(TM) and $\psi \neq 0, \pi/2$ (neither TE nor TM).

For each case, Table 1 presents the existence of the SH signal, its origin (surface source term and/or volume source term), and its state of polarization.

TABLE 1. States of polarization and corresponding origin of the 2ω signal.

Polarization of the pump beam	$2\,\omega_{TE}$	$2\,\omega_{TM}$
$\psi = (\pi/2) : \omega_{TE}$	No	Yes Origin : P^{NL}_{V}
$\psi = 0 : \omega_{TM}$	No	Yes Origin P^{NL}_{V}, P^{NL}_{S}
$\psi \neq 0, (\pi/2)$ (neither TE or TM)	Yes Origin : $P^{NL}_{S,z}$	Yes Origin : P^{NL}_{V}, P^{NL}_{S}

3.3. Summary of the theory and general remarks

The formalism developed in the previous sections is summarized on figure 2 where it is seen that this formalism corresponds to a three-step theory.

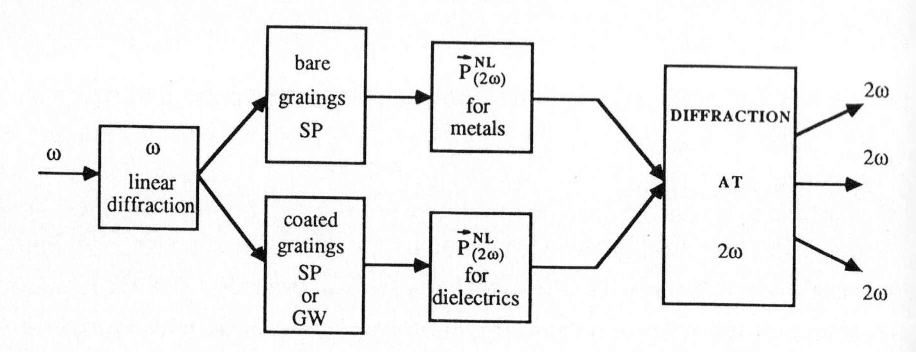

Figure 2
Successive steps of the theory of diffraction.

The following remarks are in order : some diffracted orders at ω are radiated ; the rest remains bounded to the bare or coated grating. The EM resonance, which involves normal modes (SP or GW) of the associated smooth structure, is associated with the resonant excitation of one of the bounded diffracted orders[13] at ω. This leads to the enhancement of the intensity of the resonantly excited diffracted order. For SP resonance, the pump field must be TM polarized, whereas there is no requirement concerning the polarization of the pump beam for GW resonance. Since there is a coupling between the diffracted orders at ω, the resonant excitation of a given bounded diffracted order[1-3] increases the intensity of the neighbouring diffracted ones. The increase of the intensity of the resonantly excited diffracted order and its neighbours induces an enhancement of the magnitude of the source terms at 2ω, leading to an increase in the intensity at 2ω compared to the flat case ($\delta = 0$), where no EM resonance takes place.

The integral and differential methods presented here are rigorous in the sense that the groove depth δ of the grating is not considered as a perturbative parameter, neither at ω nor at 2ω. These methods allow deriving the direction of propagation of the diffracted orders at 2ω which are not bounded to the grating[13] (whatever the geometry of diffraction may be : figures 1). In the case of SHG, one obtains :

$$\frac{d}{\lambda/2}\left[n \sin \psi_p - n_1(\lambda) \sin \theta\right] = p \qquad (10)$$

where :

$$\omega = \frac{2\pi c}{\lambda}$$

$$n_1 = \sqrt{\varepsilon_1}$$

ψ_p = diffraction angle of the diffracted order p at 2ω in the outside medium of index n at 2ω.

Equation 10 is very important since it represents the nonlinear grating equation. This equation allows determining which diffracted orders p at 2ω are radiated in one, or both, of the outside media. The direction of propagation of these orders are given by ψ_p.

4. NUMERICAL RESULTS

4.1. SP enhanced SHG

The enhancement of the SH intensity is computed with respect to the associated smooth structure, which in this case is a flat vacuum/metal interface. We first calculate the reflectivity at 2ω of a flat metallic medium.

4.1.1. SH reflectivity in the flat case

Calculation of the reflectivity at 2ω of a flat metallic medium is achieved using the results of Sec. 3.1.2. The geometry is that of Figure 1.a with $\delta = 0$.

We call nonlinear reflectivity of the metallic mirror the ratio :

$$R(2\omega) = \frac{D(2\omega)}{D^2(\omega)}$$

where $D(\omega)$ and $D(2\omega)$ are the power densities at frequencies ω and 2ω, respectively.

From the boundary conditions at the metal-vacuum interface, we get the following expression of the reflectivity at 2ω for a unit power TM polarized pump beam :

$$R(2\omega) = \frac{1}{2}\left(\frac{\mu_0}{\varepsilon_0}\right)^{3/2} \left| \frac{-2\omega\varepsilon_0\varepsilon_2(2\omega)\, J(E_x) + \alpha_{2,2} J(H_z) + 2\omega P^{NL}_{v,x}}{\alpha_{2,2} + \alpha_{1,2}\varepsilon_2(2\omega)} \right|^2 \qquad (11)$$

where $\alpha_{1,2}$, $\alpha_{2,2}$ are perpendicular components of the wave vector at the SH frequency in vacuum and in the metal, respectively, and where $P^{NL}_{v,x}$ is the x-component of the volume part of $\mathbf{P}^{NL}$ calculated at $y = 0^-$.

The reflectivity of a silver mirror at the SH frequency has been measured[15,16] as a function of the angle of incidence θ using a pump wavelength of 1.06 μm. It has been found that $R(2\omega)$ calculated from Eq. (11) is 20 times higher than the measured reflectivity at 2ω.

This leads us to a slight modification of Bloembergen's expression of $\mathbf{P}^{NL}(2\omega)$ given in Eq. (3), by introducing[15-17] two phenomenological coefficients A and B :

$$\mathbf{P}^{NL}(2\omega) = A\gamma\nabla[\mathbf{E}(\omega)\,.\,\mathbf{E}(\omega)] + B\beta\mathbf{E}(\omega)\nabla\,.\,\mathbf{E}(\omega)$$

$R(2\omega)$ is still given by Eq. (11), but now it depends on A and B through the quantities $J(E_x)$, $J(H_z)$, and $P^{NL}_{v,x}$. Use of the method of Sec. 3.1.2. yields

$$\hat{n} \times J[\mathbf{H}(2\omega)] = -Bj2\omega\hat{n} \times \left[\mathbf{P}_S^{NL}(2\omega) \times \hat{n}\right]$$

instead of Eq. (6.a)

$$J[\mathbf{E}_{v,t}(2\omega)] = \frac{1}{3}(2A + B)\, J_0[\mathbf{E}_{v,t}(2\omega)] - \frac{\beta}{3\varepsilon_0}(B - A)\,[\varepsilon_2^2(\omega) - 1]\,\nabla_t[\mathbf{E}_v^{-2}(\omega)]$$

where $J_0[\mathbf{E}_{v,t}(2\omega)]$ is given by the right-hand member of Eq. 6.b. Also, $\mathbf{P}^{NL}{}_v(2\omega) = A\mathbf{P}^{NL}{}_{0,v}(2\omega)$, with $\mathbf{P}^{NL}{}_{0,v}(2\omega)$ being equal to the right-hand member of Eq. (5.a).

The two unknown parameters A and B are determined by fitting (using a least square routine) the theoretical and experimental SH reflectivities $R(2\omega)$. The best agreement[15,16] is obtained for A = 2.2 and B = 0.57, with the corresponding values of $\varepsilon_2(\omega)$ and $\varepsilon_2(2\omega)$ for silver, measured by ellipsometry[15,16] :

$\varepsilon_2(\omega) = -36 + 2j$, $\varepsilon_2(2\omega) = -7{,}9 + 0{,}1j$

Note that the computation of the enhancement of the SH intensity as compared to the flat case is performed using these two values of A and B. That is, in computing the enhancement at 2ω there are no unknown parameters.

4.1.2. SP enhanced SHG : numerical results and comparaison with experiments

The computation of the SH enhancement has been done for a TM polarized pump field exciting a SP resonance on a sinusoidal, trapezoidal or triangular silver grating[10].

The results are summarized in Table 2.

TABLE 2. Computed (th) and measured (exp. when available) optimum groove depth and optimum SP induced enhancement of the intensity of a given diffracted order p at 2ω for different grating periodicities[16] and profiles. λ = 1.064 μm.

	OPTIMUM GROOVE DEPTH Å		OPTIMUM ENHANCEMENT	
PROFILE d Å	Th. p	Exp p	Th. p	Exp p
Sinusoidal 5556	250 -1	300 -1	47 -1	36 -1
Sinusoidal 15300	900 +1	1000 +1	2370 +1	2500 +1
Trapezoidal 15300	670 0		~ 105 0	
Triangular 15300	780 +1		7784 +1	

As predicted by theory, the data show the existence of an optimum value δ_{opt} for which the enhancement E is the greatest : $E(\delta = \delta_{opt}) = E_{opt}$. This value of δ_{opt} depends on the number p of the diffracted order at 2ω, and also on ω. Indeed in the case of a sinusoidal silver grating, (a) for d = 0.5556 μm, $\delta_{opt}(\omega)$ = 110 Å, whereas $\delta_{opt,p=0}(2\omega)$ = 600 Å and $\delta_{opt,n=-1}(2\omega)$ = 250 Å ; and (b)[18] for d = 1.53 μm, $\delta_{opt}(\omega)$ = 300 Å, whereas $\delta_{opt,p=0}(2\omega)$ = 1700 Å and $\delta_{opt,p=+1}(2\omega)$ = 900 Å.

The value of E_{opt} depends strongly on d, as shown in Table 2. An increase in d from 0.5556 μm to 1.53 μm leads to a strong increase in E_{opt}, from 36 to 2500. The reason can be understood as follows : when d increases, the ratio λ/d decreases. Thus, the diffracted orders at ω become closer and closer, i.e., more coupled to each other and especially to the resonantly excited order. Of course, the number of propagating diffracted orders, and therefore the radiation losses, increases with d, inducing a less efficient EM resonance at ω.

4.2. GW enhanced SHG

The numerical results[19] refer to a ZnS coated grating with a sinusoidal profile, where the substrate is silver and the upper medium is vacuum (Figure 1). Since ZnS belongs to the $\bar{4}3$ m symmetry class, the pump beam must be TM polarized to generate a nonlinear polarization at 2ω. This nonlinear polarization has a single component along the z axis,

$$P_z^{NL}(2\omega) = 2\varepsilon_0 \chi_{xyz}(2\omega) E_x(\omega) E_y(\omega)$$

which gives rise (Eqs. 7) to a TE polarized diffracted signal at 2ω. The enhancement at 2ω, due to the GW resonance at ω, is computed with respect to the associated smooth structure (deduced from the modulated structure by letting $\delta \to 0$).

The GW resonance occurs at ω for the -1 diffracted order, the corresponding incidence angle being $\theta = \theta_{res}$ = 28.92°. The optimum value of the groove depth at ω is $\delta_{opt}(\omega)$ = 321 Å.

As for SP enhanced SHG, there exists an optimum value δ_{opt} linked with the GW resonance at ω for which the peak enhancement takes its greatest value E_{opt}. Table 3 summarizes the main results.

TABLE 3. Computed values of δ_{opt} and E_{opt} in the case of GW enhanced SHG.

Diffracted order at 2 ω	δ opt	E_{opt}
0	335 Å	7.2×10^3
+1	333 Å	1.7×10^5

According to the periodicity, 0.400 μm, the value of $E_{opt} = 1.7 \times 10^5$ obtained for GW enhanced SHG must be compared to the value of 47 resulting from SP resonance (See Table 2).

As predicted in Ref. (20) when enhanced SHG is desired, it is worth using the GW rather than the SP resonance. The fact that GW resonance leads to a much greater increase in the SH efficiency than does the SP resonance can be easily understood. Indeed, with GW, the greatest part of the pump field lies in the nonlinear guiding layer, whereas, with SP, the EM field extends much farther into the vacuum than into the metal (which is the nonlinear medium).

5. CONCLUSION

New electromagnetic theories have been developed and implemented numerically to study gratings in nonlinear optics. It has been established that gratings have a great potentiality in increasing various NL effects and particularly SHG : they act as field amplifier in nonlinear optics when EM resonance takes place.

Several theoretical predictions have already been confirmed by experiments. This is a proof not only that the numerical implementation is correct, but also of the validity of the new boundary conditions derived at the frontier between a metal and vacuum.

The enhancement strongly depends on the groove depth δ, periodicity d, groove shape and the nature (SP or GW) of the EM resonance. It seems that profiles having cusps are better than the sinusoidal ones. In any case, the optimum groove depth corresponds to very weak modulation (~ 6%) and may lead to strong enhancements (~ 10^5).

REFERENCES

1) D. Maystre, Integral Methods, in Electromagnetic Theory of Gratings, R. Petit, Ed., pp. 63-100, Springer-Verlag, New York (1980)

2) D. Maystre, General study of grating anomalies from electromagnetic surface modes, in Electromagnetic Surface Modes, A.D. Boardman, Ed., pp. 661-724, Wiley, New York (1982)

3) M. Nevière, P. Vincent and R. Petit, Rev. Optique 5, 65 (1974)
P. Vincent, Differential methods, in Electromagnetic Theory of Gratings, R. Petit, Ed., pp. 101-121, Springer-Verlag, New York (1980)

4) R. Petit, A tutorial introduction, in Electromagnetic Theory of Gratings, R. Petit, Ed., pp. 1-50, Springer-Verlag, New York (1980)

5) N. Bloembergen, Nonlinear Optics, Benjamin, New York (1965)

6) N. Bloembergen, R.K. Chang, S.S. Jha and C.H. Lee, Phys. Rev. 174, 813-822, (1968)

7) J.E. Sipe, V.C.Y. So, M. Fukui and G.I. Stegeman, Phys. Rev. B 21, 4389-4402, (1980)

8) J.E. Sipe and G.I. Stegeman, Nonlinear optical response of metal surfaces, in Surface Polaritons, Electromagnetic Waves at Surfaces and Interfaces, V.M. Agranovich and D.L. Mills, Eds., pp. 661-701, North Holland, Amsterdam (1982)

9) D. Maystre, M. Nevière and R. Reinisch, Appl. Phys. A 39, 115-121, (1986)

10) R. Reinisch, M. Nevière, H. Akhouayri, J.L. Coutaz, E. Pic, Opt. Engineering 27, 961 (1988)

11) L.S. Schwartz, Mathematics for Physical Sciences, Addison-Wesley, Reading, Mass. (1966)

12) H.R. Jensen, K. Pedersen and D. Keller, Proceedings of the Int. Conf. on nonlinear optics NLO'88, Ireland 1988

13) R. Reinisch and M. Nevière, Phys. Rev. B 28, 1870, (1983)

14) M. Nevière, P. Vincent, D. Maystre, R. Reinisch and J.L. Coutaz, J. Opt. Soc. Am. B5, 330-336 (1988)

15) M. Nevière, J.L. Coutaz, D. Maystre, E. Pic and R. Reinsich, CLEO '86-IQEC'86 (San Francisco, Sept. 1986), p. 68

16) J.L. Coutaz, D. Maystre, M. Nevière and R. Reinisch, J. Appl. Phys. 62, 1529-1531, (1987)

17) D. Maystre, M. Nevière, R. Reinisch and J.L. Coutaz, J. Opt. Soc. Am. B 5, 338-346, (1988)

18) The characteristics of silver corresponding to all the experiments performed with d = 1.53 μm periodicity gratings lead to :

$\varepsilon_2(\omega) = -43.6 + 2j$

$\varepsilon_2(2\omega) = -8.4 + 0.1j$. The best fit is obtained for A = 2.9 and B = 0.4245

19) H. Akhouayri, M. Nevière, P. Vincent and R. Reinisch, in Proc. 14th Congress of the International Commission for Optics (Québec), pp. 239-240 (1987)

20) R. Reinisch and M. Nevière, Phys. Rev. B26, 5987 (1982)

Scattering in Volumes and Surfaces
M. Nieto-Vesperinas and J.C. Dainty (Editors)

INTRINSIC INSTABILITIES OF LASER-IRRADIATED SURFACES

J.E. SIPE and H.M. VAN DRIEL

Department of Physics and Ontario Laser and Lightwave Research Centre, University of Toronto, Toronto, Ontario, M5S 1A7 Canada

We are all used to employing what might loosely be called "one-dimensional models" to describe phenomena in optics which really exist in a three-dimensional space. The simplest such model is the plane wave approximation of a beam incident on a surface, but the description of the operation of a laser in terms of the amplitude of a single mode falls in the same category. Although in the latter example the laser beam need not be uniform in the directions perpendicular to its propagation, depending on which mode is excited, still the field variation in that so-called "transverse" direction is fixed, and does not enter as a set of dynamical variables. Another such example is the propagation of light in a single mode in either a waveguide or an optical fibre. Even within such one-dimensional models, physical systems in optics can exhibit complicated dynamical behavior if nonlinearities are present, the onset of chaos in a homogeneously-broadened, single mode laser being a notable example.

But recently, interest has turned to even more complicated systems where the kind of one-dimensional analysis outlined above breaks down. Although a system may have, for example, complete symmetry about a propagation axis, the solutions of the dynamical equations which describe it need not exhibit that symmetry. As some control parameter is varied an initially stable, one-dimensional like solution may suffer the onset of transverse instabilities. The resulting complicated dynamical behavior of the system then becomes much more interesting, from a physicist's point of view, albeit perhaps more annoying and problematic from the point of view of an engineer. Current work on such transverse instabilities [1] includes studies of transverse soliton formation, conical emission, pattern formation and pattern dynamics in semiconductor diode arrays, transverse laser instabilities, pixellation cross talk, and a host of other problems including the subject of this paper: structure formation and development when a single, intense laser beam interacts with a surface.

In studying such phenomena, exciting opportunities appear for the "scattering theorist." In many scattering problems in optics, such as the scattering of light from a disordered material, or from a diffraction grating, the inhomogeneities that are doing the scattering are prescribed -- either as a deterministic or stochastic structure -- and fixed. It is essentially a static geometry that confronts the theorist. But in the study of transverse instabilities, the calculation of how light scatters from the inhomogeneities is only part of the problem. In a nonlinear medium, one must then ask how such scattering will feed back into the development of the scattering structure itself. The full problem thus involves not only electrodynamics, but the study of the material response to electromagnetic fields and its resulting evolution. So, at least in the full problem, the scattering theorist is confronted by a dynamic geometry, and indeed the dynamics of that geometry becomes part of the problem to be solved.

However, a modest beginning can be made by studying more-or-less stable transverse structures that develop, trying to understand why they must appear, and perhaps under what conditions they become unstable and develop into different structures. Such is the nature of the investigation reported here, work carried out at the University of Toronto with our students John Preston (now at the University of Illinois) and Ken Dworschack. Some of the work has been reported previously, as indicated by the references below. In this brief sketch we try to pull together and summarize the main results of the work, and contrast the kind of analysis necessary with other scattering problems in optics.

The experimental geometry is shown in Fig. 1 [2], and is in principle exceedingly simple. The beam from a 10.6 μm CO_2 continuous-wave laser is focused down, to typically a 500 μm to 800 μm spot size, on a silicon-on-sapphire film. Initially the silicon is essentially transparent to the radiation, but as the sapphire absorbs some of it and heats up, it heats up the silicon and free electron-hole pairs are thermally excited. These can then absorb the infrared radiation, heating up the silicon directly and bringing it to its melting threshold. The process can be initiated, and thermal "run-away" effects eliminated, by using a heating stage to thermally excite the electron-hole pairs. The control parameters are the angle of incidence and polarization of the laser beam, the focus spot size, and of course the intensity of the laser beam. Since liquid silicon has a much lower emissivity in the visible that solid silicon near the melting temperature, the state of the silicon film can be monitored easily with the use of a microscope; essentially all the infrared radiation is either reflected, or absorbed by the silicon or the sapphire, and so none impinges on the microscope optics. The results are found to be qualitatively independent of the exact thickness of the silicon film used, at least in the range from about 0.5 μm to 5 μm, and there are indications that many of the qualitative results reported below could be seen at the surface of a bulk sample of silicon.

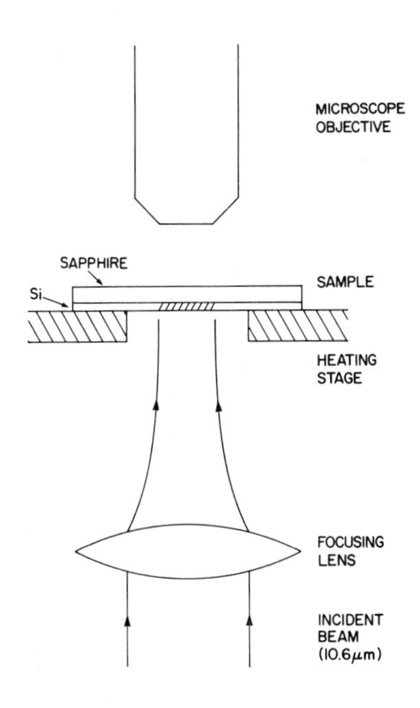

Fig. 1: The experimental geometry for studying CW laser-induced surface structures on silicon-on-sapphire films.

To set the stage, it is useful to recall that the study of structure formation at the surface of materials illuminated single, intense beams of *pulsed* laser irradiation has been studied since 1965 [3], with bursts of enthusiam at different intervals in the past quarter century [4]. Ordered and disordered structures can be formed, usually with characteristic sizes on the order of the wavelength of the light used. At normal incidence, a linearly polarized beam tends to lead to the formation of permanent ordered structures -- "ripples" -- which run perpendicular to the polarization of the light beam. The effect is quite universal [5], occurring on metals, insulators, and semiconductors, over a wide range of incident wavelengths and pulse lengths. And although the initiation of the structure formation is fairly well-understood [6], the detailed interpulse and intrapulse feedback effects, as well as the surface tension effects which lead to the permanent structure formation after melting occurs locally, have not been completely analyzed. The charm of the geometry Fig. 1, an experimental protocol pioneered by Biegelsen et al. [7], is that since a CW laser is used stable solid/liquid structures can be identified. These are qualitatively similar to the structures which result in the pulsed experiments, but because they are time independent they are much easier to analyze and understand in detail.

Even if one were unaware of the work with pulsed laser beams, a moment's thought would lead to the expectation that something interesting would happen in the geometry of Fig. 1 as the silicon is brought to its melting point. For as the laser intensity is increased, eventually too much energy is being deposited for the uniform solid silicon to remain solid. Yet since at the melting point the reflectivity of uniform liquid silcon (90%) is much greater than that of uniform solid silicon (30%), if the film were to melt uniformly it would then not be absorbing enough energy to remain liquid! The system suffers from a sort of physical analog of the psychological phenomenon that "the grass is always greener on the other side of the fence." One might guess that only a thin layer of silicon would melt, for if the molten layer were much less than the skin depth of the metallic, liquid silicon the reflectivity would not rise to the full liquid reflectivity. This is not observed: whenever and wherever the film melts, it is found to melt straight through to the sapphire substrate. In fact, a stability analysis [8] shows that a thin melt layer would be extremely unstable to the growth of spatial inhomogeneities at a host of wavelengths. What the silicon actually does is to melt through to the substrate but only in certain regions, so that the average reflectivity of the inhomogeneous structure is somewhere between the solid and liquid values.

A typical structure that is observed is shown in Fig. 2, resulting from a spot size of 500 μm at an intensity of 2.7 kW/cm^2. Alternating strips of solid and liquid appear, with a spacing of 10.6 μm if the laser beam is normally incident, and oriented perpendicular to the polarization of the laser beam. These can be thought of as the CW analog of the localized melting which leads to the formation of "ripples" in the pulsed experiments. But here other morphologies can be observed as well: increasing the intensity to 3.2 kW/cm^2 leads to the "period-doubling" shown in Fig. 3, and at a lower intensity of 1.7 kW/cm² in a larger spot size of 700 μm the "disordered" melt structure shown in Fig. 4 is observed. The experimental results can be collected and displayed in a "stability diagram" shown in Fig. 5. For any spot size, as the intensity is increased the film eventually converts from uniform solid to uniform liquid. But for small spot sizes it does this by passing through first a phase where liquid lamellae appear in the solid matrix, and then settling into an ordered structure which is quite stable over a range of intensities until its spacing finally doubles as the intensity is increased. Higher orders gratings can also be seen as the material becomes more and more liquid. For larger spot sizes, on the other hand, the film passes through a series of "amorphous" structures, changing its topology from liquid-in-solid to solid-in-liquid as the fraction of the film which is liquid increases from zero to one more gradually than for the smaller spot sizes.

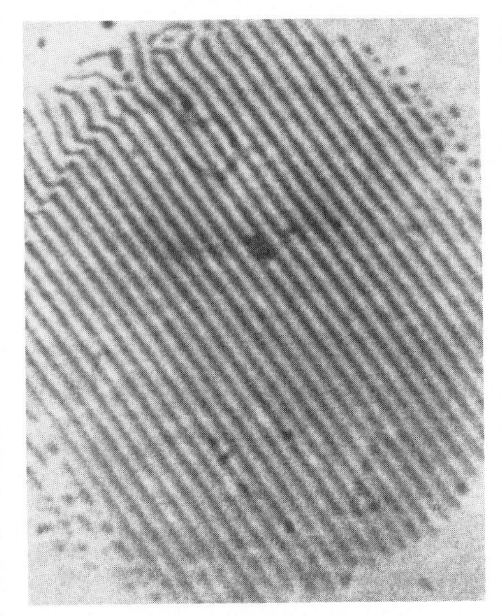

Fig. 2:A laser generated melt pattern which exhibits long range order. The laser intensity is 2.7 kW/cm², and the spot size is 500 μm. The strips are perpendicular to the laser polarization and the periodicity is equal to 10.6 μm.

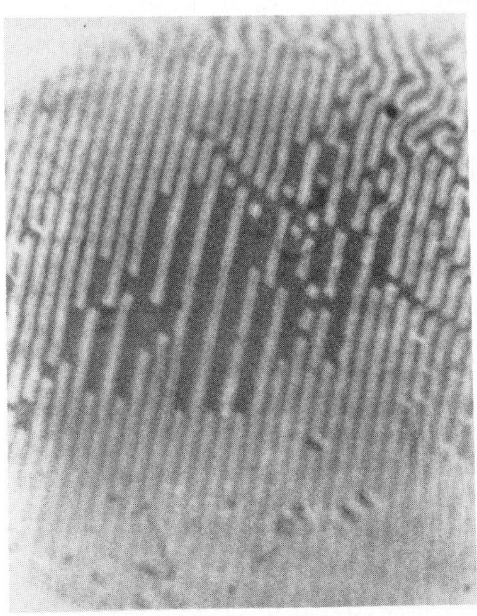

Fig. 3: In the central portion of the irradiated region, the simple grating which appeared in Fig. 2 has been replaced by a doubled grating structure. The laser intensity is 3.2 kW/cm² and the spot diameter is 520 μm.

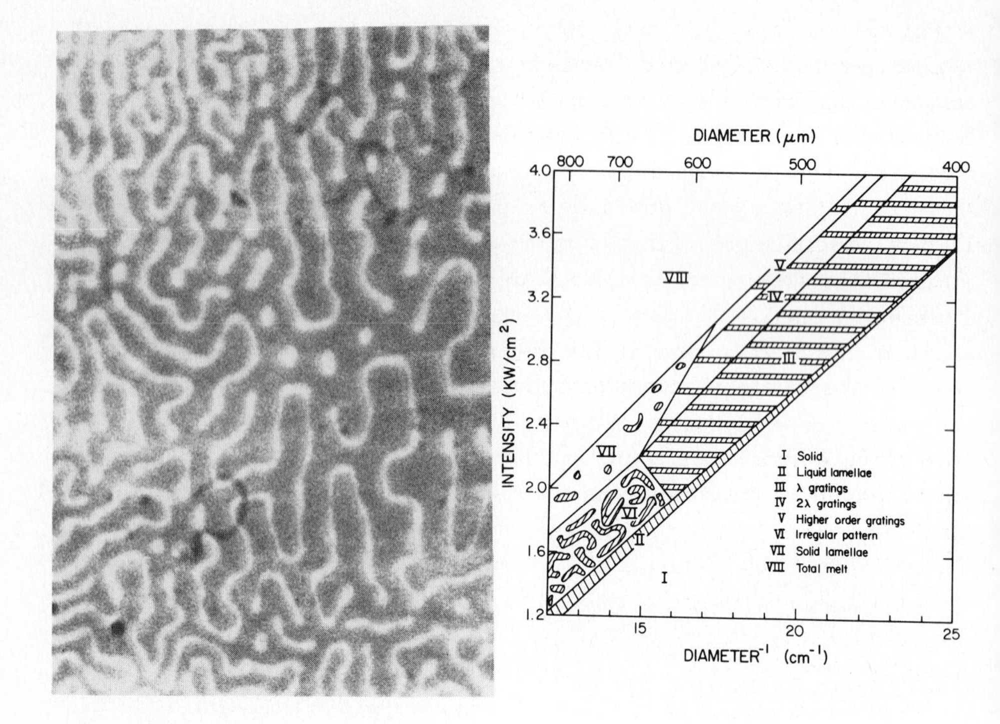

Fig. 4: An example of a disordered melt structure. The laser intensity is 1.7 kW/cm², the spot size is 700 μm. The average spacing between solid regions is slightly greater than the incident wavelength.

Fig. 5: The stability diagram for Si films on sapphire under laser irradiation at 10.6 μm.

There is some evidence that the "disordered" or "amorphous" morphologies should be thought of as metastable structures that have not been able to evolve to more stable "ordered" structures, much as a solid can be quenched into an amorphous form if it is cooled so quickly that it cannot settle into its more stable, crystalline form. One might expect this here, since a characteristic time for one of our structures to evolve could be estimated by taking the energy content in the system and dividing by the rate at which energy is flowing through the film. That rate is smaller for larger spot sizes, since the transverse temperature gradients are less, and so the "evolution time" for the larger spot sizes is longer, possibly long enough that the structures cannot settle into a more stable, ordered form. Perhaps a more convincing argument comes from experiment: Hysteresis effects can be seen at the boundaries of the regions indicated in Fig. 5, in that by slightly changing the intensity and moving from one region to another an amorphous structure can be "encouraged" to convert to an ordered structure, but ordered structures seem able to maintain their stability if the intensity is changed back. Thus, we are led to (very) tentatively think of the ordered structures as the ultimately stable ones, and attempt to understand their formation and stability first.

Now in a scenario where the film goes from uniform solid to uniform melt via a series of ordered structures with an increasing fraction of liquid, the global energy balance of the system can be understood, since the average reflectivity of the structure can vary from that of the uniform solid to that of the uniform melt. Yet it is clear that much more of an explanation is required. In the first place, the spacing of the strips and their orientation with respect to the polarization of the laser beam must be understood. Perhaps more importantly, the local energy balance of the system is still not explained: If geometrical optics were applied, the liquid strips would be highly reflecting, and the solid strips highly absorbing. Unless the liquid were undercooled and/or the solid superheated, such structures could not persist. But geometrical optics is manifestly the wrong way to tackle this problem, since the spacing of the strips is equal to the wavelength of light. Very strong interference can occur between the incident laser beam and the scattered light, and this can affect where energy is preferentially deposited in the structure. The physical optics of these structures must therefore be understood if one is to attempt to understand their stability.

In principle, of course, this is straightforward. The dielectric constants of molten silicon and solid silicon at the melting point are known, and for a given angle of incidence and polarization, and a given structure, the electromagnetic problem is well-defined. But it is not an easy one. While the thickness of the solid silicon is much less than or on the order of the wavelength of light, the skin depth of the molten, metallic regions is on the order of a few hundred Angstroms, much less than the wavelength of light; the structures themselves have characteristic sizes on the order of the wavelength of light; the dielectric constants of the solid and molten regions are vastly different. So none of the usual approximate techniques can be applied, and one seems to be driven to the computer to get estimates of the fields and thus the absorption in the different regions. While this can provide an answer, it must be done yet again for each new proposed structure. And it is easy for the underlying physics, which must be leading to the qualitatively simple structures observed, to be lost in a wash of code and output.

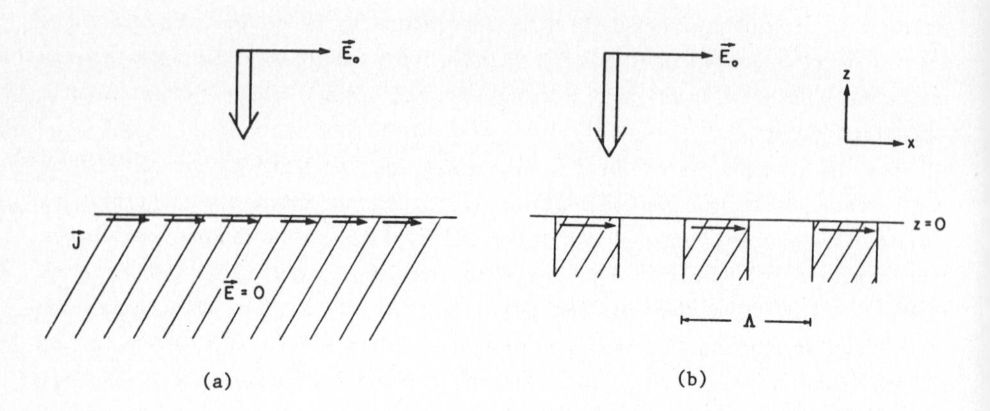

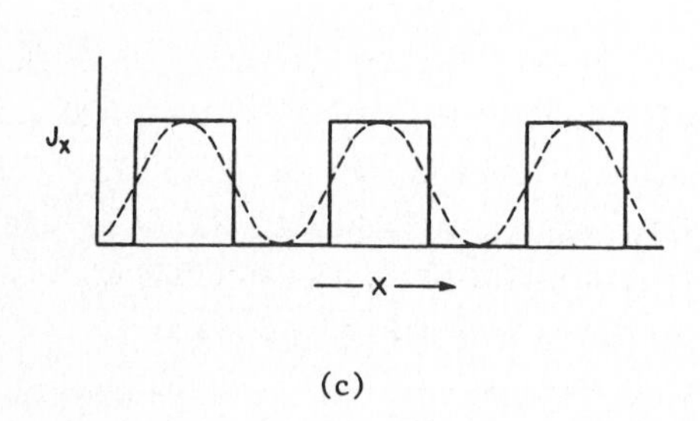

Fig. 6: a) The shielding of the bulk metal by surface currents. In reality, J extends roughly within a skin depth of the metal. b) A simple model for the surface currents in a periodic metal structure. c) Solid line: the model of the surface current in Fig. 6b. Dashed line: the approximation by two Fourier components, one with wavenumber zero and the second with wavenumber $2\pi/\Lambda$.

Luckily, it turns out to be possible to construct a simple picture of the electrodynamics of this problem, which both leads to an understanding of why the ordered structures observed are stable, and can serve as the basis for a detailed, semi-quantitative calculation of the absorption in the different parts of the structures [9]. To illustrate how this picture can be built up we look first at the simple geometry shown in Fig. 6a, where a plane wave is normally incident on a half-space of metal, such as molten silicon. The incident field induces currents in the metal which are essentially contained within a skin depth of the surface. The magnitude and orientation of these currents are just such that the total electric field -- that is, the field from the currents plus the incident field $\mathbf{E}_0$ -- essentially vanishes deep within the metal. This way of looking at the electrodynamics at a metal surface, which can be quantified in the so-called "extinction theorem" [10], leads to a simple physical argument which explains how interference effects lead to significant modifications of the energy deposition at a composite surface.

In Fig. 6b we show such a structure, where the shaded regions indicate metal (molten silicon) running along the y-direction, separated by semiconductor (solid silicon) strips, the structure having a periodicity of Λ; for the moment we neglect the presence of an underlying substrate. If a plane wave is incident, currents will be induced within a skin depth of the metal-vacuum and metal-semiconductor interfaces. In the simplest approximation we neglect the latter, and assume that the current density at $z = 0-$ is as sketched as the solid curve in Fig. 6c -- that is, it is uniform within the metallic regions, and points only in the x-direction. We also neglect polarization currents within the semiconductor regions.

Now, if we approximate the solid curve in Fig. 6c by the dotted sinusoidal curve, we consider J_x as consisting of a spatially uniform component with wavenumber zero, and a component with wavenumber $2\pi/\Lambda$. In general both of these components lead to radiated fields, such that their sum when added to $\mathbf{E}_0$ gives a total electric field $\mathbf{E} = 0$ deep inside the metallic regions.

Consider first the component at $2\pi/\Lambda$. In the special case of $\Lambda = \lambda$, where λ is the wavelength of light in vacuum, the source is phased to generate radiation fields propagating in the $\pm x$ directions. However, since the current density $\mathbf{J}$ points in the x-direction, no field can be radiated in that direction since, if it were, the generated electric field would point in the x-direction and it would be a longitudinal electromagnetic wave; such radiated fields do not exist. Thus for $\Lambda = \lambda$ the incident field $\mathbf{E}_0$ must be cancelled deep inside the metal by only the field generated by the spatially uniform component of J_x.

But since both the incident field and the field generated by the spatially uniform component of J_x do not depend on x, if the resulting total field **E** is zero deep within the metal, it also vanishes deep within the dielectric. Of course, the metal will absorb energy in the region within a skin depth of $z = 0$, which is where the induced current exists. However, to first approximation there is no electric field in the semiconductor and thus no absorption there.

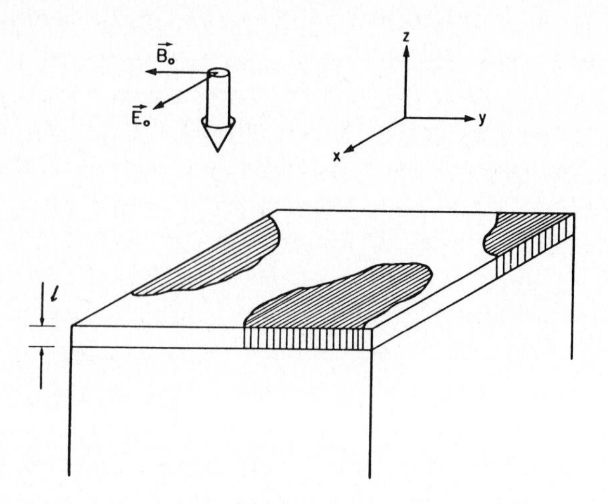

Fig. 7: The general geometry of a plane wave incident on a metal-semiconductor composite supported by a substrate. The z=0 plane is at the centre of the film.

In short, the semiconductor has been "shielded" from the incident field by interference effects between the incident field and the field generated by the induced currents in the metal, and no energy will be absorbed by the semiconductor. Note that the simple argument we have given here holds only for Λ near λ, and for $\mathbf{E}_o$ directed perpendicular to the direction along which the strips run. Yet, it shows how interference effects can completely invalidate the picture of energy absorption one would construct from geometrical optics: In that limit, as mentioned above, rays of light incident on the metal would be largely reflected, while those impinging on the semiconductor would largely proceed through and ultimately be absorbed. The geometrical optics prediction that the semiconductor regions absorb more than the metallic regions is completely invalidated by the "interference shielding" present in the periodic structure, and precisely the reverse occurs.

From this argument it is clear how the simple pattern shown in Fig. 2 can be stable: The incident energy is redistributed by the structure preferentially into the metallic liquid, which then can remain molten at a slightly higher temperature than the semiconductor. The structure, as it were, "stabilizes itself".

It is fairly simple to quantify the physical picture sketched above to construct an approximate theory to calculate the energy deposition in an arbitrary, metallic-semiconductor composite above a substrate (Fig. 7). We begin with exact expressions [11] for the electric and magnetic fields at points $z > -l/2$,

$$\mathbf{E}(\mathbf{r}) = \mathbf{E}_e(\mathbf{r}) + \int d\mathbf{r}'\, G_E(\mathbf{r},\mathbf{r}')\mathbf{P}(\mathbf{r}')$$

$$\mathbf{B}(\mathbf{r}) = \mathbf{B}_e(\mathbf{r}) + \int d\mathbf{r}'\, G_B(\mathbf{r},\mathbf{r}')\mathbf{P}(\mathbf{r}') \tag{1}$$

where $\mathbf{P}(\mathbf{r})$ is the total induced polarization field (due to both metal and semiconductor regions) in the film. The substrate is taken into account implicitly in the Green functions G_E and G_B, so the integrals in (1) extend only over the film; the fields $\mathbf{E}_e(\mathbf{r})$ and $\mathbf{B}_e(\mathbf{r})$ are the fields that would exist in the absence of the film.

We now posit an approximate form for $\mathbf{P}(\mathbf{r})$ in terms of a set of parameters. In the metallic regions, we set $\mathbf{P}(\mathbf{r}) \simeq \mathbf{P}_m(z)$,

$$\mathbf{P}_m(z) = \mathbf{Q}_t\, \delta\left(z-\frac{l}{2}\right) + \mathbf{Q}_b\, \delta\left(z+\frac{l}{2}\right) \tag{2}$$

where $\mathbf{Q}_t$ and $\mathbf{Q}_b$ have only x and y components, and are taken independent of (xy); they must be determined. In the semiconductor regions we put $\mathbf{P}(\mathbf{r}) \simeq \mathbf{P}_s$, where $\mathbf{P}_s$ is a vector with only x and y components, is independent of $\mathbf{r}$, and yet to be determined. Then, introducing $\mathbf{R} \equiv (x,y)$ and characterizing the surface structure by two functions $m(\mathbf{R})$ and $s(\mathbf{R})$,

$$m(\mathbf{R}) = 1 \text{ if } \mathbf{R} \text{ in a metallic region}$$
$$= 0 \text{ if } \mathbf{R} \text{ in a semiconductor region}$$

$$s(\mathbf{R}) = 1 - m(\mathbf{R}) \tag{3}$$

our parameterization of the polarization in the film can be written

$$\mathbf{P}(\mathbf{r}) = m(\mathbf{R})\mathbf{P}_m(z) + s(\mathbf{R})\mathbf{P}_s. \tag{4}$$

Putting eq. (4) in eq. (1) yields

$$m(\mathbf{R})\mathbf{E}(\mathbf{r}) = m(\mathbf{R})\mathbf{E}_e(\mathbf{r})$$
$$+\int d\mathbf{r}' \, G_E(\mathbf{r},\mathbf{r}')m(\mathbf{R})m(\mathbf{R}')\,\mathbf{P}_m(z')$$
$$=\int d\mathbf{r}' \, G_E(\mathbf{r},\mathbf{r}')m(\mathbf{R})s(\mathbf{R}')\mathbf{P}_d \tag{5}$$

and a similar equation for $m(\mathbf{R})\mathbf{B}(\mathbf{r})$.

Now since the total $\mathbf{E}$ and $\mathbf{B}$ must vanish deep in the metal, we must have $m(\mathbf{R})\mathbf{E}(\mathbf{r}) = m(\mathbf{R})\mathbf{B}(\mathbf{r}) = 0$ for z in the film and more than a few skin depths from the interfaces at $z = \pm l/2$; outside the metallic regions the function $m(\mathbf{R})$ guarantees that these terms are zero. Not surprisingly, our assumed form (4) for $\mathbf{P}(\mathbf{r})$ is too simple to allow these conditions to be satisfied for all $\mathbf{R}$. To deal with a general structure, we think of an ensemble of characteristically similar structures and, as the best we can do, choose the parameters in eq. (4) so that we satisfy those conditions on average:

$$\langle m(\mathbf{R})\mathbf{E}(\mathbf{r})\rangle = 0,$$
$$\langle m(\mathbf{R})\mathbf{B}(\mathbf{r})\rangle = 0, \tag{6}$$

for points near $z = 0$. In practice, the ensemble averages which appear, such as $\langle m(\mathbf{R})\rangle \equiv f_m$ and $\langle s(\mathbf{R})\rangle \equiv f_s$, can be identified as spatial averages over the surface structure; then the ensemble corresponds to the different structures which result from rigidly displacing the given structure in the xy plane from its original location. Since $m(\mathbf{R}) + s(\mathbf{R}) = 1$, the fractions metallic and semiconductor obviously satisfy $f_m + f_s = 1$; using these relations all the two point averages, such as $\langle m(\mathbf{R})m(\mathbf{R}')\rangle$ and $\langle m(\mathbf{R})s(\mathbf{R}')\rangle$, can be related to one correlation function, which we take as

$$c(\mathbf{R}-\mathbf{R}') \equiv \frac{\langle m(\mathbf{R})m(\mathbf{R}')\rangle - f_m^2}{f_m f_s} \tag{7}$$

Note that $c(\mathbf{R})$ has been defined so that it is unity at $\mathbf{R} = 0$; further, if there is no long range order we see that $c(\mathbf{R}) \to 0$ as $|\mathbf{R}| \to \infty$.

From eq. (6) it is then possible to determine $\mathbf{Q}_t$, $\mathbf{Q}_b$, and $\mathbf{P}_s$ in terms of f_m and $c(\mathbf{R})$, and once these parameters are found it is easy to calculate the absorption in the metal and the semiconductor [9]. Thus the fraction of melt f_m and the correlation function $c(\mathbf{R})$ become the "order parameters" of the system; once they are specified, the absorption in the different regions can be calculated (approximately!), and the stability of a structure can be investigated.

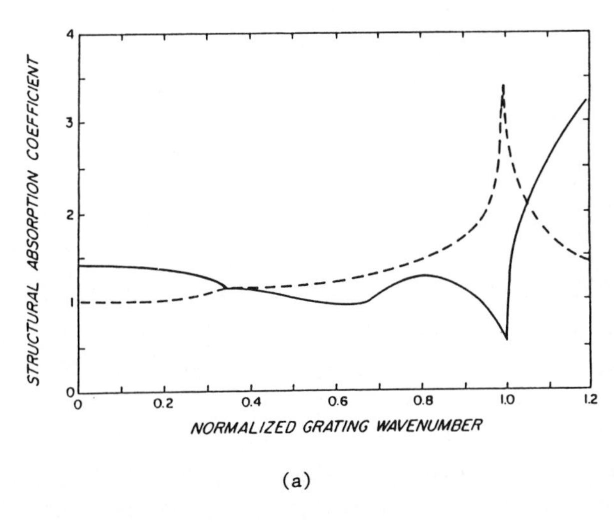

(a)

Fig. 8: a) The structural absorption coefficients as a function of the normalized grating wavenumber (= λ/Λ) for a grating structure with 50% of the surface molten and with the strips oriented perpendicular to the incident polarization. The solid line refers to the solid, the dotted line to the melt.

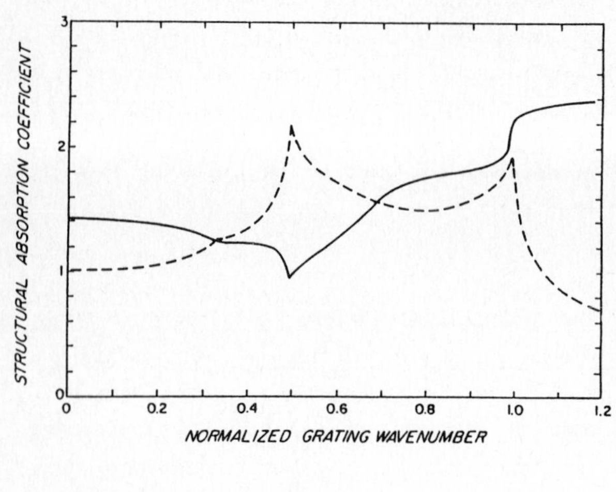

(b)

Fig. 8 b) As in Fig. 8a, but with 75% of the surface molten.

To display the results, it is convenient to introduce "structural absorption coefficients" for the metal and the semiconductor. These are defined to be the factors by which the geometrical optics expressions for absorption must be multiplied by to give the correct absorption. In Fig. 8a we show the structural absorption coefficients for a simple periodic structure with $f_m = 1/2$, as a function of λ/Λ; the structure is oriented with respect to the polarization of the laser as in Fig. 6. Note that there is maximum redistribution of energy deposition from the semiconductor to the metal at $\lambda/\Lambda \cong 1$, in agreement with the simple argument given above. Qualitatively, we can thus expect such structures to be stable. If the same calculation is made for the structure oriented with the strips running parallel to the laser polarization, by the way, the energy is directed primarily into the semiconductor. Thus, we would expect such structures to be unstable, and indeed they are not observed experimentally.

In Fig. 8b the parameters are the same as in Fig. 8a, except the fraction of metal $f_m = 3/4$. Here the maximum redistribution occurs at $\lambda/\Lambda \cong 0.5$; this corresponds to the "period doubled" grating, and the energy redistribution is relying on the spatial second harmonic fourier component of the grating. On the basis of Fig. 8 one can then see the reason for the higher order gratings to appear: As the fraction of melt is finally driven higher by the increasing laser intensity, the film finds that the higher order gratings are more efficient in redistributing the energy primarily into the molten regions, and thus those higher order gratings become more stable.

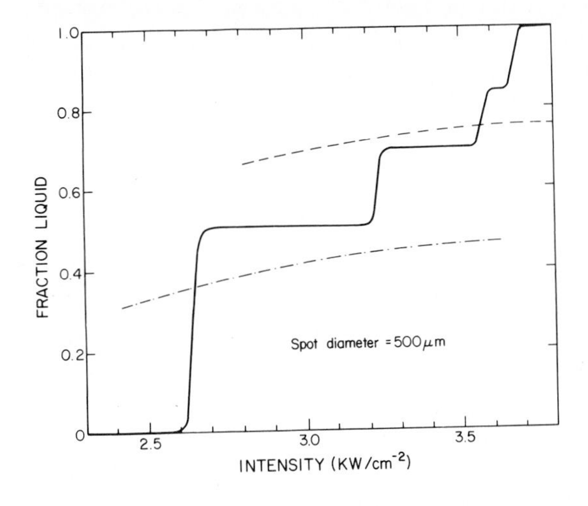

Fig. 9: A comparison between the observed molten fraction of the surface (solid line) in the small spot size regime, and simple theoretical predictions for the simple grating (dashed-dot line) and the doubled grating structure (dashed line).

To quantify this one must consider the energy deposition and heat flow in the composite film. A very simple model for this predicts how the primary and period-doubled gratings should increase their liquid fraction as the laser intensity is increased [12]; the results are in good agreement with experiment (see Fig. 9). Further, an extention of the calculation to a structure which can develop from a primary to a period-doubled grating gives a good estimate of when the primary grating should "convert" to the period-doubled structure [12].

But this begins to take us beyond the limits of the electromagnetic theory of the problem, and into the larger problem of the dynamics of pattern formation in these films; thus it is an appropriate place for this brief sketch to conclude. Suffice it to say that there are a host of other interesting problems concerning this system. For example, in the limits where there are very few inclusions of one material in another (liquid in solid at low intensities, solid in liquid at high intensities), the inclusions, or "lamellae", seem to behave as nearly-free "particles". They move, for example, in response to temperature gradients. By "move" we do not mean that there is any actual motion of material in the film, but rather that material melts and solidifies so as to move the solid/melt interfaces of the "particle" in such a way that the centre of the inclusion moves, and the inclusion to good approximation maintains its size. Both the size of these inclusions, and also how they "interact" with other inclusions during "collisions," can be semi-quantitatively understood by means of a simple extension of the theory presented above [12]. Further, for a laser beam at non-normal incidence, in the ordered regimes the spacing of the gratings observed differs from the spacing observed at normal incidence, and depends on the polarization of the laser beam; these results have been reported and understood using a generalization of the theory outlined here [13].

In conclusion, these results demonstrate that the simple problem of an intense laser beam interacting with a surface is a much richer phenomenon that one might have guessed. The richness results from the extremely nonlinear nature of the equations describing the coupled fields in this problem, the dielectric constant being essentially a step-function of the temperature at the melting point, and from the interference that can arise because the laser is a coherent source. They also indicate the kind of exciting phenomena that confront scattering theorists as they go beyond the calculation of scattered fields from a given scattering source, and consider the coupled problem of the scattering and the evolution of the source itself.

REFERENCES:

1. See an upcoming special issue of the Journal of the Optical Society of America (B) to be devoted to transverse instabilities.

2. J. Preston, H.M. van Driel, and J.E. Sipe, Phys. Rev. Lett. *58*, 69, (1987).

3. M. Birnbaum, J. Appl. Phys. *36*, 3688 (1965).

4. See, e.g., J.E. Sipe, Jeff F. Young, J.S. Preston, and H.M. van Driel, Phys. Rev. *B27*, 1141 (1983); Jeff F. Young, J.S. Preston, H.M. van Driel, and J.E. Sipe, Phys Rev. *B27*, 1424 (1983); Jeff F. Young, J.E. Sipe, and H.M. van Driel, Phys. Rev. *B30*, 2001 (1984); and references cited therein.

5. H.M. van Driel, J.E. Sipe, and Jeff F. Young, Phys. Rev. Lett. *49*, 1955 (1982).

6. J.E. Sipe, H.M. van Driel, and Jeff F. Young, Can. J. Phys. *63*, 104 (1985).

7. D.K. Biegelson, R.J. Nemanich, L.E. Fennel, and R.A. Street in *Energy Beam - Solid Interactions and Transient Thermal Processing*, ed. J.C. Fan and N.M. Johnson (North-Holland, New York, 1984), pg. 383; W.G. Hawkins and D.K. Beigelson, Appl. Phys. Lett. *42*, 358 (1983).

8. J.E. Sipe, Jeff F. Young, and H.M. van Driel, in *Laser-Controlled Chemical Processing of Surfaces*, ed. A.W. Johnson and D.J. Ehrlich (North-Holland, New York, 1984), pg. 415.

9. J.S. Preston, J.E. Sipe, H.M. van Driel, and J. Luscombe, "Optical absorption in metallic-dielectric microstructures," Phys. Rev. B (in press).

10. See, e.g., M. Born and E. Wolf, *Principles of Optics*, 6th ed., (Pergamon, New York, 1980) pg. 100.

11. See ref. 9; The Green functions can be easily derived from, e.g., the formalism presented in J.E. Sipe, Journal of the Opt. Soc. Am. B4, 481 (1987).

12. J.S. Preston, H.M. van Driel, and J.E. Sipe, "Pattern formation during cw laser induced melting," Phys. Rev. B (in press).

13. K.F. Dworschak, J.E. Sipe, and H.M. van Driel, "Laser induced melt morphologies in Si films at non-normal incidence," in preparation.

Scattering in Volumes and Surfaces
M. Nieto-Vesperinas and J.C. Dainty (Editors)

BRILLOUIN SCATTERING FROM THIN FILMS

G.I. Stegeman, J.A. Bell*, W.R. Bennett§, G. Duda†, C.M. Falco, U.J. Gibson, B. Hillebrands$^{\Delta}$, W. Knoll†, L.A. Laxhuber°, Suk Mok Lee, J. Makous, F. Nizzoli#, C.T. Seaton, J.D. Swalen§, G. Wegner† and R. Zanoni

Optical Sciences Center, University of Arizona
Tucson, AZ 85721

Recent experiments on Brillouin scattering from a variety of thin films, including metallic superlattices, Langmuir-Blodgett films and rough ZnS films, are described.

1. INTRODUCTION

Light scattering from thermally excited acoustic phonons, better known as Brillouin scattering, has proven to be a powerful tool for measuring the velocity of sound waves in bulk media and films, and at surfaces.[1] The acoustic phonon spectrum for surfaces, interfaces and thin films is particularly rich, and includes pure wave modes such as Rayleigh, Sezawa, Stonely and Love waves, and "leaky" guided shear and longitudinal waves.[2] If the acoustic velocities of a sufficient number of these modes can be determined by Brillouin scattering, the elastic constants and other properties of thin films can be evaluated and interesting thin film physics can be studied.

The acoustic phonon velocities depend on the elastic constants and densities of the sample materials, as well as on the sample geometry.[2] The various modes, all of which we studied by means of Brillouin scattering, are:

1) For bulk, or "infinite," media, those with volumes much larger in all dimensions than the acoustic decay distance, there are three normal acoustic modes, two of primarily transverse nature, and one primarily longitudinal.[3] For isotropic materials, and for high symmetry planes and directions in crystalline media, one or more of these modes are pure modes consisting of a longitudinal (compressional), and two orthogonally polarized transverse (shear) waves.

† Max-Planck-Institut Fur Polymerforschung, Postfach 3148, D-6500 Mainz, West Germany
§ Almaden Research Center, K34/802 (D), 650 Harry Rd., San Jose, CA 95120
* Boeing Electronics Co., P.O. Box 24969, MS 7J-27, Seattle, WA 98124-6269
° Romerstrasse 15, 80000 Munich 40, West Germany
Dipartimento di Matematica e Fisica, Universita' di Camerino, 62032 Camerino, Italy.
Δ 2, Physikalisches Institut, RWTH Aachen Templergraben 55, 5100 Aachen, Federal Republic of Germany

2) At the surface of a semi-infinite medium, there exists a Rayleigh mode which contains both shear and compressional components.
3) At the interface between two media, Stoneley waves can propagate under certain restrictive conditions on relative densities and elastic constants.
4) For thin films, there exist various film-guided modes, each of which contains transverse and longitudinal components in both the film and the substrate. For the simplest case of isotropic media, there is always a Rayleigh mode which is polarized in the saggital plane (the plane containing the surface normal and acoustic wavevector). For films sufficiently thick and with $V_T > V_{T'}$, where V_T and $V_{T'}$ are the film and substrate shear wave velocities, respectively, a discrete number of Sezawa modes occurs, also polarized in the saggital plane.
5) Under the conditions of 1.4, Love waves polarized parallel to the surface occur. Both Love and Sezawa modes give rise to standing waves in the film. Sezawa and Love modes also exist for anisotropic media, but their properties are more complex.
6) Under special conditions, "leaky" modes, which primarily are standing shear and/or longitudinally polarized waves in the film, exist with long propagation distances because they couple only weakly to bulk modes in the substrate.[2,4]

The key point is that in both "infinite" media and thin films there exist acoustic modes whose velocities can be measured by Brillouin scattering. The number of these modes is sufficient to allow an evaluation of the elastic constants of a medium.

In Brillouin scattering, the scattering geometry determines which acoustic wavevector will be probed. The phonon frequency (Ω), which is equal to the frequency shift of the scattered (ω_s) from the incident light (ω_i), is measured interferometrically. For scattering from bulk modes, the total wavevector is conserved in the scattering process, i.e.,

$$\mathbf{q} = \mathbf{k}_s - \mathbf{k}_i \quad , \tag{1}$$

where $\mathbf{q}$, $\mathbf{k}_s$, and $\mathbf{k}_i$ are the acoustic, scattered optical and incident optical wavevectors respectively. There are three "infinite medium" acoustic modes; therefore, at most, three upshifted ($\omega_s = \omega_i + \Omega$) and three downshifted ($\omega_s = \omega_i - \Omega$) frequency components are obtained for optically isotropic media. For scattering from thin films, or at surfaces, only the wavevector parallel to the surface must be conserved, that is,

$$\mathbf{q}_p = \mathbf{k}_{s,p} - \mathbf{k}_{i,p} \tag{2}$$

When the wavevector normal to the surface is also conserved, this usually leads to a maximum in the scattering cross-section. For thick films, with $q_p h >> 1$, scattering occurs from many Sezawa (or Love) modes which, under insufficient experimental resolution, can appear as a continuum in the spectrum, peaking at the Bragg condition given by Eqn. (1). What is important is that the structure and angular dependence of the spectrum differ for scattering from bulk media versus scattering from thin film or surface guided waves. It is therefore possible to identify spectral components resulting from scattering from different modes, to measure their dispersion relations and hence to evaluate their elastic constants.

We have used this approach to study the elastic properties of a variety of thin film systems, and report our findings in this paper. For metallic thin films, we observed scattering from Stoneley waves. For metallic superlattices we were able to observe Rayleigh, Sezawa and Love modes and so were able to evaluate all of the corresponding elastic constants. We fabricated Langmuir-Blodgett (LB) films from organic molecules one monolayer at a time using dipping techniques, and we investigated their highly anisotropic acoustic properties. Our experiments on ZnS films, in which the film roughness strongly affects the phonon spectrum, also are discussed in the body of this paper.

2. EXPERIMENTAL DETAILS

Here we briefly discuss the Brillouin scattering apparatus, a schematic of which is shown in Fig. 1. The key component in all of our experiments is the tandem multi-pass Fabry-Perot interferometer. This device consists of two three-pass Fabry-Perots, each with one of its mirrors mounted on the same scanning stage.[5,6] Each interferometer is a very narrow bandwidth interference filter, whose peak transmission wavelength is tuned by changing, piezoelectrically, the spacing between the plates. Peak transmission for each wavelength incident on the Fabry-Perot device occurs whenever a standing wave arises inside the cavity at that wavelength, and therefore recurs whenever the cavity spacing changes by a half wavelength. For two interferometers, the transmission is a convolution of the transmission functions of the individual etalons. The interferometer spacings are set to some ratio m/n where $m > n$ are integers. Thus the transmission peaks of the two cavities overlap only every m orders of the first cavity to produce the main transmission peak (and requires a spacing change of $m\lambda/2$). "Ghost" peaks occur whenever the separate cavities achieve a resonance condition, and their transmission is typically reduced by two to three orders of magnitude from the main transmission peaks. This pattern of primary and secondary maxima is repeated for every wavelength of light incident onto the multi-pass interferometer. Because the interferometer effectively has six passes through the cavities, the rejection ratio (the ratio of minimum to maximum transmission for one incident wavelength) for the complete interferometer typically is 10^{-9}. This property allows very weak Brillouin scattered components to be identified, except those in the immediate vicinity of the transmission peaks for stray light at the incident light frequency.

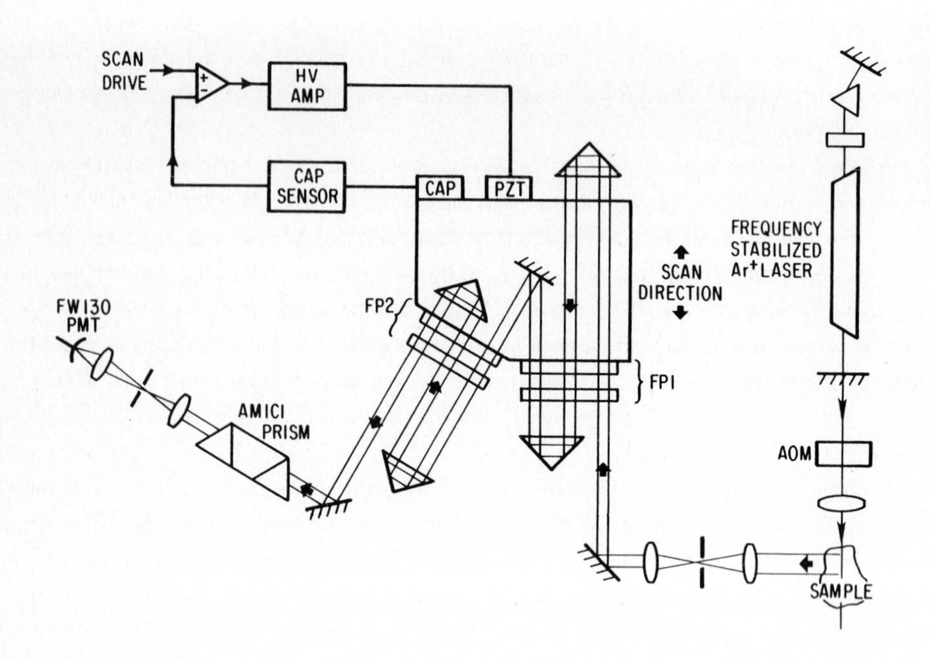

Figure 1. Schematic of the Brillouin scattering apparatus.

The Brillouin scattering apparatus also consists of a light source, the argon ion laser, and a detection system.[6] A low noise photomultiplier with a dark current of less than 1 count per second was used and experimental data was stored in a computer. Because of mechanical vibrations and thermal effects in the interferometer, the plates' parallel orientation can deteriorate in time, leading to a loss of resolution. These problems, as well as instabilities in the laser, can cause the spectrum to drift relative to the zero voltage spacing. A sensor-feedback system was used to actively tune the tandem Fabry-Perot and to collect the Brillouin spectrum.

3. METALLIC FILMS

A number of experiments were performed on metallic films fabricated by means of a special multi-target DC sputtering system. The important feature of this sputtering system is that the sputtering rate is controlled to within 0.3%, making it possible to fabricate multilayered films of precise thicknesses.[7] This capability is crucial in making the metallic superlattices we studied by Brillouin scattering.

A typical spectrum obtained from a Mo/Ta superlattice is shown in Fig. 2.[8] This superlattice consists of alternating films of Mo and Ta, with equal thicknesses, with a total film thickness of $\simeq$ 0.35 μm. The large number, about fifteen, of acoustic guided modes observed results from the large value of $q_p h$, and from the small shear wave velocity of metals. We measured the dispersion relations for the superlattices by varying the incidence angle, scattering angle and the total film thickness. These relations are shown in Fig. 3. Although both of the constituent metals in single crystal form are bcc, the films consist of small crystallites with their [001] axis normal to the surface and with random orientation in the surface plane. These films exhibit a net hexagonal symmetry. Using a least squares fit to the data, we obtained the elastic constants shown in Table 1 and the theoretical curves shown in Fig. 3. The agreement between experimental data and calculations based on a crystallite structure with either stress or strain continuous between crystallites is surprisingly good.

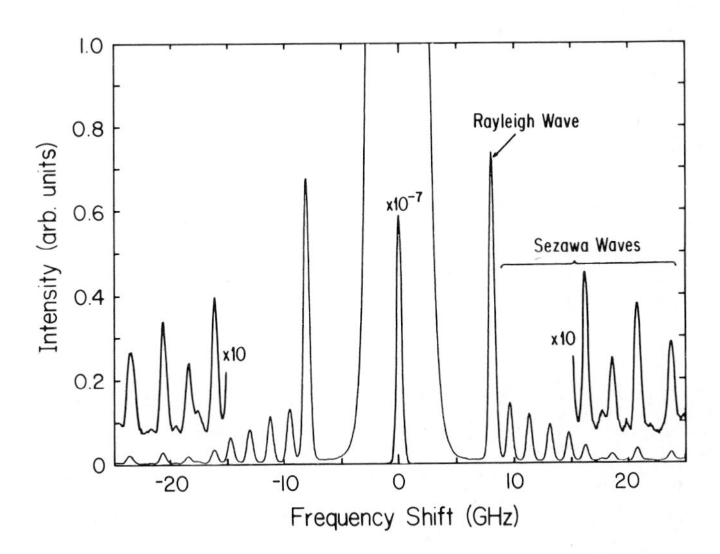

Figure 2. The Brillouin spectrum obtained on backscattering from a Mo/Ta metallic superlattice with Λ = 171 Å and 50% Mo.

Table 1. Comparison of the measured and calculated elastic constants for two metallic superlattices.*

Film material		**Stiffness (10^{10} N/m²)**				
		c_{11}	c_{12}	c_{13}	c_{33}	c_{44}
Mo/Ta	Bulk	35.2	15.9	15.7	34.1	8.7
50% Mo; Λ=171Å	Fit	37.2		14.9	34.0	8.15
Mo/Ta	Bulk	35.8	16.0	15.8	34.6	8.9
54% Mo; Λ=160Å	Fit	37.8		15.0	33.2	7.86
Cu/Nb	Bulk	20.7	12.0	12.1	20.9	4.3
60% Cu; Λ=91Å	Fit	21.7	13	12.8	22.0	3.37

**Voigt estimates are used.*[6] *It is assumed the Reuss averaging approach changed the calculated elastic constants by 1-2%.*[6]

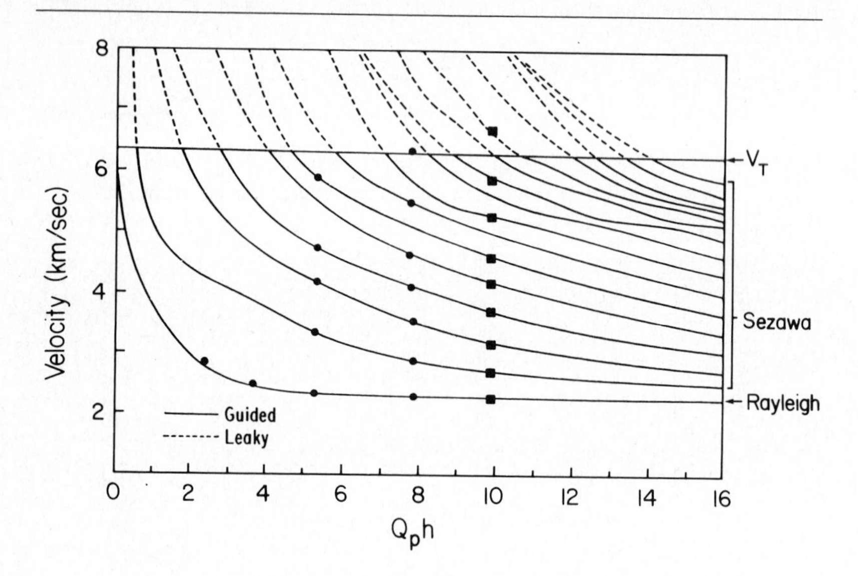

Figure 3. The best-fit dispersion curve of acoustic velocity versus q_ph for Rayleigh and Sezawa modes supported by a Mo/Ta superlattice.

The current interest in the elastic properties of metallic superlattices stems from the anomolous softening of the c_{44} shear elastic constant when the film repeat wavelength $\Lambda \simeq 25$ angstroms.[9] The variations we measured for the elastic constant c_{44} are shown in Fig. 4 for Mo/Ta films and for Cu/Nb films.[8,10] No significant variation with Λ was found for the other elastic constants. It appears that the magnitude of the softening in c_{44} is related to the mismatch between the film's two constituent metal lattices. This mismatch is 4.5% for Mo/Ta and 9.4% for Cu/Nb. In addition to a larger lattice constant mismatch for Cu/Nb, this latter material also has a symmetry mismatch between the fcc Cu and the bcc Nb. Further work to understand the origin of these anomalies is called for.

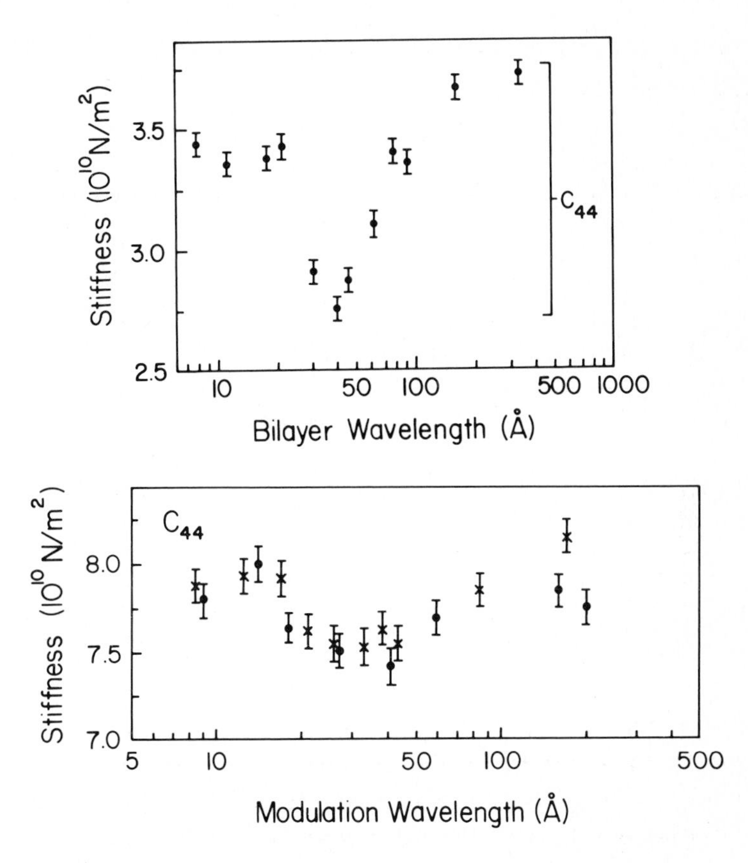

Figure 4. The measured variation in the C_{44} elastic constant for (top) Cu/Nb and (bottom) Mo/Ta superlattices versus the bilayer (modulation) wavelength.

Our experiments were all performed with regards to reflection from the air-superlattice surface. The primary scattering mechanism was the acoustically produced surface ripple.[2] In order to observe Love waves, and hence to measure the remaining elastic constants, the light was incident through the saphire substrate along the surface normal and the depolarized spectrum was measured on reflection at 66° from the normal.[11] Since Love waves do not ripple the film surface, the elastooptic interaction occurs within the weak evanescent tail of the Love waves in the substrate. The spectra were very weak and required long data acquisition times. The dispersion curves, and their best least squares fit, are shown in Fig. 5. As in Table 1, the agreement between calculation and experiment is good for c_{12}.

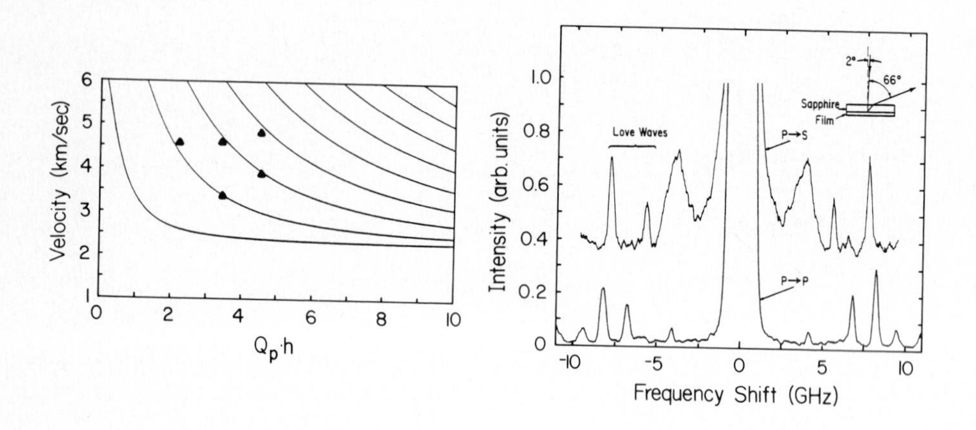

Figure 5. The calculated and measured dispersion curves for Love waves in a Cu/Nb superlattice film, 318 nm thick, with a 9.1 nm bilayer periodicity.

The fortuitous combination of the elastic constants and densities of molybdenum and a fused silica substrate results in the existence of a Stoneley wave at the molybdenum-fused silica interface. Although we did not have available an interface between two semi-infinite media, it was possible to make Mo films thick enough so that the lowest order Sezawa mode degenerates, effectively, into a Stoneley wave, localized at the film-substrate interface.[12] The Mo film thickness was varied and the scattering intensities of the two lowest order acoustic modes, Rayleigh and lowest order Sezawa, were monitored by means of light incidence from the air and from the substrate. For thin films, the two scattering intensities are comparable. For thick films, the Brillouin intensity of the Rayleigh mode was strong and that of the first Sezawa mode was weak when probed from the air side. When measured from the substrate side, the Brillouin intensity of the Rayleigh mode was weak and that of the Sezawa mode was strong. Typical spectra are shown in Fig. 6. Calculations for the thick film show that the Rayleigh mode is localized at the film-air surface, and the first Sezawa (or Stoneley) mode is localized at the film-substrate interface. The corresponding dispersion relations are shown in Fig. 7, along with calculations for which it was assumed that the elastic constants are the same throughout the film. The excellent agreement between experiment and theory indicates that the elastic properties of the film are identical at both film interfaces.

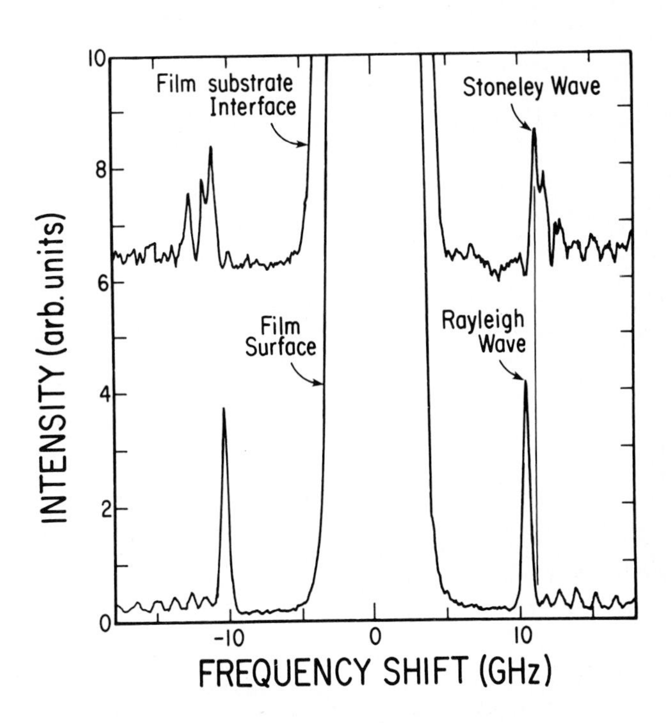

Figure 6. Brillouin spectrum obtained on reflection from the air (lower) and glass (upper) sides of a Mo film, 800 nm thick.

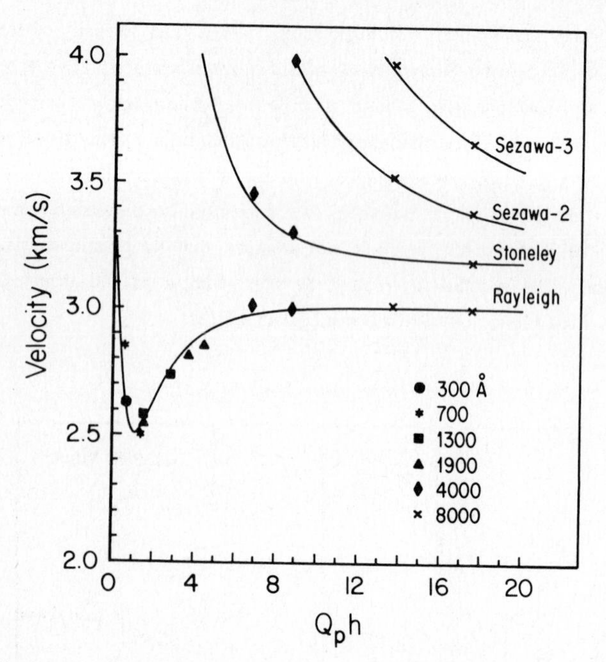

Figure 7. Experimental and best-fit calculated velocity dispersion curves for saggital plane polarized guided acoustic modes versus normalized film thickness for Mo films on fused silica. The first Sezawa mode degenerates into a Stonely wave localized at the glass-Mo interface for large values of $q_p h$.

4. LANGMUIR-BLODGETT FILMS

Langmuir-Blodgett (LB) films are composed of organic molecules and are fabricated one monolayer at a time with a Langmuir trough.[13] The classical LB molecule has a $(CH_2)_n$ chain as its backbone. Often, functional groups are attached to the chain to give the molecule specific desired properties such as photosensitivity, and two groups, one hydrophilic and the other hydrophobic, are attached to either end of the chain. The hydrophilic group is attracted to water molecules while the hydrophobic group is repulsed from the water surface. When these molecules are spread on the water surface and compressed, they form a monolayer with their backbones approximately normal to the water surface. When an appropriate substrate is dipped back and forth through this surface (surface density is maintained via a barrier with a surface-pressure-sensitive feedback system), a multi-monolayer film is formed, as shown in Fig. 8. Under carefully controlled conditions, it is possible to build films up to micron thicknesses. Many of these films are "artificial" in the sense that they do not ordinarilly occur in nature. Furthermore, their elastic properties are expected to be highly anisotropic due to weak forces, probably ionic, between the head and tail groups, which bond the layers together. These films are fragile and highly susceptible to damage via shear stresses.

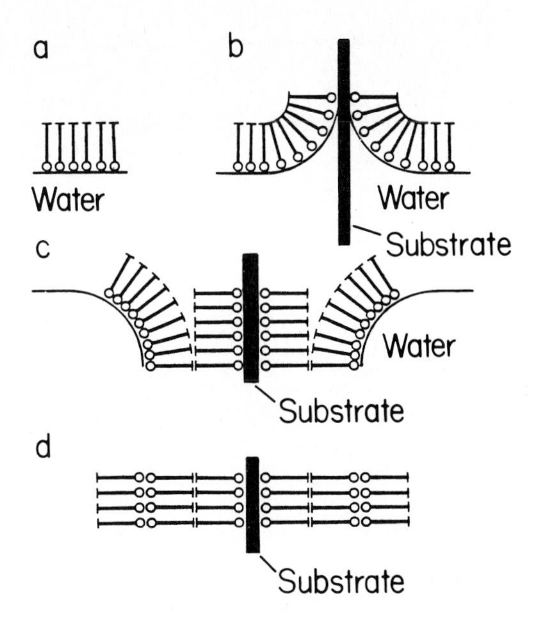

Figure 8. The dipping procedure used to fabricate Langmuir-Blodgett films.

Our films were fabricated primarily with Vickers Langmuir 4 troughs, modified to allow for drainage of the films after every dip and to operate at controlled temperatures down to 1°C. The films exhibited the classic surface pressure versus molecular area curves required for high quality films.[13] Details of the fabrication conditions can be found in References 6 and 14 through 16. The LB films were deposited onto Mo substrates. This substrate was chosen because: 1) it is metallic and therefore a) has a high thermal conductivity and avoids overheating in the laser-illuminated film region, and b) the scattering from bulk phonons inside the substrate is minimized; 2) it is readily available with a high quality surface finish as a mirror for CO_2 laser aplications; and 3) the surface oxide layer forms slowly and is at most 50 angstroms thick, resulting in only a small perturbation in the acoustic mode spectrum.

The first LB film system we studied consisted of films of cadmium arachidate [CdA] molecules and varied in thickness from 10-400 monolayers.[14] In the representative spectrum shown in Fig. 9, peaks due to scattering from Rayleigh, Sezawa and leaky waves are observed. The spectrum changes with scattering angle and film thickness, allowing measurement of the dispersion relations (velocity versus $q_p h$). Using a separate backscattering experiment at normal incidence to measure c_{33}, the least squares fit yielded the elastic constants listed in Table 2. The values of the c_{11} and c_{33} coefficients are typical of values reported for polymers. However, the shear elastic constant c_{44} is anomolously small, indicative of the weak inter-plane binding forces.

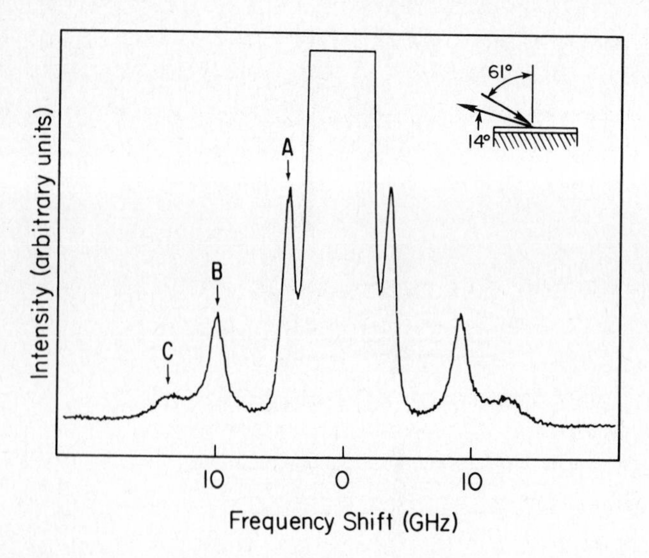

Figure 9. The Brillouin spectrum obtained from a fifty-one-monolayer-thick CdA film on a molybdenum substrate. The peaks in order of increasing frequency shift are due to scattering from the Rayleigh mode (A), a Sezawa mode (B) and a continuum of leaky waves (C).

Table 2. Measured elastic constants of the LB films (in GPa), glass, and for other materials with small ratios of $c_{55}^2/c_{11}c_{33}$.*

Material	c_{11}	c_{33}	c_{12}	c_{13}	c_{44}	c_{66}
SiO_2	78.5				31.2	
PZT-2	135	113	67.9	68.1	22.2	
Polyethylene	3.4				0.26	
CdA	11	21		10	0.4	
ODF	6.2	7.7		3.5	1.1	
ODF*	3.8	3.3		1.2	0.4	
GD-10	4.9	11.0	3.5	3.8	0.78	0.71
GD-14	5.0	10.5	3.4	3.4	0.71	0.8

**Measurements are in units of GPa. Here $c_{55} = c_{44}$. The ODF* data is for the microcracked film.*

We studied other LB multilayer films by means of Brillouin scattering.[15,16] The measured elastic constants are summarized in Table 2, along with the elastic constants of glass and two materials exhibiting some of the smallest known ratios $c_{55}^2/c_{11}c_{33}$ ($c_{44} = c_{55}$). The polyglutamates, GD-10 and GD-14, are different in structure from ODF and CdA since their long molecular axes lie in the plane of the film and not approximately orthogonal to it.[16] This property allowed us to evaluate all of the elastic constants for this film system for the first time.

The second noteworthy film system we studied is ODF, for which elastic constants were measured before and after polymerization.[15] It was widely expected that polymerization, which produces cross-linking bonds between adjacent molecules, would increase the elastic constants. In fact, for this material system we found a decrease in all of the elastic constants on polymerization. Subsequently, the electron micrograph of the film, shown in Fig. 10, revealed microcracks resulting from a large shrinkage of the film on polymerization. Additional Brillouin scattering experiments on a composite LB film of ODF and ω-ODF, with ratios chosen to cause no net shrinkage on polymerization, exhibited the expected increase in the elastic constants on polymerization. These results indicate that LB films can be strengthened on polymerization, provided that the polymerization process does not produce shrinkage or expansion of the film volume.

Figure 10. An electron micrograph of a polymerized ODF LB film showing the microcracks resulting from polymerization.

5. ROUGH ZnS FILMS

ZnS films were deposited on silicon substrates under conditions which usually result in very poor film quality. The Rayleigh wave at the film-air interface penetrates less than 1000 angstroms into the material and usually has its maximum amplitude right at the surface. This mode, then, is sensitive to surface quality. In fact, as shown in Fig. 11, we found that we could not detect the presence of a Rayleigh wave by means of light scattering. Also shown in Fig. 11 are calculations of the expected spectrum based on both a single crystal ZnS film and on an amorphous ZnS film. The peak with the smallest frequency shift corresponds to the Rayleigh mode. These results indicate that this mode is heavily damped by the rough film surface. SEM examination of the film surface revealed many microcracks. We did, however, detect Brillouin scattering from Sezawa modes, which fill the complete volume of the film and thus are less heavily damped by a rough surface. The experimental frequency shift, however, is smaller than those calculated for cubic or isotropic films, indicating a net softening of the elastic constants of the films.

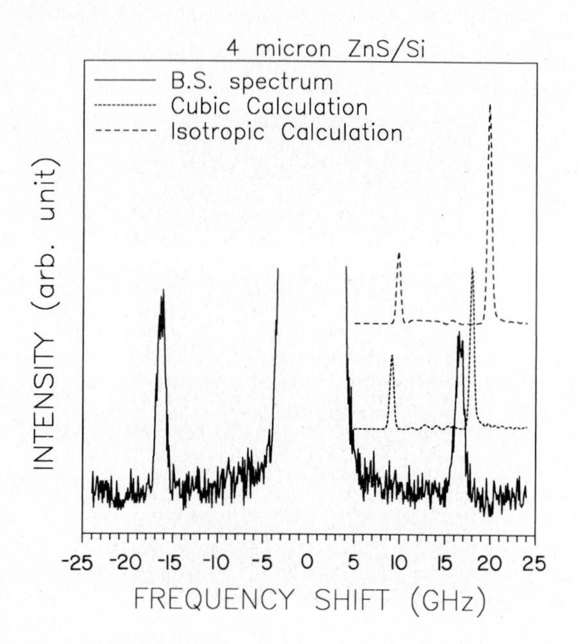

Figure 11. The Brillouin spectrum for a ZnS film, 4000 nm thick, deposited on a silicon substrate. Only one Sezawa mode is observed. Also shown are theoretical spectra calculated for the smallest Rayleigh wave velocity for the crystalline ZnS (solid line) and isotropic ZnS (dashed line).

6. SUMMARY

We have shown that Brillouin scattering is a valuable tool for measuring the elastic properties of thin films, and for studying the physics reflected in the films' mechanical properties. For metallic films, Stoneley waves were used to verify that the mechanical properties of a tantaluum film were identical at the film-air and film-substrate interfaces. For metallic superlattices, Brillouin scattering from Rayleigh, Sezawa and Love waves was used to evaluate all of the film elastic properties and to study the anomolous softening of the c_{44} elastic constant due to lattice mismatch for constituent film thicknesses of 10's of angstroms. For Langmuir-Blodgett films, very small shear elastic constants due to weak interlayer bonding forces were observed, and the effects of polymerization on the film mechanical properties were investigated. Finally, the Brillouin spectrum of very rough ZnS films was measured and interpreted.

The research at the University of Arizona was supported by the AFOSR URI program in thin films (F4962-86-C-0123). FN acknowledges research support from Consiglio Nazionale delle Ricerche.

REFERENCES

1) J.R. Sandercock in: Light Scattering in Solids III, eds. M. Cardona and G. Guntherodt, (Springer-Verlag, Berlon, 1982), p.173

2) G.W. Farnell in: Physical Acoustics, Principles and Methods, Vol. VI, eds. W.P. Mason and R.N. Thurston (Academic Press, New York, 1970); G.W. Farnell and E.L. Adler in: Physical Acoustic, Principles and Methods, Vol. IX, eds. W.P. Mason and R.N. Thurston (Academic Press, New York, 1972) chap. 2

3) B.A. Auld in: Acoustic Fields and Waves in Solids, Vol. I, (John Wiley & Sons, New York, London, Sydney, Toronto, 1973).

4) B. Hillebrands, S. Lee, G.I. Stegeman, H. Cheng, J.E. Potts and F. Nizzoli, Phys. Rev. Lett., 60 (1988) 832.

5) S. M. Lindsay and J. R. Sandercock, Rev. Sci. Inst. 52 (1981) 1478.

6) Many details can be found in R. Zanoni, *Brillouin Scattering From Langmuir-Blodgett Films*, PhD dissertation (University of Arizona, 1987); J.A. Bell, *Brillouin Scattering From Metal Superlattices*, PhD dissertation (University of Arizona, 1987).

7) C.M. Falco, W.R. Bennett and A. Boufelfel, in: Dynamical Phenomena at Surfaces, Interfaces and Superlattices, eds. F. Nizzoli, K.H. Rieder and R.F. Wallis (Springer-Verlag, Berlin, 1985) p. 35.

8) J.A. Bell, W.R. Bennett, R. Zanoni, G.I. Stegeman, C.M. Falco and F.Nizzoli, Phys. Rev. B Rapid Commun. 35 (1987) 4127.

9) A. Kueny, M. Grimsditch, K. Miyano, I. Banerjee, C.M. Falco and I.K. Schuller, Phys. Rev. Lett. 48 (1982) 166.

10) J.A. Bell, W.R. Bennett, R. Zanoni, G.I. Stegeman, C.M. Falco and C.T. Seaton, Solid State Comm. 64 (1988) 1339.

11) J.A. Bell, W.R. Bennett, R. Zanoni, G.I. Stegeman, C.M. Falco and C.T. Seaton, Appl. Phys. Lett. 51 (1987) 652.

12) J.A. Bell, W.R. Bennett, R. Zanoni, G.I. Stegeman, C.M. Falco and C.T. Seaton, Appl. Phys. Lett. 52 (1988) 610.

13) Recently reviewed by G. G. Roberts, Advances in Physics 34 (1985) 475.

14) R. Zanoni, C. Naselli, J. Bell, G.I. Stegeman and C.T. Seaton, Phys. Rev. Lett. 57 (1986) 2838.

15) S. Lee, B. Hillebrands, G.I. Stegeman, L.A. Laxhuber and J.D. Swalen, J. Chem. Phys., in press.

16) F. Nizzoli, B. Hillebrands, S. Lee, G.I. Stegeman, G. Duda, G. Wegner and W. Knoll, Phys. Rev. B, in press.

AUTHOR INDEX

SUBJECT INDEX